Nelson Mathematics

Nelson Mathematics 7

Series Authors and Senior Consultants
Marian Small • Mary Lou Kestell

Senior Author
David Zimmer

Assessment Consultant
Damian Cooper

Authors
Bernard A. Beales • Maria Bodiam • Doug Duff
Robin Foster • Cathy Hall • Jack Hope • Chris Kirkpatrick
Beata Kroll Myhill • Geoff Suderman-Gladwell • Joyce Tonner

NELSON

NELSON

Nelson Mathematics 7

Series Authors and Senior Consultants
Marian Small, Mary Lou Kestell

Senior Author
David Zimmer

Authors
Bernard A. Beales, Maria Bodiam,
Doug Duff, Robin Foster,
Cathy Hall, Jack Hope,
Chris Kirkpatrick, Beata Kroll Myhill,
Geoff Suderman-Gladwell,
Joyce Tonner

Assessment Consultant
Damian Cooper

Associate Vice President of Publishing
David Steele

Senior Publisher, Mathematics
Beverley Buxton

Acquisitions Editor, Mathematics
Colin Garnham

Executive Managing Editor, Development and Testing
Cheryl Turner

Senior Program Manager
Shirley Barrett

Program Manager
Tony Rodrigues

Developmental Editors
Christa Bedwin, Santo D'Agostino,
Lee Geller, Betty Robinson,
Bob Templeton

Developmental Consultant
Lynda Cowan

Editorial Assistant
Megan Robinson

Executive Managing Editor, Production
Nicola Balfour

Senior Production Editor
Debbie Davies-Wright

Copy Editor/Proofreader
Paula Pettitt-Townsend

Senior Production Coordinator
Sharon Latta Paterson

Production Coordinator
Franca Mandarino

Creative Director
Angela Cluer

Art Director
Ken Phipps

Art Management
Allan Moon

Illustrators
Kyle Gell, Kathy Karakasidis,
Steve MacEachern, Allan Moon,
Val Sanna

Cover Design
Ken Phipps

Cover Image
Martin Barraud/Stone/Getty Images

Design
Ken Phipps, Peggy Rhodes

Composition Team
Kyle Gell, Kathy Karakasidis,
Allan Moon, Val Sanna

Photo Research and Permissions
Vicki Gould

Set-up Photos
Dave Starrett

COPYRIGHT © 2005 by Nelson Education Ltd.

Printed and bound in Canada
9 10 11 12   22 21 20 19

For more information contact Nelson Education Ltd.,
1120 Birchmount Road, Toronto, Ontario, M1K 5G4. Or you can visit our Internet site at nelson.com

National Library of Canada Cataloguing in Publication Data

Nelson mathematics 7 / Marian Small ... [et al.].

Includes index.
ISBN 13: 978-0-17-626912-8
ISBN 10: 0-17-626912-6

1. Mathematics—Textbooks.
I. Small, Marian

QA107.2.N44 2004
510   C2004-902634-8

Advisory Panel

Michael Babcock
Teacher
Enterprise Public School
Limestone District School Board
Enterprise, Ontario

Beth Bond
Grade 8 Teacher
Toniata School
Brockville, Ontario

Mark Cassar
Vice Principal
Holy Cross Catholic School
Malton, Ontario

Donna Commerford
Retired Principal
Burlington, Ontario

Ron Curridor
Vice Principal
York Catholic District School
    Board
Maple, Ontario

Elizabeth Fothergill
Mathematics Literacy Consultant
Waterloo Catholic District School
    Board
Waterloo, Ontario

Lee Jones-Imhotep
Teacher, Literacy Coordinator
Lawrence Heights Middle School
Toronto District School Board
Toronto, Ontario

Peter Martindale
Teacher
Hamilton-Wentworth District
    School Board
Hamilton, Ontario

Lee McMenemy
Elementary Coordinator
Algoma District School Board
Sault Ste. Marie, Ontario

Kevina Morrison
Intermediate Math Teacher
Highbush Public School
Durham District School Board
Pickering, Ontario

Wayne Murphy
Department Head of Mathematics
Ajax High School
Durham District School Board
Ajax, Ontario

Barbara Nott
Teacher
Rainbow District School Board
Sudbury, Ontario

Kathy Perry
Teacher
Peel District School Board
Brampton, Ontario

Leonora Scarpino-Inglese
Teacher
Holy Cross Catholic School
Malton, Ontario

Silvana F. Simone
Mathematics Instructor
Ontario Institute for Studies in
    Education of the University of
    Toronto
Toronto, Ontario

Susan Stuart
Assistant Professor
Nipissing University
North Bay, Ontario

Christine Suurtamm
Assistant Professor, Mathematics
    Education
Faculty of Education, University
    of Ottawa
Ottawa, Ontario

James Williamson
Teacher
Nipissing-Parry Sound Catholic
    District School Board
North Bay, Ontario

T. Anne Yeager
Mathematics Teacher
Department Head of Special
    Education
Orangeville District Secondary
    School
Upper Grand District School
    Board
Orangeville, Ontario

Rod Yeager
Independent Mathematics
    Education Consultant
Department Head of Mathematics
    (retired)
Orangeville District Secondary
    School
Upper Grand District School
    Board
Orangeville, Ontario

Reviewers

Equity Reviewer

Mary Schoones
Educational Consultant/Retired
    Teacher
Ottawa-Carleton District School
    Board
Ottawa, Ontario

Literacy Reviewer

Kathleen Corrigan
Consultant
Simcoe County District School
    Board
Midhurst, Ontario

Contents

● Guided Activity   ● Direct Instruction   ● Exploration

● Guided Activity   ● Direct Instruction   ● Exploration

● Guided Activity   ● Direct Instruction   ● Exploration

● Guided Activity   ● Direct Instruction   ● Exploration

Chapter 9: Fraction Operations 303

Chapter 10: 3-D Geometry 345

● Guided Activity ● Direct Instruction ● Exploration

● Guided Activity   ● Direct Instruction   ● Exploration

CHAPTER 1

Factors and Exponents

▶ GOALS

You will be able to

- determine factors, greatest common factors, multiples, and least common multiples of whole numbers
- use exponents to show repeated multiplication
- calculate square roots of perfect squares
- use order of operations with whole numbers

Annual Garage Sale

Ever since last year's community garage sale, Kyle has been saving $2 coins, $5 bills, and $10 bills. Now he has $30 worth of each. He is looking forward to buying some items at this year's sale.

? **What prices can Kyle pay for exactly, using only $2 coins, $5 bills, and $10 bills?**

A. What can Kyle pay for using only $2 coins?

B. What can he pay for using only $5 bills?

C. What can he pay for using only $10 bills?

D. Which prices require Kyle to use more than one type of bill or coin? Explain why.

E. Suppose that the price of another item is a whole number ending in the digit 0. Can Kyle use only one type of bill or coin to pay this amount? Explain. Can he use more than one type of bill or coin? Explain.

F. Write a price that is greater than $100 and that someone can pay using only one type of bill or coin.

G. Write a price that is greater than $100 and that someone cannot pay using only one type of bill or coin.

Do You Remember?

1. A bag of marbles can be divided evenly among two, three, or four friends.

   a) How many marbles might be in the bag?

   b) What is the least number of marbles that can be in the bag?

   c) How many marbles would there be if there are between 30 and 40 marbles in the bag? How many marbles would each friend get? Use a diagram or another strategy to show your answer.

2. Suppose that you have three different lengths of linking cubes, as shown below. Assume that you have as many of these lengths as you need, but you may not take them apart.

   Can you make each length below using only one colour? If it is possible, show more than one way.

   a) 25 cubes   d) 29 cubes

   b) 20 cubes   e) 30 cubes

   c) 18 cubes   f) 32 cubes

3. Find all the possible whole-number lengths and widths of rectangles with each area given below. You might draw on centimetre grid paper or use another strategy.

   a) 12 cm²   c) 20 cm²

   b) 17 cm²   d) 24 cm²

4. A **factor** is a whole number (not including zero) that divides into another whole number with no remainder. Explain why 2 is a factor of 6. Name another factor of 6.

5. 16 is a **multiple** of 2 and of 8. Why is 16 also a multiple of 1, 4, and 16?

6. 2 and 3 are **prime numbers**, but 1 and 6 are not.

   a) List the next four prime numbers after 3.

   b) How many factors does a prime number have?

   c) The number 853 is a prime number. What are its factors?

7. A whole number greater than 1 that is not a prime number is a **composite number**. List three composite numbers.

8. Which numbers below are prime and which are composite? Explain your answer.

   a) 12   d) 23

   b) 17   e) 29

   c) 18   f) 39

9. Draw a square with each area on centimetre grid paper. Write the dimensions on your drawing.

   a) 25 cm²   c) 49 cm²

   b) 36 cm²   d) 100 cm²

10. Evaluate each expression. Show your calculations.

    a) $13 + 40 \times 3$

    b) $5 \times 12 - 3$

    c) $100 - 2 \times 8$

    d) $100 - 40 \div 2$

Using Multiples

▶ **GOAL**

Identify multiples, common multiples, and least common multiples of whole numbers.

Learn about the Math

Some Grade 7 students are planning a hot-dog sale at a volleyball tournament. Based on the last tournament, they expect to sell about 100 hot dogs. The local grocery store sells wieners in packages of 12 and buns in packages of 8.

? **How many packages of wieners and buns should the students buy if they want to make about 100 hot dogs, with no wieners or buns left over?**

A. Calculate the number of wieners contained in 1, 2, 3, 4, and 5 packages by listing **multiples** of 12.

12, 24, ▨, ▨, ▨

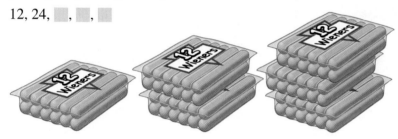

B. Calculate the number of buns contained in 1, 2, 3, 4, and 5 packages by listing multiples of 8.

8, 16, ▨, ▨, ▨

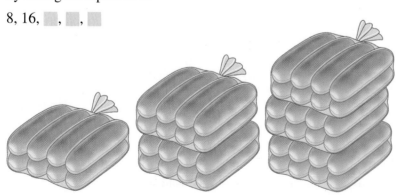

C. Continue your lists from steps A and B for at least 12 multiples of each number. Circle the multiples that are common to both lists.

D. Write the circled multiple that is the least number. This is the **least common multiple**, or **LCM**.

E. Explain how your answer in step D will help you solve the problem about how many packages of wieners and buns the students should buy.

Reflecting

1. Why might the hot-dog problem be easier to solve if wieners were in packages of 6 and buns were in packages of 12?

2. How does making a list of the multiples of different numbers help you determine which number is the LCM?

Work with the Math

Example 1: Finding the least common multiple

What is the LCM of 5 and 8?

Sandra's Solution

5, 10, 15, 20, 25, 30, 35, ④⓪, 45, …
8, 16, 24, 32, ④⓪ …
The LCM of 5 and 8 is 40.

I listed multiples of each number.
Then I circled the least number common to both lists.

Example 2: Finding the LCM of three numbers

Stephen is training for the three events in a triathlon. He runs every second day, swims every third day, and rides a bicycle every fifth day. How many times during the month of April will he have to practise all three events on the same day?

Ravi's Solution

2, 4, 6, 8, 10, 12, 14, 16, 18, 20, 22, 24, 26, 28, ㉚
3, 6, 9, 12, 15, 18, 21, 24, 27, ㉚
5, 10, 15, 20, 25, ㉚
Only one time in April will Stephen have to practise all three events on the same day.

I listed the multiples of 2, 3, and 5 to 30.
The LCM is 30.
On April 30th, Stephen will have to practise all three events.

Example 3: Using common multiples

Yuki plans to make roll-ups for a party.
She rolls a slice of meat with a slice of cheese.
Each meat package has 10 slices.
Each cheese package has 12 slices.
How many packages of each should she buy
so that there are no leftovers?

Yuki's Solution

10, 20, 30, 40, 50, (60) 70, 80, 90, 100, 110, (120) 130, ...

12, 24, 36, 48, (60) 72, 84, 96, 108, (120) ...

I listed multiples of 10 and 12.

I circled the common multiples.
60 is the LCM.

60 ÷ 10 slices of meat = 6 packages
60 ÷ 12 slices of cheese = 5 packages

120 ÷ 10 slices of meat = 12 packages
120 ÷ 12 slices of cheese = 10 packages

The least amount I should buy is 6 packages of meat and 5 packages of cheese.

I could buy 60 slices of meat
and 60 slices of cheese to make
60 roll-ups. That's 6 packages of
meat and 5 packages of cheese.

Or, I could buy 120 slices of
each to make 120 roll-ups.
That's 12 packages of meat
and 10 packages of cheese.

A Checking

3. List the first five multiples of each number.

   a) 2   **b)** 5   **c)** 6

4. Continue the patterns started in question 3 to determine the LCM of 2, 5, and 6.

5. Make lists of multiples and write three common multiples for each pair of numbers. What is the LCM?

   a) 4 and 8   **b)** 3 and 5   **c)** 12 and 18

B Practising

6. Find the LCM of each set of numbers.

   a) 7 and 9

   b) 3, 4, and 6

   c) 2, 3, 5, and 20

7. Which numbers are multiples of 5? Justify each answer by dividing.

   a) 15   **b)** 10 000   **c)** 137   **d)** 1001

8. Use the information below to solve this problem:

How many packages of hamburger buns and meat patties should you buy to sell at a baseball tournament?

- Buns are sold in packages of 6.

- Meat patties are sold in packages of 8.

- You expect to sell between 80 and 100 hamburgers at the tournament.

- You do not want leftovers.

9. Which numbers are common multiples of 3 and 5? Show the steps you used.

   a) 15   **c)** 100

   b) 135   **d)** 50

10. The number 7 is a factor of 1001.

   a) Explain how you know that 1001 is a multiple of 7.

   b) How does knowing that 7 is a factor of 1001 help you to get another factor of 1001?

11. Which numbers in the box are multiples of each number below?

5	15	180
300		12
100	60	50

   a) 3   **b)** 4   **c)** 25   **d)** 10   **e)** 30

12. The number 108 is the LCM of 36 and 54. List the next multiple of 36 and 54. Explain your reasoning.

13. Is a multiple of a number always greater than the number? Use an example to support your answer.

14. Suppose that the Grade 7 students expect to sell between 100 and 150 hot dogs.

   a) How many packages of 12 wieners and 8 buns should the students buy if they don't want any leftovers? Justify your answer.

   b) How many packages of each should they buy if they expect to sell more than 300 hot dogs?

15. In the opening hour of a new store, a bell rang every 2 min and lights flashed every 3 min. If the store opened at 10 o'clock, at what times did both events happen at the same time? Explain or sketch a diagram to show the strategy you used.

16. On an automobile assembly line, every third car is green. Every fourth car is a convertible.

   a) How many cars out of the first 100 will be green convertibles?

   b) Which number of car is the first green convertible?

   c) Show how writing common multiples helped you solve this problem.

ⓒ Extending

17. What is the LCM of 1 and any other number greater than 1?

18. The number 12 is a factor of 156. How can you use this information to determine the LCM of 12 and 156?

19. A number is divisible by 6 if it is divisible by both 2 and 3. List the numbers that are divisible by 6 and are between the numbers given below. Show the steps you used.

   a) 40 and 50

   b) 120 and 130

   c) 6000 and 6020

20. You add three multiples of a number together. Will the sum be a multiple of the number? Explain why or why not. (*Hint*: Make up some examples to help you see how to answer this question.)

A Factoring Experiment

▶ **GOAL**

Identify factors of numbers.

Explore the Math

Sarah wants to put new floor tiles in her bathroom shower. The floor measures 120 cm by 120 cm. There are several different-sized square tiles that Sarah likes. She wants to know how many of each different-sized tile she will need if she does not want to cut any of the tiles.

? **How do you determine the sizes of tiles that Sarah can use for her project?**

A. Draw a model of the shower floor on small grid paper. Use 1 square on your grid to represent a 1 cm square.

B. Sarah can use square tiles that measure 2 cm by 2 cm. Colour or draw to decide how many of these tiles will cover the floor.

C. What other sizes of tiles can Sarah use? Colour a new floor diagram for each size of tile.

D. Make a table like the one below to record your findings. (The first two rows are done for you.)

Side length of one tile (cm)	Area of one tile (cm²)	Number of tiles per side	Total number of tiles
1	1	120	14 400
2	4	60	3 600

E. In which two columns will the numbers always give a product of 120 when you multiply them?

F. Describe the relationship between the numbers in the first and second columns.

G. If the floor had 20 tiles on each side, what would be the side length of each tile?

H. If there were only 4 tiles completely covering the floor, what would be the side length of each tile?

Reflecting

1. How does thinking about the factors of 120 help you decide which tiles would fit?

2. How can you use the floor width and size of tile to calculate the number of tiles needed?

3. How would your answers change if the shower floor measured 240 cm by 240 cm instead of 120 cm by 120 cm?

Curious Math   POOL-TABLE REFLECTIONS

On a pool table, a ball will bounce off the edge at the same angle that it hits the edge.

Look at this imaginary pool table with four pockets. To answer the questions below,
- always think of starting a ball at one corner
- aim at a 45° angle (*Hint*: A 45° angle is formed by drawing a diagonal of a square.)
- draw the path of the ball
- remember to bounce at a 45° angle when you hit an edge

A. Draw a 6 by 8 rectangle on grid paper. Use your ruler to draw the path of a ball. Count the number of squares the ball passes through before it falls into a pocket.

B. Copy the table below. Record your answer for step A in the first empty square. Then draw rectangles for each of the other sizes of tables. Record the number of squares the ball passes through before falling into a pocket.

Size of table	6 by 8	4 by 6	3 by 5	8 by 12	3 by 9	2 by 7	4 by 10
Number of squares							

C. Examine the data. What is the relationship between the size of the table and the number of squares the ball passes through before falling into a pocket?

D. Draw a table of a different size. Predict the number of squares the ball will pass through. Draw the path of the ball to check your prediction.

E. Predict the number of squares the ball will pass through on a 16 by 12 table. Check your prediction with a diagram.

1.3 Factoring

You will need
• linking cubes or beads
• a calculator

▶ **GOAL**

Determine factors, common factors, and greatest common factors of whole numbers.

Learn about the Math

Many First Nations people in Canada make bead designs on clothing, footwear, and belts.

Ravi's grandmother challenged Sandra and Ravi to design patterns with rectangles. They can use any red rectangles they can make with 24 beads and any blue rectangles they can make with 18 beads.

? **How can you match up 24-bead red rectangles and 18-bead blue rectangles in patterns?**

First, Ravi and Sandra decided to put together red and blue rectangles so that the side lengths matched up. To find out the possible side lengths, they factored 18 and 24.

Example 1: Finding factors by dividing

Factor 24 by dividing.

Ravi's Solution

I started listing the factors of 24 in order.
For each factor, I listed its partner—like 1 and 24.
I found the partner by dividing 24 by the first factor.

$24 \div 2 = 12$ Another pair is 2 and 12.

$24 \div 3 = 8$ Another pair is 3 and 8. Then there's 4 and 6.

I stopped when there were no more factors.
I connected the factor pairs. The picture looks like a rainbow.
This helped me keep track so I know I didn't miss any factors.

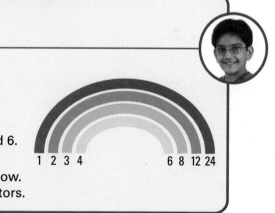

1 2 3 4   6 8 12 24

Example 2: Finding factors by rearranging rectangles

Factor 18 by making rectangles with an area of 18.

Sandra's Solution

I found all the factors of 18 by rearranging the blue beads to make all the possible rectangles.

When I finished, the lengths and widths showed all the factors.

I wrote them in order. The factors of 18 are 1, 2, 3, 6, 9, and 18.

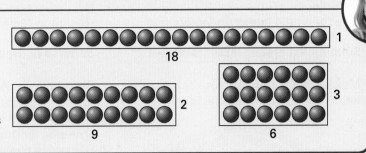

Next, Sandra and Ravi looked for factors that were common to both lists so they could see which side lengths would match up. They also wanted to know what the longest common side length would be. This is the **greatest common factor**, or **GCF**.

Ravi said, "We circled the common factors in our lists.

Factors of 24: ①, ②, ③, 4, ⑥, 8, 12, 24

Factors of 18: ①, ②, ③, ⑥, 9, 18

The common factors are 1, 2, 3, and 6.
The GCF is 6."

Sandra said, "We can match rectangle sides that are 1, 2, 3, or 6 beads long to make our patterns."

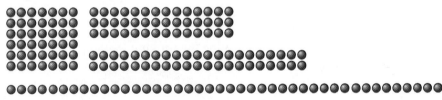

> **greatest common factor (GCF)**
>
> the greatest whole number that divides into two or more other whole numbers with no remainder; for example, 4 is the greatest common factor of 8 and 12

Reflecting

1. Think about making a factor rainbow or drawing rectangles to list all the factors of a number. How are the two methods alike? How are they different?

2. Why would you not bother to make a factor rainbow or a rectangle model for a prime number?

3. Once you determine one factor of a large number, how can you use a calculator to help you determine the other factor in the pair? Explain why it is important to keep the factors in order.

Example 3: Using models to determine common factors and the GCF

Determine the common factors and greatest common factor (GCF) of 30 and 36.

Solution A: Creating factor rainbows

Make a factor rainbow for each number, and circle the common factors.

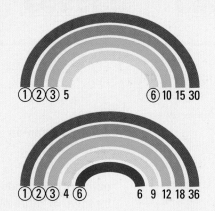

①②③ 5   ⑥ 10 15 30

①②③ 4 ⑥   6 9 12 18 36

The common widths and common factors are 1, 2, 3, and 6. The GCF is 6.

Solution B: Drawing rectangles

Draw all the possible rectangles that have an area of 30 square units. Then draw all the possible rectangles that have an area of 36 square units. (The side lengths must be whole numbers.) See which rectangle sides match.

1 × 30	1 × 36
2 × 15	2 × 18
3 × 10	3 × 12
	4 × 9
6 × 5	6 × 6

A Checking

4. Make a factor rainbow for each number. Use your rainbow to determine the number of different rectangles that can be formed with each number of beads. Sketch the rectangles.

  a) 28   **c)** 64

  b) 32   **d)** 120

5. Use your answers to question 4 to determine the common factors and GCF of each pair of numbers.

  a) 28 and 32   **c)** 28 and 120

  b) 28 and 64   **d)** 32 and 64

B Practising

6. Determine the missing factors.

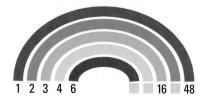

1 2 3 4 6   16   48

7. a) List the factors of 27.

  b) Use question 6 to list the factors of 48.

  c) What is the GCF of 27 and 48?

8. What is the GCF of each pair of numbers?

  a) 8 and 10   **d)** 6 and 24

  b) 3 and 12   **e)** 4 and 10

  c) 4 and 6   **f)** 2 and 5

9. a) Explain why 2 is a common factor of 12 and 24.

b) What are the other common factors of 12 and 24?

10. Which of the following numbers are factors of 120? How do you know?

a) 3　**b)** 5　**c)** 7　**d)** 12　**e)** 60

11. Determine whether or not each number is a common factor of 144 and 240. Try to use different methods, and show your work.

a) 2　**b)** 3　**c)** 10　**d)** 48　**e)** 120

12. Which number is a common factor of every number? Explain.

13. Both 2 and 5 are common factors of a number. The number is between 101 and 118. What is the number? Give a reason for your answer.

14. a) How many different ways can 50 players in a marching band be arranged in rectangle arrangements?

b) If marching bands vary from 21 to 49 players, which number of players can be arranged in the greatest number of rectangles?

c) Which number of players can be arranged in the shape of a square?

d) Explain how finding factors can help you answer parts (b) and (c).

15. Two prime numbers, 11 and 17, are multiplied together.

a) List all the factors of the product.

b) How many factors does the product have?

c) Will you get the same number of factors if you multiply two different prime numbers? Use examples to support your answer.

d) Will you get the same number of factors if you multiply a prime number by itself? Use examples to support your answer.

16. What is the GCF of two different prime numbers? Explain your reasoning.

C Extending

17. Two even numbers are multiplied together. List three factors of the product. (*Hint*: Make up an example to help you.)

18. Decide whether each statement is true or false. If a statement is false, give an example to show why it is false.

a) If the last digit of a number is even, then 2 is a factor.

b) If the last digit is a 4, then 4 is a factor.

c) If the last digit is 0, then 2, 5, and 10 are factors.

d) If the last two digits are 25, then 5 and 25 are factors.

e) If the last digit is a 6, then 6 is a factor.

19. The LCM of two numbers is 24, and the GCF is 12. What are the two numbers?

20. Determine each number.

a) the GCF of 2 and any even number

b) the LCM of 2 and any even number

c) the LCM of 2 and any odd number

d) the GCF of 2 and any odd number

e) the GCF of three different prime numbers

Exploring Divisibility

You will need
- centimetre grid paper
- a calculator

▶ **GOAL**

Use divisibility rules to identify factors of numbers.

Explore the Math

Ryan knows, without having to divide, that 6480 is divisible by 2, 5, and 10. It is divisible by 2 because it is an even number. It is divisible by 5 and 10 because it ends in 0. These are called **divisibility rules** .

Ryan wonders if there is a way to determine if 3 or 9 are factors of a number without having to divide.

divisibility rule

a way to determine if one number is a factor of another number without actually dividing

? **How can you tell if a number is divisible by 3 or 9?**

A. If you were to divide 10 into groups of 3, how much would be left over? What if you were to divide 100 or 1000 into groups of 3? How much would be left over?

B. Think of 138 as 1 hundred + 3 tens + 8 ones. Imagine dividing each of the hundreds and tens into groups of 3. How much is left over after the groups are made?

C. Combine the leftovers with the ones in 138. What is the total now? How does that relate to the sum of the digits of 138?

D. Can your answer to step C be grouped into 3s with no remainder?

E. Explain why 138 is divisible by 3.

F. Repeats steps A to E using 9 instead of 3.

G. Choose another three-digit or four-digit number. Use the sum of the digits to determine if each number is divisible by 3 or 9. Explain your reasoning.

Reflecting

1. Use the class results to help you answer these questions.

a) Are all numbers that are divisible by 3 also divisible by 9? Use an example to support your answer.

b) Are all numbers that are divisible by 9 also divisible by 3? Explain.

2. Describe a rule for deciding if a number is divisible by 3. Consider numbers greater than 1000 as well. Explain your thinking.

3. How could you adapt the rule for division by 3 to decide easily when a number is divisible by 9?

4. If 3 and 9 are both factors of 67 645 60▪, what is a possible value for the ones digit? Explain your reasoning. Divide to check.

5. Determine the least four-digit number that is divisible by 2, 3, and 5. Explain your reasoning.

6. Explain how you know that 123 123 123 900 is divisible by 2, 3, 5, 9, and 10. How do you know that it is also divisible by 6?

Mental Math

DOUBLING AND HALVING AGAIN AND AGAIN

It's easy to multiply or divide a number by 4 and 8 after you have multiplied or divided the number by 2.

Example 1: Calculate 8×15.
- Start with $2 \times 15 = 30$.
- 4×15 is double (2×15); double 30 = 60.
- 8×15 is double (4×15); double 60 = 120.

> Think of 2 × 15 as doubling 1 ten and 1 five. You get 2 tens and 2 fives.

Example 2: Calculate $360 \div 8$.
- Start with $360 \div 2 = 180$.
- $360 \div 4$ is half of 180 = 90.
- $360 \div 8$ is half of 90 = 45.

> Think of 360 ÷ 2 as half of 300 and half of 60. You get 150 and 30.

You can check the answer of 45 by doubling.
- Start with $2 \times 45 = 90$.
- 4×45 is double (2×45); double 90 = 180.
- 8×45 is double (4×45); double 180 = 360.

> Think of 2 × 45 as doubling 1 forty and 1 five. You get 2 forties and 2 fives.

1. Use doubling to multiply.

a)	4×25	d)	8×45
b)	8×25	e)	4×75
c)	4×45	f)	8×75

2. Use halving to divide.

a)	$60 \div 4$	d)	$500 \div 4$
b)	$180 \div 4$	e)	$1000 \div 4$
c)	$280 \div 8$	f)	$1000 \div 8$

3. a) Show how you can use doubling to multiply 16 by 25.

b) Show how you can use halving to divide 2000 by 16.

Powers

▶ **GOAL**

Use powers to represent repeated multiplication.

Learn about the Math

Chang's family and Sandra's family bought tickets for a charity lottery. The winner will have two options for the prize:

Option 1: Take $500 cash immediately.
Option 2: Take the prize on March 10, if it is calculated by doubling the amount each day, beginning with $2 on March 1.

? **Which option is the better deal for the lottery winner?**

Chang decides to figure out the value of Option 2 by using a table of values.

Example 1: Using a pattern to make a table of values

Which option is worth more—Option 1 or Option 2?

Chang's Solution

I made a table.

I knew that each amount was twice the amount from the day before.

I stopped calculating on March 9 because that amount is already above the $500 I'd win if I chose Option 1.

I'll take Option 2 because I get more.

Date	Amount won ($)
March 1	2
March 2	4
March 3	8
March 4	16
March 5	32
March 6	64
March 7	128
March 8	256
March 9	512

Sandra decides to calculate the value of Option 2 by multiplying by 2 over and over again.

She writes the repeated multiplication as a **power** of 2 to show how the money doubles each day.

For March 2, 2×2 or $2^2 = 4$.
For March 3, $2 \times 2 \times 2$ or $2^3 = 8$.
For March 5, $2 \times 2 \times 2 \times 2 \times 2$ or $2^5 = 32$.

power

a numerical expression that shows repeated multiplication; for example, the power 4^3 is a shorter way of writing $4 \times 4 \times 4$

Example 2: Using repeated multiplication and powers

What is the greatest amount that can be won in the lottery?

Sandra's Solution

$2 \times 2 \times 2 \times 2 \times 2 \times 2 \times 2 \times 2 \times 2 \times 2 = 2^{10}$

For Option 2, I know there will be ten 2s multiplied together. I can write this as a power of 2, which is 2^{10}.

There is a shortcut for calculating this, if you use a calculator with a repeat function.

For March 10, press 2 × followed by nine = signs to get 1024.

The greatest amount is $1024. Choose Option 2 because $1024 is greater than $500.

Reflecting

1. Why did Sandra press the ⬛ button on her calculator 9 times instead of 10 times? Does your calculator work the same way?

2. What power represents the amount of money you would win if the amount doubled every day from March 1 to 15?

3. How do you know that 2^7 must be double 2^6?

4. How does 2^{10} relate to 2^5?

Communication Tip

- The number 5^4 is read as "5 to the fourth power." It can also be read as "the fourth power of 5." Four 5s are multiplied, so you would calculate it as $5 \times 5 \times 5 \times 5 = 625$.

- The number 10^3 is read as "10 to the third power" or "ten cubed." You would calculate it as $10 \times 10 \times 10 = 1000$.

- 4^2 is "4 squared," or $4 \times 4 = 16$.

- A power has a **base** and an **exponent**. The exponent tells the number of times you repeat the base as a factor.

        3 is the *exponent* of the power.
$4^3 = 64$

4 is the *base* of the power.

Factors and Exponents **17**

Example 3: Enlarging a photograph

Simon is using photo software to enlarge a photo by doubling its length and width.

Length of original image

2 × length of original image

How many times longer will the length of the original image be if Simon doubles the length six times?

Solution: Using multiplication

$2 \times 2 \times 2 \times 2 \times 2 \times 2 = 2^6$ or $2 \boxed{\times} \boxed{=} \boxed{=} \boxed{=} \boxed{=} \boxed{=}$

The length of the original image will be 64 times longer in the enlarged photo.

A Checking

5. a) Identify the base and the exponent in 9^4.

  b) Write the power as a repeated multiplication.

  c) Calculate the product.

6. A lottery winner has two options.

  Option 1: Take $500 000 cash immediately.

  Option 2: Take the prize on March 12, if it is calculated by tripling the amount each day, beginning with $3 on March 1.

  a) Use a power to represent the amount on March 12.

  b) Use your calculator to calculate the amount on March 12.

  c) Which option is the better deal for the lottery winner?

  d) If the offer were continued, on what day would the amount be greater than $1 million?

B Practising

7. Use powers to represent each multiplication. Then calculate each product.

  a) $2 \times 2 \times 2 \times 2$

  b) $5 \times 5 \times 5$

  c) $8 \times 8 \times 8 \times 8$

  d) $10 \times 10 \times 10 \times 10 \times 10 \times 10$

  e) 100×100

8. Which calculations in question 7 can you do more quickly mentally than by using a calculator? Explain why.

9. Determine the number that is missing from each box.

  a) $2^{\blacksquare} = 32$

  d) $4^{\blacksquare} = 64$

  b) $\blacksquare^3 = 1000$

  e) $3^1 = \blacksquare$

  c) $5^6 = \blacksquare$

  f) $17^{\blacksquare} = 17$

10. Express each prize as a power. (Assume that the amount won each day is multiplied by the same number to get the winnings the next day.)

a) If you win $5 on May 1, $25 on May 2, and $125 on May 3, how much would you win on May 11?

b) If you win $10 on March 1, $100 on March 2, and $1000 on March 3, how much would you win on March 15?

c) If you win $4 on March 1, $16 on March 2, and $64 on March 3, how much would you win on March 9?

d) If you win $7 on March 1, $49 on March 2, and $343 on March 3, how much would you win on March 8?

11. a) What is the area of a 5-by-5 square?

b) Write a power to represent the area of the square.

c) Explain why the expression in part (b) is read as "5 squared."

d) Why do you think 5^3 is read as "5 cubed"?

12. Express each number as a power.

a) 36   **c)** 81

b) 49   **d)** 100

13. You know that $2 \times 5 = 5 \times 2$. Is it true that $2^5 = 5^2$? Explain. If it is not true, which is greater, 2^5 or 5^2?

14. a) How many numbers between 100 and 200 can be expressed as powers of 2? Show the steps you followed.

b) Which power of 2 is closest to 1000? Show your work.

15. Copy and fill in the missing exponent.

a) $10^{\blacksquare} = 1000$ **c)** $2^{\blacksquare} = 64$

b) $100^{\blacksquare} = 10\,000$ **d)** $3^{\blacksquare} = 243$

16. Here is a riddle:

As I was going to Halifax,
I met a man with 7 sacks,
Every sack had 7 cats,
Every cat had 7 kits,
Man, kits, cats, and sacks,
How many were going to Halifax?

a) How many cats were there? Show your work using powers.

b) How many kittens were there? Show your work using powers.

17. A chest has 8 containers. Each container has 8 boxes. Each box has 8 bottles. Each bottle has 8 quarters.

a) Write a power to describe the total number of quarters.

b) What is the value of the money in the chest?

C Extending

18. Explain how mental math can be used to calculate each power.

a) 1^{100} **c)** 2^4 **e)** 1000^2

b) 0^{50} **d)** 10^3

19. Powers can be used to express some numbers in several ways. For example, 64 can be expressed as 4^3 or 2^6 or 8^2. Write each number in two ways as powers.

a) 16   **b)** 81   **c)** 256   **d)** 1024

20. a) Describe a pattern in this list of numbers: 3, 5, 9, 17, 33, 65, …

b) Show that each number can be written as 1 added to a power of 2.

c) Use your answer to part (b) to predict the 9th and 10th numbers in the pattern.

Frequently Asked Questions

Q: How do you represent the factors of a number?

A: Think of multiplying numbers that give 20, for example,
1×20, 2×10, and 4×5.

- You can match factor pairs in a factor rainbow.
- You can draw rectangles, where the lengths and widths represent the factors of the number.
- You can write a list of all the factors.

  factors of 20: 1, 2, 4, 5, 10, 20

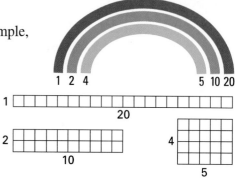

Q: How do you determine the factors of a number?

A:
- You can divide (mentally, with pencil and paper, or with a calculator) and determine divisors that leave no remainders.
- You can arrange the number of squares into rectangles.
- You can use divisibility rules to determine some factors.

Q: How do you find common factors and the greatest common factor (GCF)?

A: After writing the factors of two numbers, you circle the factors that are the same. The greatest of these numbers is the GCF.

For example, the common factors of 8 and 12 are 1, 2, and 4. The GCF is 4.

Q: How do you find common multiples and the least common multiple (LCM)?

A: To determine common multiples of 2 and 3, list some of the multiples of each and circle the numbers in both lists.

multiples of 2: 2, 4, ⑥ 8, 10, ⑫ 14, 16, ⑱ …
multiples of 3: 3, ⑥ 9, ⑫ 15, ⑱ ….

Therefore, 6, 12, and 18 are common multiples of 2 and 3. The LCM is 6.

Q: How do you use a power to show repeated multiplication?

A: The number to be multiplied is written as a base. The number of times it is to be multiplied is written as a small raised number called an exponent.

For example, the multiplication expression $2 \times 2 \times 2$ can be written as the power 2^3 because there are three 2s multiplied together.

Practice Questions

(1.1) **1.** A multiple of a number is 12. Use symbols, numbers, or words to show how you can determine three other multiples of this number.

(1.1) **2.** Write three common multiples of each pair of numbers. What is the LCM?

    a) 3 and 7   **c)** 12 and 15

    b) 15 and 30   **d)** 16 and 24

(1.1) **3.** You pick a number and then toss a die. You get a point if your number is a multiple of the number tossed.

    a) Which numbers should you pick so that you are most likely to win a point after a toss?

    b) Which numbers could you pick so that you are least likely to win a point?

(1.2) **4.** Ten is a factor of a number. What must be some other factors of the number?

(1.3) **5.** Use a factor rainbow or draw rectangles to show all the factors of each number.

    a) 100 **b)** 27   **c)** 45   **d)** 1000

(1.3) **6.** Write all the common factors of each pair of numbers. Circle the GCF.

    a) 100 and 25   **c)** 40 and 60

    b) 27 and 60   **d)** 75 and 135

7. Cookies come in packages of 12. Drinks come in packages of 3, 4, 5, 6, 8, 10, and 12. Make a table to show how many packages of each size to buy so that you have the same number of cookies and drinks. (1.3)

8. Both 7 and 13 are factors of 1001. Explain how to use this information to get two other factors. (1.3)

9. Determine the least three-digit number that is divisible by 3, 5, and 9. (1.4)

10. Determine the last two digits in the number 20▮▮ so that it is divisible by 2, 3, 5, and 9. Write the number. (1.4)

11. Identify the greater number in each pair. Show your work. (1.5)

    a) 2^{10} and 10^2 **c)** 7^3 and 3^7

    b) 3^2 and 2^3 **d)** 4^5 and 5^4

12. Explain how you know that 2^{100} is greater than 2^{75} without using a calculator. (1.5)

Square Roots

You will need
- grid paper
- a calculator

▶ **GOAL**

Determine the square roots of perfect squares.

Learn about the Math

The floor mat in gymnastics is a square with an area of 144 m².

? **How can you calculate the dimensions of the square mat?**

A. Use the **formula** *Area = length × width* to explain why the product of the dimensions of the mat is 144 m².

B. In the **equation** ▓ × ▓ = 144, ▓ represents both the length and width of the mat. Fill in the boxes with the missing numbers.

C. Why is the area of the mat a **perfect square**?

D. The mat problem can be solved by calculating the **square root** of 144. Explain why.

E. What are the dimensions of the mat? Show how you got your answer.

Ravi said, "It's easy to calculate a square root. Enter a number like 121. Then press the square root key: $\boxed{\sqrt{}}$. The number in the display, 11, is the square root."

Sandra replied, "You can check that a number is the square root if you multiply it by itself. You should get the original number again."

perfect square

the product of a whole number multiplied by itself; for example, 81 is a perfect square because it is 9 × 9

square root

a number when multiplied by itself equals the original number; for example, the square root of 81 is represented as $\sqrt{81}$ and is equal to 9 because 9 × 9 or 9² = 81

Communication Tip

The power 11² or "11 squared" represents the area of an 11-by-11 square while $\sqrt{121}$ or "square root of 121" represents the side length of a square with an area of 121.

Reflecting

1. Why are there no perfect squares between 144 and 169?

2. How did you know that the last digit of the whole-number dimensions of the floor mat must be a 2 or an 8? (*Hint*: Make a table of all the squares from 1 to 10, to see the last digits of greater squares.)

3. Suppose that your calculator does not have a square root key. How could you still use your calculator to determine $\sqrt{144}$?

Work with the Math

Example 1: Determining a square root by guessing and testing

The floor mat in rhythmic gymnastics is a square with an area of 169 m².
What are its whole-number dimensions?

Yuki's Solution

▨ × ▨ = 169	I have to find two equal factors with a product of 169.
10 × 10 or 10² = 100 Too low	The dimensions must be between 10 m and 20 m.
20 × 20 or 20² = 400 Too high	
3 × 3 = 9	I know that the last digit of the dimensions must be 3, because 9 is the last digit of 169.
13 × 13 or 13² = 169	The dimensions of the square must be 13 m by 13 m.
$\sqrt{169}$ = 13	This means that the square root of 169 must be 13.

Example 2: Using the square root key on a calculator

In artistic gymnastics, the square floor mat has an area of 196 m².
What are its whole-number dimensions?

Ryan's Solution

196 $\boxed{\sqrt{\ }}$

I entered 196, the area of the square mat, into my calculator.

Then I pressed the $\boxed{\sqrt{\ }}$ key to calculate the square root.

The dimensions of the square must be 14 m by 14 m.

I checked my answer by squaring 14.

$14^2 = 14 \times 14$
$ = 196$

$\boxed{\times}\boxed{=}$

A Checking

4. The "tatami," or mats, in judo are squares with a minimum area of 36 m² and a maximum area of 64 m².

   a) Sketch diagrams of the mats on grid paper.

   b) What are the possible whole-number dimensions of the mats? Check by multiplying.

5. Use mental math to calculate.

   a) $\sqrt{4}$ b) $\sqrt{16}$ c) $\sqrt{81}$

6. When 32 is multiplied by itself, the product is 1024. What is the square root of 1024?

7. The square root of a number is 11. What is the number?

B Practising

8. The area of a square weightlifting platform is 16 m².

   a) Sketch the platform. What are its dimensions?

   b) What is the perimeter of the platform?

9. a) Explain how you know that the square root of 225 is between 10 and 20.

   b) Will the square root of 225 be closer to 10 or 20? Explain.

   c) Guess and test to find the square root of each number.

   i) 289   iii) 2209   v) 8649
   ii) 3025   iv) 3721

10. Use mental math to determine the square root of each number.

    a) 1   d) 100
    b) 0   e) 400
    c) 25   f) 900

11. The preferred overall competition area in judo, including the mats, is a square with an area of 256 m².

    a) How do you know that the length and width of the competition area are between 10 m and 20 m?

    b) What are the possible last digits of the dimensions if the side lengths are whole numbers?

    c) Use your answers to parts (a) and (b) to predict the dimensions.

    d) Check your prediction using a calculator.

12. The number 121 is the first perfect square that is greater than 100. Calculate the following numbers. Show the steps you used to calculate each answer.

    a) all perfect squares between 121 and 200

    b) the first perfect square greater than 1000

13. Explain two different ways to calculate the square root of 225.

14. a) Explain how you know that $\sqrt{441}$ must be close to 20.

    b) Explain how you know that the last digit of $\sqrt{441}$ must be 1 or 9 if the answer is a whole number.

    c) Use your answers to parts (a) and (b) to predict $\sqrt{441}$.

    d) Use a calculator to check your prediction.

C Extending

15. Explain how you can calculate each square root mentally.
  a) $\sqrt{31 \times 31}$
  b) $\sqrt{431 \times 431}$
  c) $\sqrt{17^2}$
  d) $\sqrt{43^2}$

16. Calculate the dimensions of a square that has the same area as a 16 m by 64 m rectangle. Show your steps.

17. Calculate the area of your classroom floor. Estimate the dimensions of a square that has the same area as your classroom floor. Check your estimate.

18. a) Calculate the square root of each power of 10.
  i) $\sqrt{100}$
  ii) $\sqrt{10\ 000}$
  iii) $\sqrt{1\ 000\ 000}$

  b) Describe the pattern for the number of zeros in each square root.

  c) Use your pattern from part (b) to predict the number of zeros in $\sqrt{100\ 000\ 000}$.

19. Why might squaring a number and taking the square root of a number be thought of as opposite operations? Give an example to justify your answer.

Math Game

ROLLING POWERS

Number of players: 2 to 4

Rules

1. Roll one die to get a number that will be the base. Roll the other die to get the exponent.

2. Before the second roll, each player predicts whether the power will be greater than or less than 100.

3. The players calculate the answer.

4. Players score 1 point for each correct prediction.

5. Take turns rolling the dice.

6. The first player to reach 10 points wins.

You will need
- 2 dice
- a calculator
- paper and pencil

Prediction: less than 100

Rolled: 6 and 3

$6^3 = 6 \times 6 \times 6 = 216$

I do not score a point because I predicted incorrectly.

1.7 Order of Operations

▶ **GOAL**

Apply the rules for order of operations.

Learn about the Math

Ravi and Ryan investigated the effects of smoking on lung health. They found the following formula to estimate lung capacity, which is related to the size of your lungs and the amount of air they can hold (in millilitres).

Lung capacity = 41 × ⬚ − 18 × ⬚ − 2690

              ↑   ↑

       height in   age in years
      centimetres

? **How can you use the rules for order of operations to estimate your lung capacity?**

Rules for Order of Operations

Step 1: Perform the operations in brackets first.
Step 2: Calculate powers next.
Step 3: Divide and multiply from left to right.
Step 4: Finally, add and subtract from left to right.

> **Communication Tip**
>
> You can remember the rules for order of operations by thinking of "BEDMAS."
>
> **Rules for Order of Operations**
>
> **B**rackets
>
> **E**xponents
>
> **D**ivide and **M**ultiply from left to right
>
> **A**dd and **S**ubtract from left to right

Example 1: Underlining to show the order of operations

Ravi wants to estimate his lung capacity. He is 13 years old and 160 cm tall, so he calculates 41 × 160 − 18 × 13 − 2690. What is his lung capacity?

Ravi's Solution

$$41 \times 160 - 18 \times 13 - 2690$$
$$= \underline{41 \times 160} - \underline{18 \times 13} - 2690$$
$$= 6560 - 234 - 2690$$
$$= 3636$$

My lung capacity is 3636 mL, which is about 3.6 L.

There are no **B**rackets, **E**xponents, or **D**ivision calculations. First I had to **M**ultiply the two pairs of numbers going from left to right.

To make the calculations easier to see, I underlined the parts I do first.

Then I multiplied using my calculator.

I rewrote the calculation with the new answers.

Then I subtracted from left to right.

Example 2: Using the order of operations and mental math

Use the rules for order of operations to evaluate $(5 + 3)^2 \div (9 - 5)$.

Ryan's Solution

$(5 + 3)^2 \div (9 - 5)$	I coloured some steps to keep track.
$= 8^2 \div 4$	First I did the calculations in brackets mentally.
$= 64 \div 4$	Then I calculated the power.
$= 16$	Finally, I divided.

Reflecting

1. Press the following keys on your calculator to see if you get the correct answer for Example 1. $41 \times 160 - 18 \times 13 - 2690$ Does your calculator follow the rules for order of operations?

2. Show how to use shading or brackets instead of underlining to make sure that the operations for Example 1 are done in the correct order.

3. If you remove the brackets in Example 2, will you still get the correct answer? Evaluate $5 + 3^2 \div 9 - 5$.

4. Explain why there are four steps to the order of operations, but there are six letters to explain the order in BEDMAS.

Work with the Math

Example 3: Using the rules for order of operations

Evaluate $1028 - (12 \div 4 + 32^2)$. Rewrite the new numbers to show your work.

Solution

Identify the part you will work on at each step.
Do the math on your calculator or in your head.

32^2 or $32 \times 32 = 1024$

$12 \div 4 = 3$

$3 + 1024 = 1027$

$1028 - (12 \div 4 + \underline{32^2})$

$= 1028 - (\underline{12 \div 4} + 1024)$

$= 1028 - (\underline{3 + 1024})$

$= \underline{1028 - 1027}$

$= 1$

Factors and Exponents **27**

A Checking

5. In a recent contest for a Canadian company, the winner had 2 min to answer this skill-testing question over the telephone:

$(3 \times 50) + 20 \div 5$

a) Are the brackets necessary? Explain.

b) Calculate the answer.

c) Explain why it would be faster to do this question using mental math than using your calculator.

6. Which of these calculations will give the correct answer for $15 - 12 \div 3 \times 2 - 1$?

a) $(15 - 12) \div 3 \times 2 - 1$

b) $15 - 12 \div 3 \times (2 - 1)$

c) $15 - 12 \div (3 \times 2 - 1)$

d) $15 - (12 \div 3) \times 2 - 1$

e) $(15 - 12 \div 3) \times 2 - 1$

7. Calculate.

a) $8^2 - 15$ **d)** $(5 + 3) + 2^2$

b) $10 + 5^2$ **e)** $7 - (4 - 2^2)$

c) $16 - 4^2 + 20$ **f)** $16 \div 8 \times 2 + 1$

B Practising

8. Determine whether or not each calculation is correct. Show your work.

a) $12 \times 9 + 3 = 111$

b) $12 + 9 \times 3 = 63$

c) $5^2 + 5 - 4 \div 2 = 13$

d) $(5 + 2^2 \times 3)^2 + 1 = 290$

e) $6 + 5^2 - 4 \times 3 = 19$

f) $6 \times 9 - 10 \div 2 + 1 \times 3 = 52$

9. Use the formula Ravi and Ryan found and your height and age to estimate your lung capacity.

10. a) Explain why the word instructions below do NOT match the numerical expression.

> Square 6, add 5 to this number, and divide the sum by 2.

> $6^2 + 5 \div 2$

b) Use brackets so that the numerical expression matches the words.

c) What is the correct answer for the word instructions?

11. Explain what errors were made in these calculations. Rework them to show the corrections.

a) $4^2 - 5 \times 3$
$16 - 5 \times 3$
$11 \times 3 = 33$

b) $20 \div 10 + 2 \times 3^2 + 2$
$2 + 2 \times 3^2 + 2$
$4 \times 3^2 + 2$
$4 \times 9 + 2$
$36 + 2 = 38$

c) $50 - 5 \div 5 - 3 \times 2 + 6$
$45 \div 5 - 3 \times 2 + 6$
$9 - 3 \times 2 + 6$
$6 \times 2 + 6$
$12 + 6 = 18$

12. Evaluate each expression. Show all your work.

a) $3 \times 5 + 10^3 \times 3$

b) $10 + (3^2 - 1) \div 2 - 4$

c) $4 + 3 \times 6 \times (4 + 3^3)^2$

d) $5^3 + 4 \div 2 - 5 \times 6$

e) $4 \times (3 - 1 \times 3) + 8 \div 2^3$

13. Which expressions do not need brackets? Explain your reasoning. Calculate each answer.

a) $(3 \times 5) \times 5$

b) $4 + (3^2 + 5) \times 3$

c) $(4^2 \times 3) \div 2 + 1$

d) $(3 + 5^2)^2 \times 3$

e) $100 \div 10^2 \times 3 + (2 - 1)$

14. The boiling point of water is 100°C (Celsius) or 212°F (Fahrenheit). Try calculating the Celsius temperature using each formula below. Which formula is the correct one for converting Fahrenheit temperatures to Celsius? Show your work.

$C = 5 \times F - 32 \div 9$
$C = 5 \times (F - 32) \div 9$

15. Suppose that you were in charge of designing a skill-testing question for a lottery.

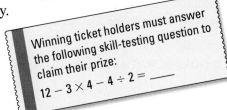

Winning ticket holders must answer the following skill-testing question to claim their prize:

$12 - 3 \times 4 - 4 \div 2 = $ _____

a) Where would you place brackets to make sure that the lottery winner gets the correct answer according to the rules for order of operations?

b) What other possible answers might people come up with for this question, if they did not know the correct order of operations?

16. Show how you could get two different answers when evaluating each expression, if there were no rules for order of operations.

a) $125 \div 5 \div 5$ **b)** $12 - 9 - 2$

ⓒ Extending

17. Write an expression that matches each instruction.

a) Add 5 to 8, square this number, and multiply by 6.

b) Add 5 to 8 squared, multiply by 6, and subtract 3.

c) Subtract 2 from 10, multiply this number by 4, and divide by 2.

d) Divide 10 by 2, multiply this number by 3, and subtract 1.

18. Explain how you can evaluate an expression with several brackets, such as $((3 + 2)^2)^2$.

19. Using BEDMAS, square roots are calculated at the same time as exponents. Explain how each expression can be evaluated.

a) $\sqrt{7 + 2 \times 3^2}$

b) $2 + \sqrt{2 + 4 \div 2} \times 3$

c) $\sqrt{3^2 + 4^2} \times 2$

20. Each expression has four 4s and equals 1.

$4 + 4 \div 4 - 4$
$(4 + 4) \div (4 + 4)$
$44 \div 44$

a) Evaluate the expressions to show that they all have a value of 1.

b) Make new expressions that equal each whole number from 2 to 10. For example, $4 \div 4 + 4 \div 4 = 2$.
- You may combine digits.
- You can multiply, divide, add, and subtract, and use powers, square roots, and brackets.
- There may be more than one solution.
- You must use all four 4s in each expression.

21. Copy each statement. Write operation signs in the boxes to make each statement true. Add brackets if required.

a) $12 \ \blacksquare \ 3 \ \blacksquare \ 2 = 18$

b) $100 \ \blacksquare \ (3^2 \ \blacksquare \ 1) = 10$

c) $(3^2 \ \blacksquare \ 1) \ \blacksquare \ 2 \ \blacksquare \ 1 = 6$

d) $2 \ \blacksquare \ (\sqrt{4} \ \blacksquare \ 1) \ \blacksquare \ 1 = 5$

22. A contractor charges a $500 flat fee, plus $25 per square metre, to install carpet in a 10 m by 10 m classroom.

a) Write a numerical expression showing how much it will cost to carpet the classroom.

b) Evaluate your expression to calculate the total cost.

Factors and Exponents **29**

1.8 Solve Problems by Using Power Patterns

▶ **GOAL**

Use patterns to solve problems with powers.

Learn about the Math

Yuki noticed that the last digit of some powers follows a pattern. She wondered if she could predict the last digit of 2^{41} without using a calculator.

? **How can you determine the last digit of a power without using a calculator?**

① Understand the Problem

Read the problem, and restate it in your own words. Discuss your ideas with someone else and revise, if necessary.

Yuki says, "The problem is to determine the last digit of a large power. I don't need to know the complete number."

② Make a Plan

Look at the information you have, and think about what it means. Try to reorganize the information so that it will help you solve the problem.

Yuki decides, "A useful strategy is looking for a pattern. I began with smaller powers of 2 and looked at the last digit."

$2^1 = 2$ $2^2 = 4$ $2^3 = 8$ $2^4 = 16$ $2^5 = 32$
$2^6 = 64$ $2^7 = 128$ $2^8 = 256$

I think that the last digit repeats: 2, 4, 8, 6, 2, 4, 8, 6, ... The pattern might be easier to see if I arrange the powers of 2 in 4 columns."

Yuki's table looks like this:

Column 1	Column 2	Column 3	Column 4
$2^1 = 2$	$2^2 = 4$	$2^3 = 8$	$2^4 = 16$
$2^5 = 32$	$2^6 = 64$	$2^7 = 128$	$2^8 = 256$
...	...	...	...

She says, "If I keep writing the pattern, 2^9 will be in Column 1, 2^{10} in Column 2, and so on.

Column 1 shows powers with these exponents: 1, 5, 9, …
If I keep adding 4 to the exponent number, I'll get this pattern of exponents: 1, 5, 9, 13, 17, 21, 25, 29, 33, 37, 41, …

So, 2^{41} is in Column 1, which means that the last digit must be **2**."

Yuki says, "I checked each calculation when I made the table. It seems that my answer is correct."

Reflecting

1. a) Could Yuki use a calculator to determine the last digit of 2^{41}?

b) How did she know that 2^{40} would be in Column 4? Explain.

2. When using a pattern to solve a problem, what are some benefits of making a table as Yuki did? Explain.

3. a) Explain how you know that when the last digit of a power of 2 is 4, the next power of 2 will end in an 8?

b) If the last digit of a power of 2 is 6, what will be the last digit of the next power of 2?

4. a) What is the last digit of 100^{10}? Explain how you know.

b) What is the last digit of 111^{56}? Explain how you know.

Factors and Exponents **31**

Example: Using a pattern to solve a problem

Use the problem solving steps to predict the last digit of 3^{31}.

Yuki's Solution

1 Understand the Problem

I can make a pattern like I did in the last example.

2 Make a Plan

I'll determine the column that 3^{31} will be in, and use the last digit.

3 Carry Out the Plan

$3^1 =$ 3	$3^2 =$ 9	$3^3 =$ 27	$3^4 =$ 81
$3^5 =$ 243	$3^6 = 729$	$3^7 = 2187$	$3^8 = 6561$

If I continue the pattern, 3^{31} will be in the same column as 3^3. This is the third column, so the last digit is a 7.

4 Look Back I checked my work to make sure that I did it right.

Ⓐ Checking

5. Use the pattern in Yuki's table to determine the last digit of each power of 2.

a) 2^{20} **b)** 2^{30} **c)** 2^{25} **d)** 2^{100}

Ⓑ Practising

6. Determine the last digit of each power by using a strategy like Yuki's.

a) 3^{20} **b)** 4^{30} **c)** 6^{25} **d)** 7^{100}

7. Nathan calculated squares of numbers that end in 5.

a) What do you notice about the last two digits?

b) Describe how to use a number pattern to predict the value of 65^2.

$15^2 = 225$
$25^2 = 625$
$35^2 = 1225$
$45^2 = 2025$
$55^2 = 3025$

c) Continue the pattern to calculate 75^2, 85^2, and 95^2. Show your method.

8. Write three powers of 2 that are greater than 2^{100} and have a last digit of 4. Explain your thinking.

9. All the numbers from 1 to 99 are multiplied together. Use a pattern to determine the last digit of the product. Justify your answer.

10. a) Calculate each sum.
$1^2 = ?$
$1^2 + 2^2 = ?$
$1^2 + 2^2 + 3^2 = ?$
$1^2 + 2^2 + 3^2 + 4^2 = ?$
$1^2 + 2^2 + 3^2 + 4^2 + 5^2 = ?$

b) Describe the pattern.

c) Explain how you can use the pattern to predict the following sum:
$1^2 + 2^2 + 3^2 + 4^2 + 5^2 + 6^2 = ?$

Chapter Self-Test

1. When Katya factored 288, she listed these factors: 1, 2, 3, 6, 8, 9, 12, 16, 18, 32, 36, 48, 72, 96, and 288. Use a factor rainbow to decide which three factors she missed.

2. Determine all the common factors and two common multiples of 12 and 56. Show your work.

3. Anthony plans to ride 175 km on his bike, and Samantha plans to ride 250 km. They both plan to complete their journeys in a whole number of days.

   a) How many kilometres could each person travel if they both ride the same number of kilometres each day?

   b) How many days will each person have to ride for each common factor you found in part (a)? Explain.

4. Asha plans to place vacation photos of her friends on a square bulletin board measuring 72 units by 72 units.

   a) Which size of photo can be placed without any spaces or overlapping? Show your work.

   b) How many photos of that size can be placed on the bulletin board? Sketch your answer if this helps you think about the problem.

5. a) Which of these numbers are perfect squares? Why?
   i) 1   **ii)** 24   **iii)** 225   **iv)** 500

   b) Use the factor rainbow to determine if 289 is a perfect square. Explain your reasoning.

   1 17   17 289

6. a) Write a power to represent three 4s multiplied together.

   b) Calculate the power.

   c) Explain how to use your answer to part (b) to calculate 4^4.

7. The number 6^{12} is equal to 2 176 782 336. Explain how to use this answer to calculate each power.

   a) 6^{13} b) 6^{11}

8. a) If 8 is a factor of a number, explain why 2 and 4 must also be factors of the number.

   b) Explain why 8 is a factor of 8^{10}.

9. Use mental math to calculate each square root.

   a) $\sqrt{25}$ b) $\sqrt{81}$ c) $\sqrt{100}$

10. Calculate the dimensions of a square with each of these areas.

    a) 441 m²   c) 900 cm²
    b) 121 m²   d) 10 000 cm²

11. Evaluate each expression.

    a) $4 \times (5 - 2) - 2^3$

    b) $12 + 2^2 - (1 + 2) \times 3$

Chapter Review

Frequently Asked Questions

Q: What is the square root of a number?

A: It is the number that, when multiplied by itself ("squared"), equals the original number. For example, 7 is the square root of 49 because $7^2 = 7 \times 7$ or 49. The square root of 49 is written as $\sqrt{49}$.

Q: Why are numbers like 36, 49, and 64 perfect squares?

A: Perfect squares are the areas of squares with whole-number side lengths. When you take the square root of a number and the result is a whole number, the original number is called a perfect square.

Q: What is a power?

A: A power is a short way of writing repeated multiplication. For example, $4 \times 4 \times 4 \times 4 \times 4 \times 4 \times 4 \times 4 \times 4$ can be written as 4^9 because there are nine 4s multiplied together.

Q: What are the rules for order of operations?

A: These are rules for calculating with mixed operations so that every person will get the same result.

Step 1: Perform operations in **B**rackets first.
Step 2: Calculate **E**xponents (including square roots) next.
Step 3: **D**ivide and **M**ultiply from left to right.
Step 4: **A**dd and **S**ubtract from left to right.

Some people use the memory aid (or *mnemonic*) BEDMAS to remember the rules.

For example, you can follow these steps to calculate $2 + (9 - 3)^2 + 8 \div 2$:

$$2 + (9 - 3)^2 + 8 \div 2 \qquad \text{Perform operations in \textbf{B}rackets first.}$$
$$= 2 + 6^2 + 8 \div 2 \qquad \text{Then calculate \textbf{E}xponents (powers).}$$
$$= 2 + 36 + 8 \div 2 \qquad \text{\textbf{D}ivide next.}$$
$$= 2 + 36 + 4 \qquad \text{\textbf{A}dd the first two numbers from left to right.}$$
$$= 38 + 4 \qquad \text{\textbf{A}dd.}$$
$$= 42$$

Practice Questions

(1.1) **1.** Explain how to determine two numbers whose common multiple is 120.

(1.1) **2. a)** If Kyle has only $2 coins, $5 bills, and $20 bills, what is the least amount he can pay using only $2s, only $5s, and only $20s? What is the LCM of 2, 5, and 20?

b) What is the LCM of 2, 3, and 9?

(1.3) **3.** Show at least one method you can use to list the common factors of 128 and 192.

(1.3) **4.** The number 272 is a multiple of 16. Explain how to use this information to determine the GCF of 16 and 272.

(1.3) **5. a)** Draw a factor rainbow for each number.
    i) 4   **ii)** 9   **iii)** 16   **iv)** 25   **v)** 36

b) Write another number whose factor rainbow will have a similar shape to those in part (a). Explain your thinking.

(1.4) **6.** Find a number that has these properties. Show your work or give a reason for your answer.

a) a number that is divisible by 2 and 3 and is greater than 100

b) a number that has only 2 factors and is between 90 and 100

c) a number that can be divided by 2 three times in a row with no remainder, has 5 as a factor, and is greater than 40

(1.5) **7. a)** Calculate 2^5 without using a calculator.

b) Explain how you know that 2^{10} is twice 2^9.

(1.5) **8. a)** If 9 is a factor of a number, explain why 3 is also a factor.

b) Explain why 3 is a factor of 3^{10}.

c) The number 3^{12} is equal to 531 441. Explain how to use this answer to calculate 3^{11}.

9. On Monday, Zach sends an e-mail message to 4 friends. On Tuesday, each of these friends forwards the message to 4 people. On Wednesday, each of these people forwards the message to 4 other people. (1.5)

a) How many people were sent the message on Wednesday? Explain how you know.

b) Show how to use powers to describe the number of messages sent on Wednesday.

c) How many people will be sent the message on Sunday if this daily process of forwarding continues? Explain your reasoning.

10. Winnie used her calculator to calculate the square root of a number. She then found the square root of the number displayed. The calculator then showed 3. What number did Winnie first enter into her calculator? Explain your reasoning. (1.6)

11. A rectangular ice rink measures 25 m by 64 m. What are the dimensions of a square with the same area? Show the steps you used. (1.6)

12. Use the rules for order of operations to evaluate each expression. Show your steps. (1.7)

a) $2 \times 5 + 5 \div 1 + 10^2$

b) $5 + (5 - 1)^2 + 5 \times 3$

c) $2^2 + 120 \div 2 \div 2 \div 2$

13. What is the last digit of 445^{12}? Explain how you know. (1.8)

Chapter Task

Designing Interesting Numbers

Sports players have numbers on their jerseys. For example, Wayne Gretzky wore the number 99. When he retired from playing hockey on April 18, 1999, the National Hockey League (NHL) also retired his jersey number. This means that no other NHL player will be allowed to use the number 99 in the future.

A. Write at least two number sentences to describe the number 99 using brackets, exponents, square roots, and any or all of the four operations.

B. What is your favourite number? Describe at least three properties of the number using words such as factor, multiple, square, square root, prime, and divisible.

C. Pick a number between 1 and 98. Have a timed contest to see who can come up with the most number sentences or word descriptions to describe this number.
- What strategies might you use to try to win the contest?
- What number between 1 and 98 has the most properties?

D. Design a jersey number that has interesting number properties. Show the number on the front of the jersey. Show its properties on the back.

CHAPTER 2

Ratio, Rate, and Percent

▶ GOALS

You will be able to

- compare and order decimals and percents
- relate fractions, ratios, decimals, and percents
- solve problems that involve ratios, rates, and percents
- multiply and divide decimal numbers

You will need
- centimetre grid paper
- a calculator

Making Number Comparisons

In the fall, some Canadian farmers build mazes in their cornfields to attract tourists and earn extra money. The grid shows a plan for a simple maze in a square cornfield. The **ratio** of green squares (with corn) to white squares (paths without corn) is 58:42.

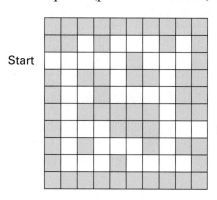

Start

Finish

? **How can you use ratios, fractions, decimals, and percents to describe a path through the maze?**

A. What fraction of the square cornfield is used for paths in the maze? Write the decimal that is **equivalent** to this fraction.

B. What **percent** of the cornfield is used for paths in the maze?

C. What fraction of the cornfield is used for corn? Write the decimal that is equivalent to this fraction.

D. What percent of the cornfield is used for corn?

E. Copy the maze and find a path from Start to Finish.

F. Count the white squares on your path. Use ratios, fractions, decimals, and percents to express the area of your path as part of the total area of the cornfield.

G. Explain why the areas of the paths are easily expressed as ratios, fractions, decimals, and percents of the cornfield's total area.

Do You Remember?

1. The number 100 has factors 2 and 50, since $2 \times 50 = 100$. List all the other factor pairs of 100.

2. Yan buys a candy bracelet that has 24 candy beads. Write each ratio, based on the picture of the bracelet.

  a) the number of red beads to the number of blue beads

  b) the number of green beads to the number of red beads

  c) the number of yellow beads to the number of white beads

  d) the number of green beads to the total number of beads

  e) the total number of beads to the number of green beads

3. Which ratios are equivalent to $4:5$?

  16:20   2:3   9:10

  12:15   90:100   32:40

4. Express each shaded area as a fraction of the area of the whole shape.

  a)    **c)**

  b)   **d)**

5. Copy the following number line, and mark each number on your number line.

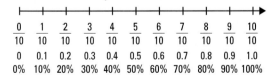

  a) 0.45   **c)** 0.82   **e)** 0.25

  b) 35%   **d)** 25%   **f)** $\dfrac{1}{4}$

6. Look at the numbers you marked on your number line in question 5. Which numbers are equivalent?

7. For each shape, write the following.

  a) the ratio that compares the number of shaded squares to the number of unshaded squares

  b) the fraction that expresses the area of the shaded squares as part of the area of the whole shape

  c) the decimal and percent that are equivalent to this fraction

  i) 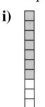   **ii)**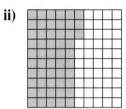

8. Use mental math to multiply.

  a) 235×0.1 **c)** 876×0.01

  b) 235×0.01 **d)** 876×0.001

9. Multiply. Check by using a calculator.

  a) 9×11 **c)** 4×52

  b) 9×0.11 **d)** 4×5.2

10. Guess and test to determine the missing number.

  a) $2 \times \blacksquare = 4$ **d)** $100 \times \blacksquare = 160$

  b) $2 \times \blacksquare = 3$ **e)** $473 \times \blacksquare = 4.73$

  c) $20 \times \blacksquare = 30$ **f)** $111 \times \blacksquare = 1.11$

11. Divide. Check by using a calculator.

  a) $67 \div 10 = \blacksquare$ **d)** $67 \div \blacksquare = 6.7$

  b) $670 \div 100 = \blacksquare$ **e)** $200 \div \blacksquare = 2.00$

  c) $30 \div \blacksquare = 3.0$ **f)** $473 \div \blacksquare = 4.73$

12. Divide. Check by using a calculator.

  a) $12 \div 12 = \blacksquare$ **c)** $5 \div \blacksquare = 1$

  b) $2.5 \div 2.5 = \blacksquare$ **d)** $5.5 \div \blacksquare = 1$

Exploring Ratio Relationships

You will need
• centimetre grid paper
• a ruler
• a red pencil

▶ **GOAL**

Explore equivalent ratios.

Explore the Math

Fawn drew a red rectangle like this one on centimetre grid paper. She wants to create a series of **similar rectangles**.

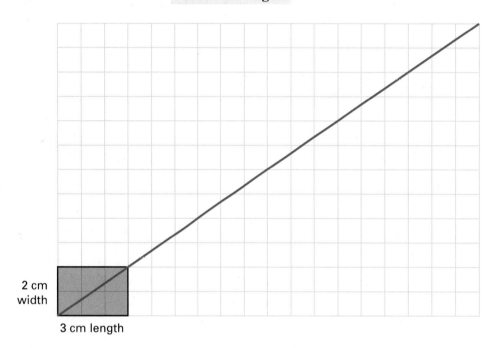

2 cm width

3 cm length

❓ How do you draw similar rectangles?

A. Use centimetre grid paper to draw a rectangle exactly like Fawn's rectangle. Use a ruler to draw a diagonal from the bottom left corner to the top right corner of your rectangle. Extend the diagonal as far as you can on the grid paper.

B. Create a new rectangle by adding 2 cm to the width and 3 cm to the length of your original rectangle. Start this rectangle at the bottom left of the first rectangle so that the two rectangles overlap.

C. Repeat step B to create four more rectangles. Each time add 2 cm to the width and 3 cm to the length of the previous rectangle. You will have six rectangles in total.

similar rectangles

rectangles that have the same shape, but not necessarily the same size

D. Copy the following table. Use your rectangles to complete the table.

	First Group of Rectangles					
	1	**2**	**3**	**4**	**5**	**6**
Width (cm)	2					
Length (cm)	3					
Width : Length						

E. On a new piece of grid paper, draw another rectangle that has a width of 2 cm and a length of 3 cm. Use a ruler to draw a diagonal from the bottom left corner to the top right corner of this rectangle. Extend the diagonal as you did in step A. Create five new rectangles by adding 3 cm to the width and 3 cm to the length each time.

F. Copy the following table. Use your rectangles from step E to complete the table.

	Second Group of Rectangles					
	1	**2**	**3**	**4**	**5**	**6**
Width (cm)	2					
Length (cm)	3					
Width : Length						

G. Choose a length and a width for a rectangle. Draw the rectangle on grid paper. Draw and extend the diagonal as in step A. Draw five rectangles that are similar to your rectangle.

Reflecting

1. Which groups of rectangles contain similar rectangles? How do you know?

2. In which groups of rectangles are the ratios of width to length equivalent? What is the connection between **ratios** of sides and whether rectangles are similar?

3. In step E, did the diagonal pass through each rectangle's top right corner? Explain why or why not.

4. If you start with any rectangle, how can you create a rectangle that is similar to it?

ratio

a comparison of two or more quantities that are measured in the same units; for example, the heights of two trees are 17 m and 28 m. The ratio can be written as 17:28 or $\frac{17}{28}$

2.2 Solving Ratio Problems

▶ **GOAL**

Compare quantities using ratios, and determine equivalent ratios.

Learn about the Math

Snowboarding is the fastest growing winter sport in Canada. A recent estimate of the Canadian male-to-female participant ratio is $3:1$. Romona's school is going on a ski trip, and 24 boys signed up to go snowboarding.

? **How many girls signed up to go snowboarding if the numbers match the Canadian ratio?**

equivalent ratios

two or more ratios that represent the same comparison; for example, $1:3$, $2:6$, and $3:9$

You can write **equivalent ratios** by multiplying or dividing each **term** in a ratio by the same number. For example,

$$\overset{\times 2}{1:3} = 2:6 \quad \text{or} \quad \overset{\times 2}{\frac{1}{3}} = \frac{2}{6} \qquad \overset{\div 4}{16:12} = 4:3 \quad \text{or} \quad \overset{\div 4}{\frac{16}{12}} = \frac{4}{3}$$
$$\underset{\times 2}{} \qquad \underset{\times 2}{} \qquad \underset{\div 4}{} \qquad \underset{\div 4}{}$$

term

the number that represents a quantity in a ratio

Example 1: Multiplying to write equivalent ratios

Determine the number of girls who signed up to go snowboarding.

Romona's Solution

$$\frac{3}{1} = \frac{15}{5}$$
$$= \frac{18}{6}$$
$$= \frac{21}{7}$$
$$= \frac{24}{8}$$

The ratio of male to female snowboarders is $3:1$ or $\frac{3}{1}$. First I multiplied each term by 5. Then I continued to write ratios that are equivalent to $\frac{3}{1}$ until I found the ratio with the term for 24 boys.

From this ratio, I can see that 8 girls signed up to go snowboarding.

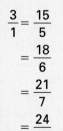

Calculate the number of girls who signed up to go snowboarding.

Paul's Solution

$3:1 = 24:\blacksquare$

$\times 8$

$3:1 = 24:\blacksquare$

$\times 8$

$3:1 = 24:8$

$\times 8$

First I wrote the Canadian ratio of boys to girls, which is $3:1$. I want to find out the number of girls when there are 24 boys, so I wrote the second ratio as $24:\blacksquare$. The ratios must be equivalent.

Since $3 \times 8 = 24$, I multiplied by 8 to get the missing term in the second ratio.

The number of girls who signed up to go snowboarding must be $1 \times 8 = 8$.

Communication Tip

Ratios can compare the pieces in this diagram in different ways: part to whole, whole to part, or part to part.

- **part to whole**

  The number of red pieces to the total number of pieces is 3 to 8, $\frac{3}{8}$, or $3:8$.

  The number of blue pieces to the total number of pieces is 5 to 8, $\frac{5}{8}$, or $5:8$.

- **whole to part**

  The total number of pieces to the number of red pieces is 8 to 3, $\frac{8}{3}$, or $8:3$.

  The total number of pieces to the number of blue pieces is 8 to 5, $\frac{8}{5}$, or $8:5$.

- **part to part**

  The number of red pieces to the number of blue pieces is 3 to 5, $\frac{3}{5}$, or $3:5$.

  The number of blue pieces to the number of red pieces is 5 to 3, $\frac{5}{3}$, or $5:3$.

Reflecting

1. How did Paul decide what **scale factor** to use to solve his **proportion**?

2. a) How can you create a ratio that is equivalent to another ratio?

   b) How can you determine whether two given ratios are equivalent?

3. Suppose that you have a proportion in which one term is missing from one of the ratios. How can you determine the value of the missing term?

4. a) How are ratios the same as fractions?

   b) How are ratios different from fractions?

scale factor

a number that you can multiply or divide each term in a ratio by to get the equivalent terms in another ratio; it can be a whole number or a decimal

$\div 4$

$\dfrac{16}{12} = \dfrac{4}{3}$ or $\dfrac{16 \div 4}{12 \div 4} = \dfrac{4}{3}$

$\div 4$

proportion

a number sentence that shows two equivalent ratios; for example,

$1:3 = 4:12$ or $\dfrac{1}{3} = \dfrac{4}{12}$

Example 3: Dividing and multiplying to write equivalent ratios

Determine the missing term in each proportion.

a) $18:9 = 8:\blacksquare$

b) $\dfrac{15}{25} = \dfrac{18}{\blacksquare}$

Solution

Divide the terms in the first ratio to write a simpler ratio.
Then multiply to calculate the value of the missing term.

a) $18:9 = 2:1$ and $2:1 = 8:4$
$\div 9 \qquad \times 4$
$\div 9 \qquad \times 4$

b) $\dfrac{15}{25} = \dfrac{3}{5}$ and $\dfrac{3}{5} = \dfrac{18}{30}$
$\div 5 \qquad \times 6$
$\div 5 \qquad \times 6$

The missing term is 4.   The missing term is 30.

A Checking

5. Use the ratios represented in each diagram.

   a) Write the two ratios as a proportion.

   b) Determine the scale factor that relates the two ratios.

   c) Calculate the missing term.

i)

2 oranges	5 apples
4 oranges	▦ apples

ii)

4 triangles	▦ squares
3 triangles	9 squares

iii)

8 stars	2 stars
▦ bells	5 bells

iv)

3 women	▦ women
5 men	15 men

6. Calculate each missing term.

   a) $\dfrac{3}{8} = \dfrac{\blacksquare}{16}$ **b)** $\dfrac{32}{24} = \dfrac{20}{\blacksquare}$

B Practising

7. Copy and shade the grid on the right so that the ratio of shaded squares to total squares is the same as the ratio in the grid on the left. Then write the proportion.

   a)

   b)

8. Write three equivalent ratios for each of the following ratios.

   a) 21 to 56   **c)** 48:36

   b) 6:54   **d)** $\dfrac{22}{55}$

9. The ratio of the number of orange sections to the total number of sections is the same for all three diagrams. Explain why.

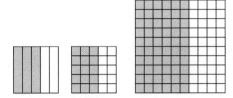

10. Determine the missing term in each proportion.

a) $\dfrac{27}{45} = \dfrac{\blacksquare}{5}$ **b)** $\dfrac{2}{6} = \dfrac{3}{\blacksquare}$

11. Measure the length or height of the animal in each scale drawing. Use your measurement and the given scale to calculate the actual length or height of each animal.

a) scale 1 : 30

b) scale 1 : 100

c) scale 2 : 1

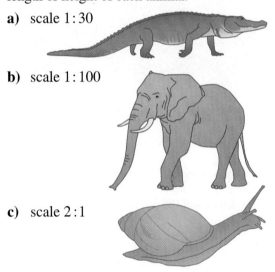

12. The height of the CN Tower is about 550 m. What is the height of the CN Tower in a scale drawing if the scale is 1 : 11 000?

13. Write each comparison as a ratio. The units for the terms must be the same.

a) 400 g to 1 kg

b) 6 cm to 7 mm

c) 200 s to 3 min

14. Katherine spends 7 h a day in school. This includes a 30 min lunch break each day.

a) Write a ratio that compares the time for her lunch break with the total time she spends at school each day.

b) Write a ratio that compares the time for her lunch breaks for a week with the total time she spends at school each week.

c) Determine the number of hours of lunch break she has in a month of school days.

15. For the numbers 2 to 100, including 2 and 100, determine each ratio.

a) the number of even numbers to the number of odd numbers

b) the number of multiples of 6 to the number of multiples of 8

c) the number of prime numbers to the number of composite numbers

16. Todd knows that the ratio of boys to girls in his class is 3 : 5. Since 12 of the students are boys, he says there must be 36 students in his class. Is he right? Explain your thinking.

17. Luca is 9 years old and 129 cm tall. Medical charts show that a boy's height at age 9 is $\dfrac{3}{4}$ of his predicted adult height. Predict Luca's adult height.

⒞ Extending

18. At Darlene's Dairy Bar, the ratio of vanilla to chocolate to strawberry ice cream cones sold is about 3 : 5 : 2.

a) Last week, 155 chocolate ice cream cones were sold. What was the total number of ice cream cones sold?

b) Two weeks earlier, a total of 380 ice cream cones were sold. How many of each flavour were sold?

2.3 Solving Rate Problems

▶ **GOAL**

Determining equivalent rates to solve rate problems.

Learn about the Math

Peter has started a new part-time job making pizza at Pizza Shack. For every 3 h he works, he earns $27.00.

? **How can you calculate Peter's earnings if he works 24 h in a week?**

A. Write the **rate** for Peter's earnings in 3 h, using 3 h as the second term.

B. Write a rate with a missing term for Peter's earnings in 24 h.

C. The rates in steps A and B are **equivalent rates**. Write a proportion and solve it.

Communication Tip

Rates are often written with a slash (/). For example, the maximum speed on a highway is 100 km/h. The slash is read as "per," meaning "for each" or "for every." The rate of 100 km/h means that, on average, you are travelling a distance of 100 km for every hour you are driving.

---

rate

a comparison of two quantities measured in different types of units; unlike ratios, rates include units

equivalent rates

rates that represent the same comparison; for example, $\frac{90 \text{ km}}{2 \text{ h}}$ and 45 km/h

Reflecting

1. a) Explain why the rates in steps A and B are equivalent.

   b) Explain how you know that Peter's rate of pay is $9/h.

2. Why is it important to include the units in a rate?

3. a) How are a rate and a ratio the same?

   b) How are they different?

4. How is solving a rate problem like solving a ratio problem?

Work with the Math

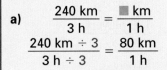
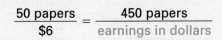
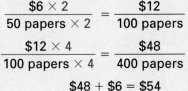
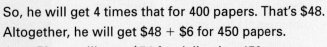

A Checking

5. Write each comparison as a rate.

   a) There was 15 mm of rain over 3 days.

   b) Four chocolate bars were on sale for $2.20.

   c) Indu saves $14.00 every week.

   d) Philip's height changed by 12 cm over 4 months.

6. Write two equivalent rates for each comparison.

   a) 5 goals in 10 games

   b) 10 km jogged in 60 min

   c) 6 pizzas eaten in 30 min

   d) 10 penalties in 25 games

B Practising

7. On a hike, Peter walked 28 km in 7 h.

   a) What was his average rate of walking?

   b) Suppose that he walked the 28 km in 8 h. What would his average rate of walking be?

8. Write a proportion for each situation. Determine the missing term in each proportion.

   a) Three trucks have 54 wheels. Six trucks have ▢ wheels.

   b) In 5 h, you drive 400 km. In 1 h, you can drive ▢ km.

   c) In 1 h, you earn $10. In 8 h, you can earn $▢.

   d) Six boxes contain 72 donuts. One box contains ▢ donuts.

   e) In 4 min, you can type 128 words. In 2 min, you can type ▢ words.

   f) Six pencils that cost $0.72 is equivalent to three pencils that cost ▢.

9. Winnie's baseball coach orders 4 pizzas for the 10 players on her team. She needs to determine the number of pizzas to order for the league's year-end party that 120 players are expected to attend. Which proportion does *not* model this situation?

   A. $\dfrac{4 \text{ pizzas}}{10 \text{ players}} = \dfrac{▢ \text{ pizzas}}{120 \text{ players}}$

   B. $\dfrac{4 \text{ pizzas}}{10 \text{ players}} = \dfrac{120 \text{ players}}{▢ \text{ pizzas}}$

   C. $\dfrac{10 \text{ players}}{4 \text{ pizzas}} = \dfrac{120 \text{ players}}{▢ \text{ pizzas}}$

   D. $\dfrac{2 \text{ pizzas}}{5 \text{ players}} = \dfrac{▢ \text{ pizzas}}{120 \text{ players}}$

10. Takumi bikes 45 km in 3 h. If he continues at the same rate, how far will he bike in 4 h?

11. Brad pays $56 for four CDs. At this rate, how many CDs can he buy with $42?

12. Tony works at a gas station. He serves 12 cars in 20 min. At this rate, how many cars can he serve in 25 min?

13. Anita earns $72 every 6 h to fix bicycles. How much money will she earn in 4 h if she is paid at the same rate?

14. Craig jogs 12 km in 1 h. If he jogs at the same rate, how long will it take him to jog 18 km?

15. If 6 kg of oranges costs $15, how many kilograms of oranges can you buy for $20?

16. The goalie for the Eagles stopped 8 shots out of every 10 shots.

a) How many shots can the Eagles goalie expect to stop if the Hawks get about 15 shots in every game?

b) The Hawks play the Eagles five times during the season. If the Hawks' rate of shots per game continues, how many shots can they expect to get against the Eagles during the season?

c) How many goals can the Hawks expect to get against the Eagles during the season?

17. Explain the difference between a ratio and a rate. Provide two examples in your explanation: a situation in which a ratio would be used and another situation in which a rate would be used.

18. Jason's mom drove at 100 km/h for a 160 km trip. How much longer would the trip have taken if she had driven at 90 km/h?

19. A grey whale's heart beats 24 times in 3 min. At this rate, how many times does it beat in a day?

C Extending

20. Dan's computer reads 68% of battery left after 1 h. If the rate of using power remains the same, about how much longer will his battery provide power?

21. Fuel efficiency is a rate that represents the number of litres of fuel required for the driving distance of 100 km. For example, the fuel efficiency for a certain model of car is $\frac{9.5\ L}{100\ km}$. To compare the fuel efficiency of different cars, you need to determine the number of litres of fuel used per 100 km for each car.

Model	Kilometres driven (km)	Litres of fuel used (L)
Chevrolet Corvette	180	25.00
Honda Civic	275	22.36
Honda Accord	126	11.45
Ferrari F40	309	40.66
Ford Escort	294	25.56
Ford Taurus	258	27.74
Porsche 911 Turbo	333	46.25

a) Which car uses the least fuel? Explain.

b) Which car uses the most fuel? Explain.

Ratio, Rate, and Percent

2.4 Communicating about Ratio and Rate Problems

▶ **GOAL**

Explain your thinking when solving ratio and rate problems.

Communicate about the Math

Alice works for a special effects company. She is building a scale model of the Lion's Gate Bridge in Vancouver, British Columbia, to be used for a movie. The distance between the vertical supports on the bridge is 473 m, and the towers are 111 m tall. She decides to make the distance between the vertical supports on the model about 5 m. What should the height of the towers be on the model?

? **How can Tien improve her solution to the problem?**

Tien's Explanation

From the actual bridge, I know the ratio
$$\frac{\text{tower height (m)}}{\text{distance between supports (m)}} = \frac{111}{473}.$$

For the model, I decided to make the distance between the supports 4.73 m.

For the model, the same ratio is
$$\frac{\text{tower height (m)}}{\text{distance between supports (m)}} = \frac{\blacksquare}{4.73}.$$

So $\frac{111}{473} = \frac{1.11}{4.73}$.

The height of the towers of the model must be 1.11 m, or 111 cm.

Nathan's Questions

Why did you write the ratio the way you did?

Why did you choose 4.73 m?

How did you calculate 1.11?

How do you know this is correct?

A. Which of Nathan's questions do you think are good questions? Why?

B. How should Tien respond to the questions?

C. What other questions do you think would be helpful?

D. Use the Communication Checklist to improve Tien's solution.

Reflecting

1. Which parts of the Communication Checklist did Tien cover well? Explain.

2. Why should you explain your thinking when solving problems?

Communication Checklist

☑ Did you identify the information given?

☑ Did you show each step of your calculation?

☑ Did you explain your thinking at each step?

☑ Did you check to see if your answer is reasonable?

Work with the Math

Example: Communicating about ratios

The Coyotes have won 8 of their first 20 soccer games. If this continues, how many games would you expect them to win out of 30 games? Explain your thinking.

Fawn's Solution

$$\frac{8 \div 4}{20 \div 4} = \frac{2}{5}$$

$$\frac{2 \times 6}{5 \times 6} = \frac{\blacksquare}{30}$$

Winning 8 out of 20 games is the same as winning 2 out of 5 games.

The team is playing 30 games. Since $5 \times 6 = 30$, the team should win $2 \times 6 = 12$ games out of 30 games.

The answer is correct because $\frac{8}{20} = \frac{2}{5}$ and $\frac{12}{30} = \frac{2}{5}$.

Also, you can use logical thinking to check the answer. If the team played half of 20, or 10, games, they would win half of 8, or 4, games. This means that they win 4 games out of every 10, or 12 games out of 30.

Miguel's Solution

$$\frac{8}{20} = \frac{\blacksquare}{30}$$

$$\frac{8}{20} = \frac{\blacksquare}{30}$$
$\times 1.5$

$\times 1.5$
$$\frac{8}{20} = \frac{12}{30}$$
$\times 1.5$

I know that the Coyotes have won 8 games out of 20 games, and there are 30 games in total.

I wrote a proportion with a missing term for the number of games they will win.

The scale factor is 1.5 because $20 \times 1.5 = 30$. So, I multiplied 8 by 1.5 to get the missing term.

The team should expect to win 12 games out of 30.

The answer is correct because the team has 10 more games to play, which is half of 20. They should win half of the number they've already won (8), which is 4.

A Checking

3. Marlene runs 4 km in 30 min. At this rate, can Marlene run 6 km in 45 min? Complete the following calculation and explanation. Use the Communication Checklist.

$$\frac{4 \text{ km}}{30 \text{ min}} = \frac{distance}{time}$$

I know that Marlene runs 4 km in 30 min. I can write a proportion to show this information.

B Practising

Use the Communication Checklist to help you answer each question.

4. Chris is having a party for 90 people. He has a punch recipe that makes 2 L of punch. This will serve 15 people. Chris thinks that he will need 12 L of punch for his party. Is Chris's reasoning correct? Explain.

5. Sam has a part-time job delivering flyers door to door. He earns a quarter for every 10 flyers he delivers. He wants to earn $45.00 to buy his mother a birthday present.

   Sam figures out that he can write a proportion to determine how many flyers he needs to deliver. He thinks, "I know that I earn $0.25 for 10 flyers, and I want to know how many flyers I need to deliver to earn $45.00."

   Complete Sam's explanation, and solve the problem.

6. Akeem measured the capacity of a glass as 270 mL and the capacity of a Thermos® as 1 L. He said that the ratio 270 : 1 compares the capacity of the glass with the capacity of the Thermos®. Is his reasoning correct? Explain.

7. A photocopier can make 1800 copies in 1 h. Rosa says that the photocopier can make 60 copies per minute. Is she correct? Explain why or why not.

8. For every 1.5 m that an iceberg rises above the water, there is 12 m of ice below the surface. If an iceberg rises 9 m above the water, is the height of the iceberg from top to bottom 72 m? Explain.

9. In a bulk-food store, peanuts in the shell are sold by their mass in grams. The rate is $0.44 per 100 g. In a grocery store, the price of a 400 g bag of peanuts in the shell is $2.20.

   Which would you buy to get the most peanuts for your money? Justify your choice.

10. These are the prices for two types of raisins in a bulk-food store.

    golden raisins ...... $0.66/100 g

    dark raisins .... $0.55/100 g

    a) A recipe calls for 500 g of raisins. How much will you save if you buy the less expensive raisins? Explain.

    b) Raj bought $3.63 worth of raisins. What was the mass in grams? Give two possible answers.

Did you know?

- It takes 21 kg of milk to produce 1 kg of butter.
- About 160 L of maple sap are needed to make 4 L of maple syrup.
- It takes 250 mL of uncooked rice to make 500 mL of cooked rice.
- It takes 2 kg of soybeans to make 5 kg of tofu.
- One cup of unpopped corn yields 36 cups of popped corn.

1. Write a proportion to describe each situation. Then answer the question.

   a) How much milk does it take to produce 25 kg of butter?

   b) How much maple sap is needed to make 60 L of maple syrup?

   c) How much uncooked rice is used to make 800 mL of cooked rice?

   d) What mass of soybeans is needed to make 3000 g of tofu?

   e) How much unpopped corn does it take to produce 72 cups of popped corn?

2. Write a proportion to describe each situation. Then answer the question.

   a) How much butter can be made with 6300 kg of milk?

   b) How much maple syrup can be made with 80 L of maple sap?

   c) How much cooked rice can be made with 1 L of uncooked rice?

   d) What mass of tofu can be made with 8000 g of soybeans?

   e) How much popped corn can be made with 8 cups of unpopped corn?

3. How are the proportions in questions 1 and 2 different? What does this show about the situations?

Frequently Asked Questions

Q: **What is the difference between a ratio and a rate?**

A: A ratio is a comparison of two quantities that are measured in the same units. A rate is a comparison of two quantities that are measured in different units.

An example of a ratio is a comparison of the heights of two students, measured in centimetres. A height comparison is 140 cm to 110 cm. Since both measurements are in centimetres, the ratio can be written as $140:110$ or $\frac{140}{110}$.

An example of a rate is the number of words a student can type in a certain number of minutes. A typing rate is $\frac{160 \text{ words}}{2 \text{ minutes}}$, or 80 words per minute.

Q: **What is a proportion?**

A: A proportion is an equation that shows two equivalent ratios or rates. Multiplying or dividing both terms in one ratio or rate by the same number produces an equivalent ratio or rate.

For example, $\frac{3}{4} = \frac{21}{28}$ is a proportion. Multiplying both terms in the first ratio by 7 results in the second ratio.

Q: **How do you solve a proportion with a missing term?**

A: You determine the missing term using what you know about equivalent ratios. For example, how do you solve the proportion $\frac{24}{48} = \frac{\blacksquare}{6}$?

Method 1
Since 48 divided by 8 is 6, the number you are looking for (the missing term) must be 24 divided by 8.
$$\frac{24 \div 8}{48 \div 8} = \frac{\blacksquare}{6}$$
The equivalent ratio is $\frac{3}{6}$. The missing term is 3.

Method 2
Since 24 is half of 48, the missing term must be half of 6, and $6 \div 2 = 3$. The missing term is 3.

Practice Questions

(2.2) **1.** Write three equivalent ratios for each ratio.

    a) 5 to 6   **b)** 28:42   **c)** $\dfrac{36}{45}$

(2.2) **2.** Write each comparison as a ratio. Remember that the units must be the same.

    a) 40 cm to 3 m

    b) 100 min to 2 h

    c) 50 g to 1 kg

(2.2) **3.** Calculate the missing term in each proportion.

    a) $\dfrac{3}{4} = \dfrac{\blacksquare}{16}$ **d)** $\dfrac{4}{12} = \dfrac{\blacksquare}{90}$

    b) $12:\blacksquare = 60:20$ **e)** $2:20 = \blacksquare:50$

    c) $\dfrac{30}{48} = \dfrac{20}{\blacksquare}$ **f)** $\dfrac{8}{24} = \dfrac{26}{\blacksquare}$

(2.2) **4.** Calculate the actual length of the boat using this scale drawing.

scale 1:1000

(2.2) **5.** On a sunny day, a streetlight that is 9 m high casts a shadow that is 6 m long. At the same time, a fence post casts a shadow that is 2 m long. How tall is the fence post?

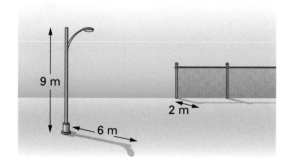

6. A quarterback on a professional football team completes, on average, 10 passes for every 15 passes he attempts. How many passes can he expect to complete if he attempts 525 passes in a season?   (2.2)

7. Determine whether each comparison is a ratio or a rate. Explain your thinking.

    a) Peter is 145 cm tall and Mei is 125 cm tall.

    b) 25 pens were on sale for $5.00.

    c) Kaj walked 6 km in 2 h.   (2.3)

8. Christine has a summer job cutting grass. She earns $400 every 2 weeks. How much will she earn over the summer holidays, which are 12 weeks long?   (2.3)

9. Rohan cycled 30 km in 2 h. How long will he take to cycle 45 km if he continues at the same rate?   (2.3)

10. 10 kg of potatoes cost $4.50. How much will 50 kg of potatoes cost at this rate? (2.3)

11. A one-week summer basketball camp promises 3 coaches for every 24 players. If 276 players register for the camp, how many coaches should the camp hire? Explain how you interpreted the calculation to decide.   (2.4)

Ratios as Percents

▶ **GOAL**
Solve problems that involve conversions between percents, fractions, and decimals.

Learn about the Math

Tynessa and Chang know that the ratio of male snowboarders to all snowboarders in Canada is 3:4. They want to express this relationship using a percent.

? **How do you rename a ratio as a percent?**

A. Does the snowboarding ratio, 3:4, compare a part to a whole or a part to another part? Explain.

B. Write a proportion that relates the ratio 3:4 to an equivalent ratio out of 100.

C. Solve the proportion to find the percent that is equivalent to 3:4.

D. What fraction represents the part of all snowboarders that is male?

E. What decimal represents the part of all snowboarders that is male?

Reflecting

1. How is relating a ratio to a percent the same as solving a proportion?

2. Why can you write the ratio of male snowboarders to all snowboarders as the fraction $\frac{3}{4}$?

3. How do you calculate the percent that corresponds to a given fraction?

4. How do you calculate the percent that corresponds to a given decimal for tenths or hundredths?

> **Communication Tip**
>
> When you write percent, think of "per hundred."
>
>
>
> A percent is used to compare a part to a whole; for example, 27% is a way of writing the ratio $\frac{27}{100}$.

Work with the Math

Example 1: Calculating a percent

A group of 20 Grade 7 students were surveyed about their favourite types of music. This table shows the results of the survey. What percent of students do not prefer rap?

Type of music	Number of students
country	1
rock	4
hip-hop	6
rap	9

Bonnie's Solution: Using a grid

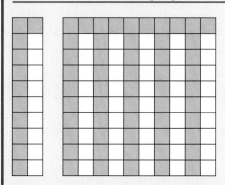

First I drew a grid of 20 squares, since 20 students were surveyed. I shaded 11 squares for the people who prefer country, rock, and hip-hop music.

Then I drew a grid of 100 squares. I figured out that 5 groups of 20 fits in my second grid. Each group of 20 has 11 squares shaded. I counted all the shaded squares and got 55.

$\frac{55}{100}$ = 55%, so this is the percent of students who do not prefer rap music.

Miguel's Solution: Using a proportion

$\frac{11}{20} = \frac{\blacksquare}{100}$

I added up the number of students who do not prefer rap and wrote the ratio of this number to the total number surveyed.

I know that a percent is a ratio out of 100, so I wrote a proportion.

$\frac{11 \times 5}{20 \times 5} = \frac{\blacksquare}{100}$

$= \frac{55}{100}$

I know that 20 × 5 = 100, so 11 × 5 = ▪ is the number I'm looking for. 11 × 5 = 55

So, 55% of the students surveyed do not prefer rap music.

Example 2: Renaming a ratio

Express the red part as a ratio, a percent, a decimal, and a fraction.

Fawn's Solution

3 : 5 = ▪ : 100

×20

3 : 5 = ▪ : 100

3 : 5 = 60 : 100

×20

I can see that 3 out of 5 squares are red.

I know that a percent is always out of 100, so I wrote a proportion using the ratio 3:5 and the ratio out of 100.

The scale factor that relates the two ratios is 20 because 5 × 20 = 100.

The red squares represent 60% of the whole shape.

Also, 60% = $\frac{60}{100}$, which is 0.60 or 0.6, and $\frac{6}{10} = \frac{3}{5}$.

Ⓐ Checking

5. What percent of each figure is shaded?

a) 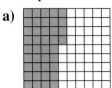   **b)**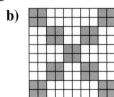

6. Suki is a goalie for her hockey team. During the last game, she made 36 saves out of 40 shots. Write this ratio as a percent.

7. Copy the table. Determine the missing equivalent fraction, ratio, decimal, or percent in each row.

	Fraction	Ratio	Decimal	Percent
a)	$\frac{1}{2}$	5:10	0.5	
b)		2:5		40%
c)	$\frac{21}{100}$			
d)			0.3	
e)				25%

Ⓑ Practising

8. Look at the figures in question 5.

a) What percent of each figure is not shaded?

b) Write each percent in part (a) as a decimal.

9. Complete each calculation.

a) $\frac{7}{20} = \frac{\blacksquare}{100}$ or $\blacksquare$%

b) $\frac{23}{50} = \frac{\blacksquare}{100}$ or $\blacksquare$%

c) $\frac{19}{25} = \frac{\blacksquare}{100}$ or $\blacksquare$%

d) $\frac{4}{5} = \frac{\blacksquare}{100}$ or $\blacksquare$%

e) $\frac{3}{4} = \frac{\blacksquare}{100}$ or $\blacksquare$%

10. Write each fraction as a percent.

a) $\frac{32}{100}$ **d)** $\frac{6}{10}$

b) $\frac{48}{50}$ **e)** $\frac{2}{5}$

c) $\frac{16}{25}$ **f)** $\frac{1}{4}$

11. The average rainfall for a region is 25 cm. If 15 cm of rain has fallen, what percent of the average rainfall has fallen?

12. a) Write each percent as a decimal.

b) Write each percent as a ratio.

i) 15%   **iii)** 76%   **v)** 98%

ii) 32%   **iv)** 7%   **vi)** 80%

13. Arrange in order from greatest to least.

a) 74%, 32%, 45%, 66%

b) 45%, 0.13, 0.68, 36%

c) $\frac{2}{8}$, 22%, $\frac{3}{4}$, 0.58, 79%, 0.06

d) $\frac{12}{48}$, 62%, $\frac{36}{60}$, 0.28, 97%, 0.46

14. Sales tax is an extra amount charged on an item, and it is expressed as a percent. Sales tax differs in the provinces. Suppose that you are buying a $1 item. How much sales tax would you pay in each province?

a) Ontario, with a sales tax of 8%

b) Manitoba, with a sales tax of 7%

c) Saskatchewan, with a sales tax of 6%

15. Andrew, Mohammed, Ian, and Tyrone receive marks of $\frac{23}{25}$, $\frac{38}{40}$, $\frac{27}{30}$, and $\frac{34}{50}$, respectively. Express their marks as percents. Then arrange their marks in order from greatest to least.

16. Milk is classified according to its fat content. One type of milk has a fat content ratio of 1 part to 50 parts. Express this ratio as a percent.

17. Lindsay won 85% of her tennis matches. What is her ratio of losses to wins?

18. A basketball team won 48 out of the 80 games they played over the season.
    a) What percent of the games did they win?
    b) If they lost 30% of their games, what percent of the games did they tie?

ⓒ Extending

19. A movie poster was selling for $5.00. Its price was increased to $7.00. By what percent was its price increased?

20. Canada's highest mountain is Mt. Logan in the Yukon Territory. The world's highest mountain is Mt. Everest in Nepal. Mt. Logan is 5951 m high and Mt. Everest is 8848 m high.
    a) What is the ratio of the height of Mt. Logan to the height of Mt. Everest?
    b) Determine the decimal equivalent of this ratio. Round your decimal to the nearest hundredth.
    c) What percent of the height of Mt. Everest is the height of Mt. Logan?

21. a) When $\frac{1}{3}$ is written as a percent such as ▓%, why is ▓ not a whole number?
    b) Name a different ratio for a part to a whole that is not equivalent to a whole number percent. Explain how you know.

Mental Math

MULTIPLYING BY TENTHS AND HUNDREDTHS

Look at the number patterns.

$34 \times 100 = 3400$ $34 \times 200 = 6800$
$34 \times 10 = 340$ $34 \times 20 = 680$
$34 \times 1 = 34$ $34 \times 2 = 68$
$34 \times 0.1 = 3.4$ $34 \times 0.2 = 6.8$
$34 \times 0.01 = 0.34$ $34 \times 0.02 = 0.68$

1. How can you use mental math to multiply a whole number by 0.1? by 0.2?

2. How can you use mental math to multiply a whole number by 0.01? by 0.02?

3. To calculate each product using mental math, multiply the whole number by 0.1 or 0.01 first.
    a) 0.2×16 d) 0.3×15 g) 0.03×120
    b) 0.7×300 e) 0.2×7500 h) 0.04×240
    c) 0.4×25 f) 0.04×21 i) 0.02×408

2.6 Solving Percent Problems

▶ **GOAL**

Solve percent problems using equivalent ratios and decimals.

Learn about the Math

Brian and his family went out to dinner to celebrate his sister's birthday. Their bill came to $80.00. It is customary to leave a 15% tip for the server.

? **What amount of tip should Brian's family leave for the server?**

A. What is 10% of $80.00?

B. What fraction of 10% is 5%?

C. What is 5% of $80.00? Explain how you found this.

D. Determine the amount of the tip.

E. You could have solved $\frac{15}{100} = \frac{\blacksquare}{80}$ to calculate the tip. Explain why. Calculate the value of the missing term.

Reflecting

1. Suppose that the bill at the restaurant had been $76.35.

   a) Why would some people round the amount to $80 before calculating the tip?

   b) Would you use this method if you had a calculator? Explain your answer.

2. What scale factor did you use to solve the proportion in step E?

3. In step E, you used a proportion to calculate the tip. In steps A to D, you did not use a proportion. Compare the strategies.

4. Explain why a 15% tip is easier to calculate than a 13% tip.

Work with the Math

Example 1: Calculating with percent

The largest oil spill in North America occurred in Prince William Sound, Alaska.
Close to 42 million litres of oil were spilled. Clean-up crews were able to recover 14% of the spill. About how many litres of oil were recovered?

Romona's Solution: Solving a proportion

$$\frac{\text{amount recovered (millions of litres)}}{\text{amount spilled (millions of litres)}}$$

$$\frac{14}{100} = \frac{\blacksquare}{42}$$

$$\frac{14 \times 0.42}{100 \times 0.42} = \frac{5.88}{42}$$

I wrote a proportion that relates the amount of oil spilled to the percent of oil recovered.

I found the scale factor that relates the two ratios. $42 \div 100 = 0.42$ so $100 \times 0.42 = 42$

I used the scale factor and a calculator to calculate the missing term in the proportion.

About 5.88 million litres of oil were recovered.

Paul's Solution: Multiplying by an equivalent decimal

amount spilled = 14% of 42 million

$$14\% = \frac{14}{100} \text{ or } 0.14$$

$$\begin{aligned} \text{amount spilled} &= 0.14 \times 42 \\ &= 5.88 \end{aligned}$$

The decimal form of 14% is 0.14. So, I can calculate 14% of 42 by multiplying 0.14×42 on my calculator.

So, 5.88 million litres of oil were recovered.

Example 2: Calculating a percent

In the student council election, Julie received 168 votes out of 240 votes. Determine the percent of the votes she received.

Bonnie's Solution

$$\frac{\text{votes for Julie}}{\text{total votes}} = \frac{168}{240}$$

$$\frac{168}{240} = \frac{\blacksquare}{100}$$

$$\frac{168 \div 2.4}{240 \div 2.4} = \frac{70}{100}$$

I wrote the first ratio, which is the number of votes for Julie to the total number of votes for all the candidates.

Since percent means out of 100, I used ■ to represent the number of votes that Julie got out of 100.

I divided 240 by 100 to determine the scale factor. The scale factor is 2.4 because $240 \div 2.4 = 100$.

I divided 168 by the scale factor using my calculator.

Julie got 70% of the votes.

Ratio, Rate, and Percent

Example 3: Calculating a number from a percent

There are 10 boys in Erin's music class. If 40% of the students in the music class are boys, how many students are in the music class?

Miguel's Solution: Using logical reasoning

Percent of class that are girls = 100% − percent that are boys

$\qquad\qquad\qquad\qquad\qquad\quad = 100\% - 40\%$

$\qquad\qquad\qquad\qquad\qquad\quad = 60\%$

I know that 40% of the students in the class are boys. So, girls must be 60% of the class.

Number of girls = 60% of class

$\qquad\qquad\quad\;\; = 40\%$ of class + 20% of class

$\qquad\qquad\quad\;\; = 10 + 5$

$\qquad\qquad\quad\;\; = 15$

If 40% of the class is equivalent to 10 boys, 40% must also be 10 girls. Since 20% is half of 40%, I need half of 10, which is 5. There are 15 girls in the class.

Total number of students = 10 + 15

$\qquad\qquad\qquad\qquad\;\; = 25$

I added the number of boys and the number of girls to get the total. There are 25 students in the music class.

Fawn's Solution: Solving a proportion

$$\frac{10}{\text{total}} = \frac{40}{100}$$

I used "total" to represent the total number of students in the class. Since percent means out of 100, I know that 40% means 40 boys out of 100. I wrote a proportion.

$$\frac{10}{\text{total}} = \frac{40 \div 4}{100 \div 4}$$

I figured out that 40 ÷ 4 equals 10, so the total must be 100 ÷ 4.

$100 \div 4 = 25$

$\quad\;\text{total} = 25$

I divided and determined that the total is 25. There are 25 students in the music class.

A Checking

5. Calculate 15% of 60 using each method.

   a) Use a mental calculation. Give only the answer.

   b) Use a proportion.

   c) Multiply by the equivalent decimal.

6. Determine each number.

   a) 25% of 60 = ▦   **c)** 20% of ▦ = 45

   b) 10% of 40 = ▦   **d)** 12% of ▦ = 54

B Practising

7. Calculate.

   a) 50% of 20   **d)** 12% of 50

   b) 75% of 24   **e)** 15% of 200

   c) 20% of 45   **f)** 44% of 250

8. Calculate.

   a) 50% of ▦ = 15   **d)** 75% of ▦ = 12

   b) 25% of ▦ = 22   **e)** 15% of ▦ = 24

   c) 10% of ▦ = 7   **f)** 44% of ▦ = 110

9. Out of a batch of 600 computers, 30 failed to pass inspection due to faulty wiring. What percent failed to pass inspection?

10. There are 12 girls with blond hair in Katya's gymnastics class. This is 25% of the entire class.

    a) Use a mental calculation to determine the total number of students in the gymnastics class. Record only the answer.

    b) Use a proportion to determine the total number of students in the class.

11. A dealer paid $6000 for a used car. The dealer wants to make a profit that is 25% of the price he paid for the car.

    a) What profit does the dealer want to make?

    b) How much should the dealer sell the car for?

12. If Sarah's mother really likes the service in a restaurant, she leaves a 20% tip. Their last bill was $110.00, and Sarah's mother left $22.00 for the tip. Explain how you know that this is 20%.

13. To sell a new chocolate bar, the manufacturer advertises "20% MORE, FOR FREE!" If a standard chocolate bar is 50 g, what size is the new bar?

14. A roast-beef sandwich contains 18 g of fat, and a slice of pizza contains 25 g of fat. If you choose to eat pizza at lunch instead of a roast-beef sandwich, what percent greater is your fat intake? Round to the nearest whole number percent.

15. Suki buys a dress on sale for 20% off the original price. She saves $40.

    a) What was the original price of the dress?

    b) How much did Suki pay for the dress?

16. A pair of jeans usually costs $80.00. The jeans are on sale at Jane's Jean Shop for 50% off. Denim Discounters offers the same jeans at 30% off, as well as a further 20% off the already discounted price. Are the jeans the same price at both stores? Justify your answer.

17. Use the circle graph to estimate the number of hours that Matthew spends on each activity in a 24 h day.

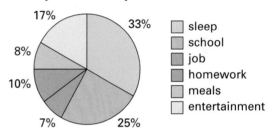

18. Last year, Jasleen was 150 cm tall. This year, her doctor measures her height and tells her that she has grown 20%. How many centimetres has she grown?

C Extending

19. A new process in a factory has increased production by 12%. If workers are now producing 30 more units per day, how many units did they produce per day before the new process was introduced?

20. The tax rate for businesses was increased by half of 1%. If a business was paying $20 000 in taxes before the increase, how much would the business pay after the increase?

21. Anthony had a 12 cm by 15 cm photo enlarged. If the dimensions of the enlargement are 24 cm by 30 cm, by what percent was the area of the photo enlarged?

Decimal Multiplication

▶ **GOAL**

Use decimal multiplication to solve ratio and rate problems.

Learn about the Math

Romona says, "I have a poster that is 0.5 m by 1.1 m. I want a copy to put above my desk. I'm going to change each dimension to 60% of its original size."

Bonnie says, "60% is $\frac{60}{100}$. That's the same as $\frac{6}{10}$ or 0.6. You can multiply 0.6×0.5 to calculate 60% of the width."

? **What is the width of the reduced copy?**

Example 1: Using a grid to model multiplication with decimals

Use a 10-by-10 grid to determine the product of 0.6×0.5.

Romona's Solution

0.5

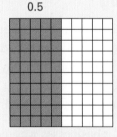

I coloured 5 columns of squares red on a 10-by-10 grid to represent 0.5.

0.5

60% = 0.6

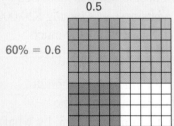

I coloured 6 rows of squares blue to represent 60% of the grid.

There are 30 purple squares in the overlap. The purple squares represent $\frac{6}{10}$ of $\frac{5}{10}$, which is $\frac{30}{100}$ of the 10-by-10 grid.

So, $0.6 \times 0.5 = 0.30$.

Example 2: Multiplying and then dividing by 10

Multiply 0.6 × 0.5 to calculate the width of the 60% reduced poster.

Bonnie's Solution

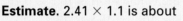

0.6 × 0.5 (0.6 × 10) × 0.5 = 6 × 0.5 = 3.0 3.0 ÷ 10 = 0.3 So, 0.6 × 0.5 = 0.30	I know how to multiply decimals by whole numbers. I multiplied one number by 10 to get a whole number. 6 × 0.5 means 6 groups of 5 tenths = 30 tenths or 3.0. Then I multiplied the whole number by the decimal. Since I multiplied by 10, now I need to divide by 10. (To divide by 10, the digits move one place to the right.) The width of the reduced copy is 0.3 m.

Example 3: Multiplying and then dividing by 100

Estimate 2.41 × 1.1. Then multiply.

Paul's Solution

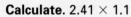

Estimate. 2.41 × 1.1 is about 2 × 1 = 2.	I rounded the numbers, and then I multiplied.
Calculate. 2.41 × 1.1 (2.41 × 100) × 1.1 = 241 × 1.1 = 265.1 265.1 ÷ 100 = 2.651 So, 2.41 × 1.1 = 2.651	I multiplied one of the numbers to make it a whole number. Since I multiplied by 100, I need to divide by 100. (To divide by 100, the digits move two places to the right.) I know my answer is reasonable because 2.651 is about 2.

Reflecting

1. Why will Romona's strategy of using a 10-by-10 grid to model multiplying tenths by tenths always have an answer in hundredths?

2. a) How did Bonnie's method allow her to use what she already knew about whole number multiplication?

  b) Why did Bonnie need to divide by 10 after she multiplied by 10?

3. a) How did estimating help Paul check his answer?

  b) Why did Paul need to divide by 100 after he multiplied by 100?

4. The length of the poster is 1.1 m. Explain how to calculate 60% of the length.

Work with the Math

Example 4: Estimating and calculating distance travelled

Stephen can ride his bike at an average speed of 15.5 km/h. At this speed, how far will he travel in each amount of time?

a) 30 min   **b)** 0.75 h

Solution

a) Estimate.

30 min is 0.5 h or half an hour. In one hour, Stephen travels 15.5 km, so he'd travel about 8 km in half an hour.

Calculate.

Distance = speed × time, so the calculation is 15.5 × 0.5.

Multiply by 10.   15.5 × (0.5 × 10) = 15.5 × 5
$$= 77.5$$

Divide by 10.   77.5 ÷ 10 = 7.75 or 7.8, rounded to one decimal place

In 0.5 h, Stephen can ride 7.8 km at 15.5 km/h.

b) Estimate.

0.75 h is a bit less than 1 h. So Stephen should travel a bit less than 15.5 km—maybe 13 km.

Calculate.

Distance = speed × time, so the calculation is 15.5 × 0.75.

Multiply by 100.   15.5 × (0.75 × 100) = 15.5 × 75

$$
\begin{array}{r}
15.5 \\
\times 75 \\
\hline
775 \\
1085 \\
\hline
1162.5
\end{array}
$$

Divide by 100.   1162.5 ÷ 100 = 11.625 or 11.6, rounded to one decimal place

In 0.75 h, Stephen can ride 11.6 km at 15.5 km/h.

Ⓐ Checking

5. Use a 10-by-10 grid to model, and then calculate.

a) 0.3 × 0.8   **b)** 0.2 × 0.7

6. Estimate each product.

a) 9.7 × 0.63   **b)** 3.75 × 5.86

7. Choose a strategy and multiply.

a) 0.2 × 3.4   **b)** 0.8 × 7.59

Ⓑ Practising

8. Use a 10-by-10 grid to model, and then calculate.

a) 0.2 × 0.6   **b)** 0.8 × 0.7

9. Estimate, and then calculate. Show the strategy you used.

a) 0.8 × 1.3   **d)** 2.6 × 1.01

b) 1.1 × 2.3   **e)** 0.02 × 1.5

c) 2.2 × 0.03   **f)** 0.35 × 10.1

10. Explain why you can multiply 0.5×0.64 mentally more easily than 0.7×0.64.

11. How would you place the digits 6, 7, and 8 so that the product is as close to 5 as possible?

$$\blacksquare.\blacksquare \times 0.\blacksquare$$

12. Calculate the distance a car will travel in 3.5 h if its speed averages 92.5 km/h.

13. A store advertises hamburger meat for sale at \$2.25/kg. How much will 3.4 kg cost?

14. A package of 12 guitar picks costs \$5.04. At that price, how much should 18 picks cost?

15. A Japanese bullet train recently maintained an average speed of 317.5 km/h for a trip that lasted 3 h 15 min. Calculate the distance covered by the train on that trip.

16. Gasoline recently cost 82.5¢/L. If Tonya's car holds 58.5 L, how much should it cost to fill the tank?

17. What is the area of a rectangle that is 3.2 m wide and 5.1 m long?

18. Suki wants her bedroom ceiling painted. The room is a rectangle with dimensions 4.2 m by 3.9 m. The label on the paint can says one can has enough paint to cover 12 m². Explain whether one can is enough for Suki to put two coats of paint on the ceiling.

19. Richard works at a clothing store. He is paid \$150 a week plus 9% of the value of his sales for the week. Last week his sales totalled \$457.85. Calculate Richard's earnings for the week.

20. The adult height of a male is about 1.19 times his height at age 12. The adult height of a female is about 1.07 times her height at age 12. Predict how tall Miguel and Romona will be as adults if Miguel is 1.5 m and Romona is 1.6 m.

C Extending

21. According to Statistics Canada, the 2002 birth rate for the country was 10.5 births for every 1000 people. In July 2002, the Canadian population was 31 361 611. Calculate the approximate number of births that took place in Canada in 2002.

22. The Information About Canada Web site indicates that about 79% of the population lives in cities. Use the July 2002 population data in question 21 to calculate the approximate number of city dwellers in Canada.

23. A used car is listed for sale at \$9000. The price of the car is reduced by 20% for quick sale. Then it is reduced by 20% of the sale price. What is the final sale price of the car?

24. Meagan said that to multiply 1.3×1.3, you can multiply 1×1 and 0.3×0.3 and the answer is 1.09. Do you agree? Explain.

2.8 Decimal Division

▶ **GOAL**

Use decimal division to solve ratio and rate problems.

Learn about the Math

Paul is selling used comic books. Miguel has $2.50 to buy some.

? **How many comic books can Miguel buy with $2.50?**

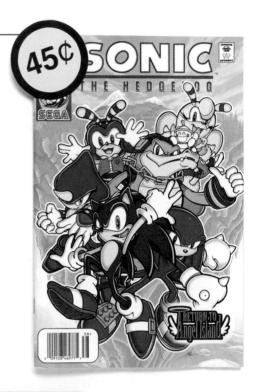

Example 1: Using grids to model division with decimals

Comic books cost $0.45 each. Use 10-by-10 grids to determine how many comic books you can buy with $2.50. How much money will be left over?

Miguel's Solution

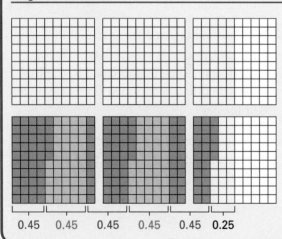

| 0.45 | 0.45 | 0.45 | 0.45 | 0.45 | 0.25 |

I used three 10-by-10 grids to represent the $2.50.

Each group of 45 squares represents the price of one comic book.

There are 5 groups of 45 squares and 25 squares left over.

I can buy 5 comic books and I'll have 25¢ left over.

Example 2: Dividing hundredths using equivalent fractions

Calculate $2.50 ÷ $0.45 to figure out how many comic books Miguel can buy.

Paul's Solution

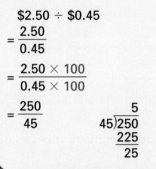

$$\$2.50 \div \$0.45$$
$$= \frac{2.50}{0.45}$$
$$= \frac{2.50 \times 100}{0.45 \times 100}$$
$$= \frac{250}{45}$$

$$\begin{array}{r} 5 \\ 45\overline{)250} \\ 225 \\ \hline 25 \end{array}$$

I know that $1 \div 4$ means $\frac{1}{4}$, so I can write $2.50 \div 0.45$ as $\frac{2.50}{0.45}$.

If I multiply 0.45 by 100 to get a whole number, I need to multiply 2.50 by 100.

The **quotient** is 5 and the **remainder** is 25.

Miguel can buy 5 comic books and have 25¢ left over.

Example 3: Dividing tenths using equivalent fractions

Estimate $91.8 ÷ 3.4$. Then divide.

Fawn's Solution

Estimate. $91.8 \div 3.4$ is about $90 \div 3 = 30$.

Calculate.
$$91.8 \div 3.4$$
$$= \frac{91.8}{3.4}$$
$$= \frac{91.8 \times 10}{3.4 \times 10}$$
$$= \frac{918}{34}$$

$$\begin{array}{r} 27 \\ 34\overline{)918} \\ 68 \\ \hline 238 \\ 238 \\ \hline 0 \end{array}$$

I rounded 3.4 to 3. I rounded 91.8 to 90 because 90 is a multiple of 3.

I can write $91.8 \div 3.4$ as $\frac{91.8}{3.4}$.

If I multiply the **divisor** 3.4 by 10 to get a whole number, I need to multiply the **dividend** 91.8 by 10.

The quotient is 27.

I know that my answer is reasonable because 27 is almost 30.

Reflecting

1. When would Miguel's strategy of using 10-by-10 grids result in a whole number quotient without a remainder?

2. Why can Paul write $2.50 ÷ $0.45 as $\frac{2.50}{0.45}$?

3. a) Why did Fawn choose 10 as the multiplier?

   b) Why is it necessary to multiply both the divisor and the dividend by the same number?

4. How did estimating help Fawn check her answer?

Work with the Math

Example 4: Dividing using measurements

Mei has 5 m of ribbon. She wants to divide it into equal pieces.

a) How many pieces will there be if each piece is 25 cm long?

b) What if each piece is 0.8 m long?

Solution

a)
$$5 \div 0.25 = \frac{5}{0.25}$$
$$\frac{5 \times 100}{0.25 \times 100} = \frac{500}{25}$$

$$\begin{array}{r} 20 \\ 25\overline{)500} \\ 500 \\ \hline 0 \end{array}$$

Change the measurements so that the units are the same.

25 cm = 0.25 m

Multiply the divisor 0.25 by 100 to get a whole number. Then multiply the dividend 5 by 100.

There will be 20 pieces, each 25 cm long, with no ribbon left over.

b)
$$5 \div 0.8 = \frac{5}{0.8}$$
$$\frac{5 \times 10}{0.8 \times 10} = 50 \div 8$$

$$\begin{array}{r} 6 \\ 8\overline{)50} \\ 48 \\ \hline 2 \end{array}$$

Multiply the divisor 0.8 by 10 to get a whole number. Then multiply the dividend 5 by 10.

There will be 6 pieces, each 0.8 m long, with some ribbon left over.

A Checking

5. Use 10-by-10 grids to model, and then calculate.

  a) $2.7 \div 0.9$

  b) $3.6 \div 0.18$

  c) $12.4 \div 0.4$

6. Estimate each quotient.

  a) $3.6 \div 0.9$

  b) $7.8 \div 1.3$

7. Choose a strategy, then divide.

  a) $2.7 \div 0.03$

  b) $4.59 \div 0.9$

  c) $0.25 \div 0.04$

B Practising

8. Use 10-by-10 grids to model, and then calculate.

  a) $3.6 \div 0.4$

  b) $2.8 \div 0.14$

  c) $2.25 \div 0.15$

9. Divide.

  a) $2.7 \div 0.4$

  b) $3.13 \div 0.02$

  c) $10.2 \div 1.5$

  d) $0.27 \div 0.04$

  e) $14.4 \div 0.12$

  f) $0.04 \div 0.02$

10. Use division to show the number of coins you would have if you had $11.50 in each type of coin.

a) dimes   c) quarters

b) nickels   d) pennies

11. Why is the result of $1.25 \div 0.01$ the same as 1.25×100?

12. Explain how you can tell without actually calculating that the remainder for $2.95 \div 0.05$ is 0, but the remainder for $2.95 \div 0.02$ is not 0.

13. Nathan has 11.4 m of rope. He wants to divide it into equal pieces. How many pieces will there be if the pieces are these lengths?

a) 80 cm long   c) 0.7 m long

b) 1.4 m long   d) half a metre

14. To the nearest hour, how long will it take to walk 10 km at each speed?

a) 4.5 km/h   b) 3.2 km/h

15. How many 0.35 L glasses can be filled from a 1.5 L bottle of water?

16. To the nearest litre, how much gasoline can you buy with $20.00 if the price for gas is 87.5¢/L?

17. Susan earned $191.25 last week. She is paid $8.50/h. How many hours did she work?

18. Calculate the average speed of a train that completed a 525 km trip in 4.7 h. Round your answer to the nearest whole number.

19. The adult height of a male is about 1.19 times his height at age 12. The adult height of a female is about 1.07 times her height at age 12. Predict how tall each of these people were when they were 12.

a) an adult male 1.8 m tall

b) an adult female 1.8 m tall

ⓒ Extending

20. Kyle is filling his brother's wading pool. The pool holds 180 L of water and the hose supplies water at 22.5 L/min. To the nearest minute, how long will it take to fill the pool?

21. When water freezes, its volume increases. For example, when 100 cm³ of water is frozen, about 109 cm³ of ice results.

a) What volume of ice will result from freezing 12 cm³ of water?

b) What volume of water must be frozen to produce 100 cm³ of ice?

22. Snails move at approximately 0.013 m/s. How long would it take a snail moving at this speed to travel the 450 km distance from Toronto to Ottawa?

23. Order the times required for each trip from least to greatest.

	Distance	Speed
Bullet train	571.5 km	317.5 km/h
Airplane	1191 km	680.5 km/h
Car	204 m	90 km/h

Tape a line on the floor, 3 m from your basket. Create a table like the one below to record your team's results.

Player	Shots in basket	Shots taken	Ratio of $\dfrac{\text{shots in basket}}{\text{shots taken}}$	Percent in basket
		10		
		10		
		10		
		10		
		40		

You will need
- a spongy ball or a crumpled piece of paper
- a measuring tape or a metre stick
- a wastepaper basket or another suitable "basket" (such as a box or a bucket)
- tape
- a calculator

Number of players: 4 per team

Rules

1. Each player on your team takes 10 shots. Record the "Shots in basket" in your table.

2. Calculate the "Percent in basket" for each player on your team. Record this in your table.

3. Calculate your team's "Percent in basket" for the 40 shots taken.

4. Compare your team's results with the results for the other teams in your class. The team with the highest "Percent in basket" is the champion.

5. What is the "Percent in basket" for your class?

Optional: Challenge other classes in your school to see which class is the "Wastepaper Basketball" champion.

Chapter Self-Test

1. a) Write the ratio of white squares to green squares.

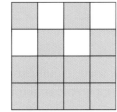

b) What fraction of the squares are white?

c) What percent of the squares are green?

2. Determine three ratios that are equivalent to $5:9$.

3. Calculate the missing term in each proportion.

a) $\dfrac{2}{11} = \dfrac{\blacksquare}{55}$ **c)** $\dfrac{9}{\blacksquare} = \dfrac{21}{28}$

b) $\dfrac{36}{42} = \dfrac{12}{\blacksquare}$ **d)** $\dfrac{6}{9} = \dfrac{\blacksquare}{15}$

4. The height of a building in a scale drawing is 9 cm. The scale is $1:600$. Explain how you would use the scale to find the actual height of the building.

5. Two large cups of coffee cost $4.00. How much money will Marianne need if she wants to buy nine large cups of coffee?

6. Nicole earned $72 in 9 h. At this rate, how much would she earn in 12 h?

7. The air you breathe is $\dfrac{1}{5}$ oxygen. What percent of the air is made up of other gases?

8. Copy and complete the table to express the equivalent forms in each row.

	Fraction	Ratio	Decimal	Percent
a)		21:60		
b)			0.18	
c)				82%

9. Calculate.

a) 40% of 35 is $\blacksquare$.

b) 21 out of 30 is $\blacksquare$%.

c) 38% of $\blacksquare$ is 95.

10. At a baseball game, the stadium is 65% full. If the stadium's capacity is 1800, how many people are at the game?

11. Braydon buys a new CD and pays 8% sales tax. If the sales tax is $1.20, calculate the price of the CD.

12. Akeem bought a used video game for $38. Later, he sold it for 40% less than he paid. For how much did Akeem sell the video game?

13. The ratio of children to adults at an art show is $22:18$.

a) What percent of the crowd are children?

b) What percent of the crowd are adults?

14. Why is calculating 10% of an amount easier than calculating 17% of the amount?

15. How is calculating 1% of an amount similar to calculating 10% of the amount?

16. Why might you calculate 50% of 212 in a different way than you would calculate 42% of 212?

17. How do you know that each answer is about 5?

a) 1.4×3.5 **c)** $65.2 \div 12.9$

b) 0.7×7.34 **d)** $2.46 \div 0.53$

18. How much greater is 27.9×4.5 than $27.9 \div 4.5$?

Frequently Asked Questions

Q: **What is a percent?**

A: A percent is a ratio out of 100. A percent is used to compare a part to a whole, so the second term refers to a whole or a total.

To determine a percent from a ratio, calculate the equivalent ratio out of 100.

For example,
$$\frac{2}{5} = \frac{\blacksquare}{100}$$
$$\frac{2 \times 20}{5 \times 20} = \frac{40}{100}$$
$$\frac{2}{5} = 40\%$$

Q: **How are ratios, fractions, decimals, and percents related?**

A: If you are given one form (ratio, fraction, decimal, or percent), you can determine the other three equivalent forms. For example, $4:5$ is $\frac{4}{5} = \frac{80}{100}$, which means 80 out of 100 or 80%, and 80 hundredths $= 0.80$ or 0.8, which is $\frac{4}{5}$.

Q: **How do you solve a problem that involves percent?**

A: Depending on the problem and numbers involved, you could draw a diagram, use the relationship between the numbers, or solve a proportion by relating to a ratio out of 100.

Q: **How can you multiply or divide with two decimals?**

A: You can use a model or you can use powers of 10.

For multiplication, multiply either factor by a power of 10 to get a whole number, multiply this whole number by the other factor, then divide the result by the same power of 10. For 0.3×4.67,
$0.3 \times 10 = 3$, then $3 \times 4.67 = 14.01$,
and $14.01 \div 10 = 1.401$.

For division, multiply the divisor by a power of 10 to get a whole number, multiply the dividend by the same power of 10, then divide. For $227.8 \div 3.4$,
write $\frac{227.8}{3.4}$, then $\frac{227.8 \times 10}{3.4 \times 10} = \frac{2278}{34}$,
and $2278 \div 34 = 67$.

Practice Questions

(2.2) **1.** Write two equivalent ratios for each ratio.

    a) 9:20   **b)** $\dfrac{4}{5}$ **c)** 21 to 3

(2.2) **2.** Express both quantities in each comparison using the same units. Write the comparison as a ratio in fraction form. Then write an equivalent ratio using whole numbers.

    a) 85¢ to $1.20   **b)** 24 kg to 80 g

(2.2) **3.** Determine each missing term.

    a) 2:7 = ■:21   **c)** 4:■ = 8:14

    b) $\dfrac{36}{9} = \dfrac{72}{■}$ **d)** $\dfrac{2}{12} = \dfrac{■}{48}$

(2.2) **4.** The average height of an ostrich is 255 cm. Suppose that you want to make a scale drawing of an ostrich. The scale is 30:1. What height will you make the ostrich in your drawing? Explain your thinking.

(2.3) **5.** A car travels 180 km in 3 h. At this rate, how far will the car travel in 5 h?

(2.5) **6.** Copy and complete the table to express the equivalent forms in each row.

	Fraction	Ratio	Decimal	Percent
a)		24:30		
b)			0.36	
c)				86%
d)	$\dfrac{3}{3}$			
e)			0.05	

(2.5) **7.** The ratio of red cars to black cars in the school parking lot is 6:8. Vanessa determines that $\dfrac{6}{8} = 75\%$. She concludes that 75% of the cars in the parking lot are red. Is she correct? Explain.

8. The following circle graph shows the ingredients that are used to make a sausage and mushroom pizza. The percent of each ingredient, by weight, is given.

    a) Determine the fraction, by weight, of each ingredient.

    b) Determine the decimal, by weight, of each ingredient.

    c) If you used fractions or decimals to represent the data, would the shape of the circle graph change? Explain. **(2.5)**

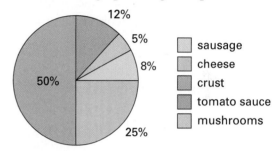

9. Determine each missing number. **(2.6)**

    a) 25% of 84 = ■

    b) 24% of 200 = ■

    c) 10% of ■ = 5

10. A movie theatre has sold 75% of its seats for the 7:00 p.m. show. If the theatre has 440 seats, how many tickets are sold? **(2.6)**

11. Last week, Raj earned $58. He spent $42 and saved the rest. Michael earned $83 and saved $20. Who saved the greater percent of his earnings? **(2.6)**

12. Draw a model to calculate. **(2.7)**

    a) 0.2×0.9 **b)** $3.2 \div 0.4$

13. Calculate. **(2.8)**

    a) 1.4×5.3 **c)** $6.3 \div 2.1$

    b) 0.9×3.28 **d)** $6.93 \div 0.33$

Chapter Task

Ball Bounce-ability

Have you ever dropped a ball to see how high it would bounce?
Do different types of balls bounce better than others?

? **How can you determine which type of ball bounces the best?**

A. With a partner, choose three different types of balls. Each ball
should be a different size and material; for example, a
basketball, golf ball, tennis ball, or soccer ball.

B. Create a table to record your data from steps C and D.

C. Select the first ball. One person will be the dropper, and the
other person will be the measurer.
- Find a height that is comfortable for the dropper to
release the ball. Measure and record this height.
- The dropper lets the ball fall to the floor.
- As the ball bounces back up, the measurer
notes the maximum *bounce height*. (This is
the distance that the ball bounces up from
the floor on the first bounce.)
- The dropper then measures the distance
from the floor to the maximum bounce
height.
Complete 5 or 10 trials, and record the average
in your data table. Use the same drop
height each time.

D. Repeat step C for the other two balls.

E. Determine the ratio of bounce height : drop
height for each ball.

F. For each ball, what is the ratio for comparing
the bounce height with the drop height? What
is the percent for this?

G. Write a short report about your findings.
Discuss the results of your experiment,
and rank the bounce-ability of each ball.

Task Checklist

- ☑ Did you measure and record all the required data?
- ☑ Did you show all your calculations?
- ☑ Did you explain your thinking?
- ☑ Did you include enough detail in your report?
- ☑ Did you discuss the reasons for the conclusion you made?

Math in Action

Rock Band Manager

Keith Porteous manages such highly successful music acts as 54/40. He lives in Vancouver, British Columbia, and has been a rock band manager for over 15 years.

Keith is paid a commission for his work. The **commission** is a percentage of the artists' earnings. "The whole music business is based on percentages," says Keith. "We call the percentages 'points.' If you earn 10%, you make 10 points." How does Keith earn his points? "I talk on the phone all day," Keith joked in a recent interview. Actually, his "talk" leads to touring and recording opportunities for the acts he represents.

Problems, Applications, and Decision Making

1. At one concert, a band earns $10 000. Keith makes 20 points. How much money does Keith earn?

2. The profit from a concert is $12 750. A manager makes 10.5 points. How much money does the manager earn?

3. Suppose that you are the manager of a rock band. You earn $3750 for a concert. The concert takes in $25 000. How many points do you make?

4. List some advantages and disadvantages of being paid by commission.

Keith Porteous explains how the concert profit is divided: "Let's say each concert ticket costs $15. The seating capacity of the concert hall is 1000 people. So you multiply $15 by 1000 seats. Then you multiply the product by 10%, which is the agent's commission. What's left over you multiply by 20%, which is the manager's commission. The rest belongs to the artists."

5. Write a formula to work out how much money the musicians earn.

6. Why is the order of operations important? Could the formula be written any other way? Explain.

7. How much money do the musicians earn? How much money does the manager earn? How much money does the agent earn?

8. What would happen to everyone's earnings if the ticket price were increased to $20? Calculate the earnings.

9. Suppose that the concert is moved to a hall that holds 1200 people. Use a calculator to find the earnings of the band, the manager, and the agent if the ticket price is $15.

10. A band plays in a large stadium and earns $270 000. The manager's share is 20% of the band's earnings. Estimate how much money the manager earns. Check your estimate.

Advanced Applications

Keith explains how the money from T-shirt sales is divided: "If T-shirts are sold at a concert, the concert hall takes 15% of the total sales, before sales taxes. What's left is called the net. The merchandisers who make and distribute our T-shirts usually have an agreement with the artists to keep 67% of the net. Out of what's left, the manager gets 20%. The band gets the rest."

11. Write a formula to calculate how much money the band receives from T-shirt sales.

12. A band plays a concert in Ontario. If T-shirt sales are $7500, find out how much money the concert hall gets, how much money is paid in GST and PST, and how much money the merchandisers, manager, and band get. What if T-shirt sales are $12 750?

"If you think of the profit from a song as being 100%, the publishing expenses are 50% of the profit," Keith explains. "The songwriter gets the other 50%. Of this, 50% is for composing the music and 50% is for writing the lyrics to the song. If the lyrics are co-written, the writers divide the writing portion."

13. The publishers of a song make $100 000. How much money does the lyricist earn? How much money does the person who composed the music earn?

14. The profit from a song is $250 000. The lyrics were co-written by two people. How much money does each person earn?

CHAPTER 3

Data Management

▶ GOALS

You will be able to

- collect data from a variety of sources
- organize data
- display data in the most appropriate form
- describe data by comparing mean, median, and mode
- draw conclusions from collected or displayed data

Typical Names

I think most girls have longer names than boys. There's Stefania and Katarina, Emil and Hans.

But Cho and Ginny are short names. Christopher is a long name.

? **How can you determine if boys' first names are generally shorter than girls' first names?**

A. Survey your class. Record the number of letters in each person's first name on a tally chart like the one below.

Number of letters in first name	Boys	Girls
2	I	
3	III	
4	⊬⊬	

B. Display your results using a bar graph. Remember to include the following:
- a title for the graph
- a title for the vertical axis
- a title for the horizontal axis
- a key to indicate "Boys" and "Girls"

C. Calculate the mean number of letters in each set of names. Round each mean to the nearest whole number.
a) boys' first names
b) girls' first names
c) everyone's first names

D. Based on your survey, which first names are shorter: boys' names or girls' names? Write a conclusion, using your data for support.

E. Do you think the results of the survey would be the same in other classes in your school? Do you think they would be the same in other classes in your province or in Canada? Why or why not?

Do You Remember?

1. Use the broken-line graph to answer the following questions.

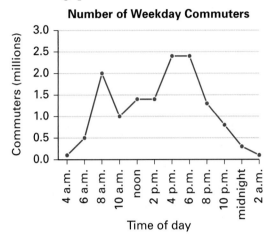

Number of Weekday Commuters

Source: Statistics Canada, General Social Survey, 1998

a) What information is shown on the horizontal axis?

b) What are the units for the vertical axis?

c) About how many people were commuting at 10 a.m.?

d) Between which 2 h interval were the greatest number of people commuting?

2. Use the data to answer the questions below.

How we usually get to school	Number
walk	80
school bus	32
car	46
bicycle	22
public transit	8
other	12

a) If each student chose only one answer, how many students were surveyed?

b) How many more students use a school bus than public transit?

c) What percent of students arrive by car?

d) What percent of students are driven (by car, bus, or public transit)?

3. Yan surveyed 20 students in his class about their favourite sport to play. He constructed this circle graph to represent his survey results. Use Yan's circle graph to complete the table below.

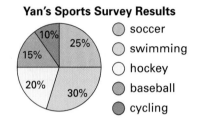

Yan's Sports Survey Results

- soccer
- swimming
- hockey
- baseball
- cycling

Favourite sport to play	Tally	Number of students	Fraction	Percent
soccer	IIII		$\frac{5}{20}$	25%
swimming		6		
hockey				20%
baseball			$\frac{3}{20}$	
cycling	II			

4. Andrea raised money for cancer research by collecting pledges from her family and friends. Her pledge sheet is shown below.

Sponsor	Pledge ($)
Mom and Dad	50
Grampa	10
Alan	5
Indu	5
Uncle Ray	10
Carlos	2
Mei	5
Cheryl	5
Steve	3
Mrs. Khan	20

a) What is the mean of the pledges?

b) What is the median pledge?

c) What is the mode?

Data Management

Collecting Data

▶ **GOAL**

Make inferences and convincing arguments that are based on primary and secondary data.

You will need
- a paragraph of text
- a calculator
- a ruler
- a strip of heavy paper 20 cm long and marked in 5 cm intervals
- tape

Explore the Math

Raj and Braydon noticed the arrangement of the letters on their computer keyboards. Raj said, "I wonder if the letters are arranged this way on purpose."

QWERTY keyboard   home row

? **Where are the most frequently used letters of the alphabet found on a standard keyboard?**

A. Find a paragraph of text to read, in any book, newspaper, or magazine. Each person in the class should choose a different source.

B. Use a tally chart to record the **frequency** of each letter of the alphabet in your paragraph. This is your **primary data**.

frequency

the number of times that an event or item occurs

primary data

information that is collected directly

Letter	Tally	Frequency
A		
B		

C. The following table shows the letter frequencies per thousand letters in English. How do the letter frequencies in your primary data compare with the letter frequencies in this **secondary data**?

secondary data

information that is collected by someone else

Letter Frequencies in English (per 1000 letters)					
A 73	B 9	C 30	D 44	E 130	F 28
G 16	H 35	I 74	J 2	K 3	L 35
M 25	N 78	O 74	P 27	Q 3	R 77
S 63	T 93	U 27	V 13	W 16	X 5
Y 19	Z 1				

D. Copy this table. Complete the blank "Count" column by adding your data.

Letters in each row	Combined frequency of letters in row (using secondary data from table)		Combined frequency of letters in row (using primary data you collected)	
(QWERTY keyboard)	Count	Percent	Count	Percent
row above home row: QWERTYUIOP	540	$\dfrac{540}{1000}$ or $\dfrac{54}{100}$ = 54%		
home row: ASDFGHJKL	299	$\dfrac{299}{1000}$ or $\dfrac{30}{100}$ = 30%		
row below home row: ZXCVBNM	161	$\dfrac{161}{1000}$ or $\dfrac{16}{100}$ = 16%		
Total	**1000**	**100%**		**100%**

E. Complete the blank "Percent" column by expressing the fraction $\dfrac{\text{count for all letters in row}}{\text{total for all letters in paragraph}}$ as a percent, rounded to the nearest whole number.

F. Create a circle graph to display your data. Follow these steps.

Step 1: Use a strip of paper 20 cm long. Mark every 1 cm. Label the marks by counting by 5s.

Step 2: Join the ends of the strip to make a circle. (Make sure there is no overlap.)

Step 3: Trace around the outside of the strip. Put a point in the centre of the circle, and join it to 0%.

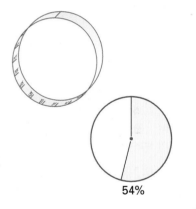

Step 4: Make each sector of the circle graph by joining the centre point to the approximate value on the strip.

54%

Reflecting

1. a) Compare the two "Percent" columns in the table.

b) Are your percents similar to the percents for other students? Explain.

2. Braydon predicted that the most frequently used letters would be on the home row of the keyboard. Compare this with your data or graph.

3. Where are the four most commonly used letters?

4. What other times might someone collect data to design something?

Data Management **83**

Avoiding Bias in Data Collection

▶ **GOAL**

Understand different ways to collect data and analyze bias in data-collection methods.

Learn about the Math

At Fleury Public School, many students seem to arrive at class late. The student council has proposed starting school 1 h later. Ms. Chan, the student council advisor, wants the council members to gather some data to support their proposal.

? **How should the council members collect their data?**

The council members start by brainstorming ways to gather the data they need.

Peter: "I think we should try an experiment. Half the school could start at the regular time, and the other half could start 1 h later. We could see if there are fewer students in the second group who are late for school."

Jasleen: "We should do some research. If we find other schools that have a later start than us, we can ask them what the good and bad points are."

Ms. Chan: "But what works for another school may not work here."

Heather: "Let's survey all the students in Grade 7 to find out their opinions. We could use actual quotes to support our proposal."

Ms. Chan: "That's a good idea, Heather, but you might get **biased results** . We have students from kindergarten to Grade 8 in this school. The Grade 7 students may not have the same sleep patterns as younger students do. They may not be a good **sample** of the entire **population** ."

Zach: "Why don't we take a **census** ? We could make a questionnaire and give it to every student in the school."

Ms. Chan: "I think most students will want to start later, but I'm not sure if their parents or teachers will. We have to get these opinions, too."

---

biased results

when the results of a survey of one group are not likely to apply to another group selected from the same population

sample

a part of a population that is used to make predictions about the whole population

population

the total number of individuals or items

census

the counting of an entire population

Reflecting

1. Whose ideas involve collecting primary data, and whose ideas involve collecting secondary data? Explain.

2. Whose suggestions are most likely to provide biased results for the student population? In what way would the data be biased?

3. Whose suggestions are most likely to provide unbiased results? Explain.

4. Would Ms. Chan's suggestions create or remove a source of bias? Explain.

Work with the Math

Example: Analyzing sources of bias in a survey

The student council decides to conduct a survey to see if the school community is in favour of changing the start time. How should the survey be set up? How might the results be biased?

Students' Solutions

Idea for Survey	Potential for bias
Jasleen: "We could arrive early one morning and survey the first 100 people—students, teachers, and parents—who walk by."	This survey would be biased in favour of those who prefer an early start. The first 100 people would not be those who have trouble arriving on time.
Heather: "At morning recess, we could phone 100 families selected randomly from the home phone number list."	This survey would be biased against parents who work outside the home, assuming that you don't leave a message that is answered.
Peter: "We could go down to the office after morning announcements and survey the first 100 people who walk into the office."	This survey would be biased in favour of those who prefer a late start. Many of the people in the office at this time would be arriving late and signing in.

Ⓐ Checking

5. The following students suggest other ways to conduct the survey. Would the results in each case be biased? Justify your answers.

a) Zach: "Ask 10 students, 10 teachers, and 10 parents. Then combine their answers."

b) Samantha: "Use the office student list, and survey the first 100 names."

c) Takumi: "Set up a desk in the front foyer, and let anyone who wishes complete the survey."

6. Explain whether each source is an example of primary data or secondary data.

   a) telephone interviews

   b) data from an encyclopedia

   c) information from a newspaper

Ⓑ Practising

7. Explain how each source could involve primary data or secondary data.

   a) completed questionnaires

   b) votes at an election

   c) observations of bird behaviour at a park

8. A survey may be biased in favour of or against different parts of a population. For each situation below, describe the groups for which the survey is likely to show bias.

9. For each survey question below, describe the sample you would use to avoid bias. Explain why your sample would avoid bias.

   a) What is the most common family size in your community?

   b) What are the most popular television shows for families in Ontario?

   c) What is the favourite type of music for people in your community?

   d) Do more people in your province prefer hockey or soccer?

   e) What is the best day of the week for students?

   f) If a movie is shown at one time only at the theatre, should it start at 4:00 p.m., 7:00 p.m., or 10:30 p.m.?

	Situation	Group that survey is likely biased in favour of	Group that survey is likely biased against
a)	A company selects every 100th name in the telephone book to call between 10:00 a.m. and 2:00 p.m., and doesn't leave messages. The question is "How many teenagers live in your home?"		
b)	Every 10th person who walks by a particular intersection in downtown Toronto is asked, "Should hunting be banned in Ontario?"		
c)	Every third person who enters a large toy store is asked, "Should the old age pension be increased?"		
d)	On Saturday mornings, from 9:00 a.m. to 11:00 a.m., every 10th family entering the local zoo is asked, "How much time, on average, do you spend with your children?"		
e)	An Internet survey is conducted to find out about computer use in Canadian households.		

10. The following names have been suggested for a new kind of cookies: Chocolatines, Mocha Chews, and Chocolicious. Explain why you would or would not use each method for collecting data to decide the best name.

a) census   **c)** questionnaire
b) interview   **d)** survey

11. Ms. Chan says that a census is the most accurate way to find out information about an entire population because each person in the population has an opportunity to respond. Why do you think a survey is used more often than a census?

12. When designing a survey, why should you be aware of bias? Explain one way to eliminate it.

ⓒ Extending

13. A town council needs to determine whether to spend money on building a new sports arena or a new library for the town. The council members want to make sure that they have the support of most of the 15 000 residents of the town. Decide on the best method to collect the data by ranking the following methods from best to worst. Then explain advantages and disadvantages of each method.

a) telephone survey

b) census

c) mail-in questionnaire

d) presentations by citizens

e) research

f) door-to-door interviews

Mental Math

MULTIPLYING AND DIVIDING BY 10, 100, AND 1000

You can multiply or divide a whole number or a decimal number by 10, 100, or 1000 by thinking about place value.

For example, to multiply 21.34 by 100:
tens become thousands,
ones become hundreds,
tenths become tens,
hundredths become ones.

Thousands	Hundreds	Tens	Ones	Tenths	Hundredths
		2	1 .	3	4

$$\times\, 100$$

Thousands	Hundreds	Tens	Ones	Tenths	Hundredths
2	1	3	4 .	0	0

1. What happens to each digit in a number when the number is multiplied or divided by 10, 100, or 1000?

2. Calculate.

a) 20×100 **c)** 10×0.425 **e)** 0.035×1000 **g)** 10.05×1000

b) $4200 \div 10$ **d)** $14.55 \div 10$ **f)** $120.6 \div 100$ **h)** $1250 \div 1000$

3.3 Using a Database

You will need
• a database program

▶ **GOAL**

Use a database to sort and locate data.

Learn about the Math

Marianne coaches a girls' softball team. She is using a **database** to store data about the players. She is looking at each player's **record** to determine who is the best batter on the team.

Player #	Last name	First name	Phone #	Position	Times at bat	Singles	Doubles	Triples	Home runs
1	Jones	Julie	555-9864	right field (RF)	22	5	4	2	1
2	Fernandez	Josie	555-3342	third base	18	1	0	1	0
3	Kong	Anita	555-8872	first base	18	5	2	0	0
4	Bown	Sally	555-9394	catcher (C)	18	3	3	0	0
5	Ali	Olivia	555-0032	pitcher (P)	23	4	1	0	0
6	Barrett	Marcie	555-9923	left field (LF)	18	5	2	1	0
7	Francis	Beth	555-7730	shortstop (SS)	20	2	2	1	0
8	Miller	Trina	555-7291	centre field (CF)	23	5	5	3	5
9	Miller	Pat	555-7291	second base	25	7	6	3	5

Record: |◄ ◄ | 4 | ► ►| ►*| of 10

- The orange row is a single record: all the data about one player.
- The yellow column is a single **field**: one type of data about all the players.
- The green cell is the position. The number in it is the **entry**. This entry is in a numeric field because it uses numbers.

? **How can you use this database to find the best batter on the team?**

A. In a database, each field should serve a particular purpose, or have a reason for being included. What fields are shown above?

B. A field can be a text field, a numeric field, a date field, or a memo field. Which type of field is each field in Marianne's database?

C. Which field has Marianne used to **sort** the information above?

D. State the name that would appear first if you sorted the information in the database by each of the following.
- **a)** last name
- **b)** first name
- **c)** times at bat
- **d)** phone number

E. Marianne wants her best batter to be up third. How should she sort the database to find her best batter? Which player should she choose to bat third? Why might someone disagree?

---

database

an organized set of information, often stored on a computer

record

all the data about one item in the database; for example, one player (see orange row)

field

a category used as part of a database; for example, last name (see yellow column)

entry

a single piece of data in a database; for example, home runs for one player (see green cell)

sort

order information from greatest (or first) to least (or last); a database can be sorted by fields

---

Reflecting

1. a) Why would it not make sense for Marianne to rank her players by sorting the database using "Phone #"? Why is it useful to have this field anyway?

b) What other fields might a coach find useful? Explain.

2. Marianne's database contains information for only 9 players. Suppose it is changed to hold the data for all 15 players on the team. How might the revised database be used?

3. Who might have information about you or your family in a database? Give two examples.

Work with the Math

Example 1: Sorting a database for letter frequencies

Indira decides to create a database for the information about the QWERTY keyboard and letter frequencies in lesson 3.1. Her database is shown. How can Indira sort her database to find out how many of the 10 most frequent letters are typed with the left hand?

Letter	Frequency in 1000 Letters	Row on keyboard	Finger used to type	Hand used to type
Q	3	top	pinky (fifth)	left
W	16	top	ring (fourth)	left
E	130	top	middle	left
R	77	top	index	left
T	93	top	index	left
Y	19	top	index	right
U	27	top	index	right
I	74	top	middle	right
O	74	top	ring (fourth)	right
P	27	top	pinky (fifth)	right
A	73	home	pinky (fifth)	left
S	63	home	ring (fourth)	left
D	44	home	middle	left
F	28	home	index	left
G	16	home	index	left
H	35	home	index	right
J	2	home	index	right
K	3	home	middle	right
L	35	home	ring (fourth)	right
Z	1	bottom	pinky (fifth)	left
X	5	bottom	ring (fourth)	left
C	30	bottom	middle	left
V	13	bottom	index	left
B	9	bottom	index	left
N	78	bottom	index	right
M	25	bottom	index	right

Record: 1 of 26

Indira's Solution

Letter	Frequency in 1000 letters	Row on keyboard	Finger used to type	Hand used to type
E	130	top	middle	left
T	93	top	index	left
N	78	bottom	index	right
R	77	top	index	left
I	74	top	middle	right
O	74	top	ring (fourth)	right
A	73	home	pinky (fifth)	left
S	63	home	ring (fourth)	left
D	44	home	middle	left
H	35	home	index	right

Record: 10 of 10

I sorted by the field "Frequency in 1000 letters." I used the "sort" button and chose "descending" instead of "ascending" to list the letters from greatest to least frequent. I see the letters in descending order of frequency. Six of these letters are typed with the left hand.

Example 2: Sorting a school database

Mrs. Cooper, the school secretary, keeps a student information database on the school computer. Part of her database is shown.

Student #	Last name	First name	Grade	Room	Home phone #	Emergency contact	Allergy	Bus #
4158	Brown	James	7	214	555-4411	555-2214	none	441
5532	Adams	Melissa	7	214	555-3392	555-9932	peanut	414
3327	Adams	Mark	5	116	555-3392	555-9932	none	414
7225	Chu	Frank	6	116	555-6248	555-8454	penicillin	walk
4233	Goring	Mandy	8	222	555-7302	555-9745	peanut	walk
3329	Hardy	Melissa	8	214	555-8780	555-4123	none	441
7764	Yan	Ye	5	116	555-9834	555-3381	none	414
7833	Walker	Mark	6	118	555-4487	555-2526	nuts	414

Record: 1 of 8

Mrs. Cooper wants to create a school phone directory of these eight names alphabetically by last name and then first name. If she sorts the database, which name will be third?

Mrs. Cooper's Solution

Student #	Last name	First name	Grade	Room	Home phone #	Emergency contact	Allergy	Bus #
3327	Adams	Mark	5	116	555-3392	555-9932	none	414
5532	Adams	Melissa	7	214	555-3392	555-9932	peanut	414
4158	Brown	James	7	214	555-4411	555-2214	none	441
7225	Chu	Frank	6	116	555-6248	555-8454	penicillin	walk
4233	Goring	Mandy	8	222	555-7302	555-9745	peanut	walk
3329	Hardy	Melissa	8	214	555-8780	555-4123	none	441
7833	Walker	Mark	6	118	555-4487	555-2526	nuts	414
7764	Yan	Ye	5	116	555-9834	555-3381	none	414

Record: 1 of 8

First I sorted the last names alphabetically. Then I sorted the first names alphabetically.

Now the third name is James Brown.

A Checking

4. State which name will appear last if Mrs. Cooper sorts the database by the following information.

 a) grade, from highest to lowest

 b) grade, from lowest to highest

 c) room number, from greatest to least

 d) room number, from least to greatest

5. Which kind of field (text, numeric, date, or memo) should you choose if you want to sort the following data?

 a) e-mail addresses

 b) math marks

 c) shoe sizes

 d) comments about student effort

 e) days when deliveries are expected

 f) ingredients in cereal

B Practising

6. So far, Anthony has input the following data in a database of all the videos his family owns.

Code	Title	Studio	Year released	Running time (min)	Notes
1	Grease	Paramount	1977	110	has interviews with the stars
2	Shrek	Dreamworks	2001	93	features Mike Myers doing Shrek's voice
3	The Little Mermaid	Disney	1989	83	was my first video
4	Casper	Universal	1995	101	produced by Steven Spielberg
5	Agent Cody Banks	MGM Studios	2003	103	stars Frankie Muniz

Record: 1 of 5

 a) Identify which fields are numeric, text, date, or memo.

 b) If Anthony sorts his database by "Year released" (from most recent to least recent), which movie will be third?

7. The following database lists information about different countries in the world.

Country	Continent	Area (square km)	Population (millions)	Population density (people/ square km)
Australia	Australia	7 687 000	19.4	2.7
Brazil	S. America	8 512 000	174.4	20.6
CANADA	N. America	9 976 000	31.6	3.5
China	Asia	9 597 000	1273	137.7
Congo	Africa	342 000	53.6	23.7
Iceland	Europe	103 000	0.3	2.7
Jamaica	N. America	11 000	2.7	248.2
Niger	Africa	1 267 000	10.4	8.2
Singapore	Asia	620	4.3	6693.4
U.S.A.	N. America	9 373 000	278.1	30.7

Record: 1 of 10

a) How is the data sorted in this database?

b) If you sorted this database by population, which country would appear first?

c) If Iceland appeared last, which field would you have used to sort?

d) If Singapore appeared first, which field would you have used to sort?

e) If you sorted this database in increasing order by area, what would be the first five countries?

8. As a school fundraiser, students are selling chocolates. This is a sample order form.

The ABC Chocolate Bar Company
Order Form

Student's name: _____   Homeroom: _____   Phone number: _____

Customer's name:	Chocolate Almond Bars @ $2.50 each		Chocolate Mint Wafers @ $1.50		Milk Chocolate Megabars @ $2.00		Sales summary	
	Quantity	Cost	Quantity	Cost	Quantity	Cost	Quantity	Cost
Jennifer Jones	4	$ 10	2	$ 3	4	$ 8	10	$ 21

a) How is an order form the same as a record in a database?

b) Explain how you could use the information in the order form to create a database for the school.

c) Use a computer to create a sample database from the order form.

9. Not all databases are computerized. You use a database every time you look up a phone number in the phone book. What databases would you use to find the following information?

a) what to order in a restaurant

b) how to spell a word

c) tomorrow's weather conditions

d) an alternate word for "good"

e) the area of a country

10. Katya wants to get some information for a project on probability. She goes to the school library and uses the library's database to find the following information.

Call #	Author	Title	Publisher	Date published	Status	Description	Category
519.2 Cus	Cushman, Jean	Do You Wanna Bet?	Clarion Books	1991	checked out	story about two boys who find out that the most ordinary events are dependent on probability	probability
123.302 Rip	Ripley, Robert	Book of Chance	Collins	1982	checked in	thousands of extraordinary facts about winners and losers	probability
519.2 Hol	Holland, Bart	What Are the Chances?	Johns Hopkins University Library	2002	checked in	voodoo deaths, office gossip, and other adventures in probability	probability

Record: 3 of 4

Identify which field Katya should use for each situation.

a) if she wants the most recent book

b) before she looks for the book on the shelves

c) when she is trying to find the book on the shelves

d) if she wants to find other books by the same author

❻ Extending

11. Describe how each worker might use a database.

a) a family doctor   **c)** a police officer

b) a veterinarian   **d)** a real estate agent

Using a Spreadsheet

You will need
- a spreadsheet program
- a calculator

▶ **GOAL**

Use a spreadsheet, and understand the difference between a spreadsheet and a database.

Learn about the Math

Heidi is organizing a pizza lunch for her school. Each class has given her a form that shows the total number of slices of pizza and drinks.

Heidi uses a **spreadsheet** program on a computer. She enters the data for each class in the **cells**, as shown below. She uses a **formula** to calculate the total number in a column.

	A	B	C	D	E	F	G	H
1	Class	Pepperoni	Mushroom	Cheese		Orange	Cola	Root Beer
2	8A	15	7	5		7	9	11
3	8B	13	12	2		9	14	4
4	8C	8	14	7		6	12	11
5	7A	10	8	12		2	15	13
6	7B							
7								
8	Total	46	=sum(C2:C7)	=sum(D2:D7)		=sum(F2:F7)	=sum(G2:G7)	=sum(H2:H7)
9	Number of whole pizzas	=B8/8	=C8/8	=D8/8	Number of cases of pop (12 cans per case)	=F8/12	=G8/12	=H8/12

spreadsheet

an orderly arrangement of numerical data using rows and columns; computerized spreadsheets can use formulas to perform calculations with the data

cell

the intersection of a column and a row, where individual data entries are stored; for example, cell B2 shows the entry in row 2 and column B

❓ How can you determine the total order for the school?

A. What information is found in column G? in row 3?

B. What data is in cell F4? in cell H5? in cell B8?

C. Describe how to enter the following data for class 7B: Cheese 7, Pepperoni 13, Mushroom 10, Root beer 5, Cola 12, Orange 13.

D. Each formula shown in red is hidden on the screen. It is automatically replaced by the calculated result after you enter a formula. This has already happened in cell B8. What formula was entered to give this result? Calculate the other results that will appear in row 8.

E. Each cheese pizza has 8 slices. How is this information used in row 9? Calculate the results that will appear in row 9.

F. If class 8C wants to change its order to 10 mushroom pizzas, how will this change the totals? What is the only change you have to make to the spreadsheet to change the order for the school? Explain.

formula

calculations made within a cell using data from other cells; formulas may vary depending on the spreadsheet program used; for example, the formula for cell C8 is =sum(C2:C7), which tells the program to add the numbers in column C from row 2 to row 7

Reflecting

1. As soon as Heidi entered a value into cell B6, the number in B8 changed. Why did this happen?

2. How can Heidi use the spreadsheet to determine the total order for the school?

3. Heidi noticed that the spreadsheet looked like a database in some ways. What can you do on a spreadsheet that you cannot do on a database, even if they both contain the same information?

Work with the Math

Example 1: Scheduling employees

Shane's employer uses the following spreadsheet to schedule the employees.

a) How many hours will Shane work this week?

b) Use the graphing button to create a bar graph that compares the hours the employees will work this week.

	A	B	C	D	E	F	G	H	I
1	Employee	Sunday	Monday	Tuesday	Wednesday	Thursday	Friday	Saturday	Total hours
2	Meghan	0	7	7	0	8	8	0	=sum (B2:H2)
3	Ahmed	4	0	0	4	7	7	0	=sum (B3:H3)
4	Shane	8	8	4	0	0	4	4	=sum (B4:H4)
5	Natasha	7	0	0	7	0	0	7	=sum (B5:H5)
6	Tien	0	4	8	8	0	0	8	=sum (B6:H6)
7	Dustin	4	4	0	0	4	4	0	=sum (B7:H7)
8	Tim	0	0	4	4	4	0	4	=sum (B8:H8)
9	Total	=sum (B2:B8)	=sum (C2:C8)	=sum (D2:D8)	=sum (E2:E8)	=sum (F2:F8)	=sum (G2:G8)	=sum (H2:H8)	=sum (I2:I8)

James's Solution

a) After I entered the formula in column I, the program automatically added all the data in row 4 and showed 28 in cell I4. So Shane will be working 28 hours this week.

b)

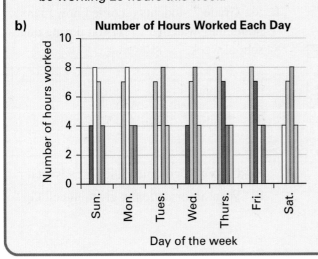

Number of Hours Worked Each Day

To create a bar graph, I highlighted all the hours worked, but not the totals. Then I chose the bar graph option in my spreadsheet program. When the screen opened, I selected "Column Graph." Then I selected "Titles" to add a title for the graph and for each axis.

This is my legend:

□ Meghan ■ Ahmed □ Shane □ Natasha
■ Tien ■ Dustin □ Tim

Create a new column, J, to calculate how much each employee in Example 1 will earn if the salary is $8.00 per hour.

Solution

	J2		= =I2*8							
	A	B	C	D	E	F	G	H	I	J
1	Employee	Sunday	Monday	Tuesday	Wednesday	Thursday	Friday	Saturday	Total hours	Total pay ($)
2	Meghan	0	7	7	0	8	8	0	30	240.00
3	Ahmed	4	0	0	4	7	7	0	22	176.00
4	Shane	8	8	4	0	0	4	4	28	224.00
5	Natasha	7	0	0	7	0	0	7	21	168.00
6	Tien	0	4	8	8	0	0	8	28	224.00
7	Dustin	4	4	0	0	4	4	0	16	128.00
8	Tim	0	0	4	4	4	0	4	16	128.00
9	Total	23	23	23	23	23	23	23	161	1288.00

To calculate the employees' salaries, create column J. Highlight the column and press the $ button to change the amounts to dollars. Then enter the formula J=I*8. This formula multiplies the total number of hours, as listed in column I, by $8. Remember that the * key means multiply.

A Checking

4. a) If Shane's hours are changed from 0 to 9 on Wednesday (cell E4), which cells will change automatically?

b) What will be the new values in these cells?

c) Explain how you would change column J if all the employees got a 50¢ per hour raise.

B Practising

5. The following spreadsheet shows the pet populations in selected countries.

	A	B	C	D	E	F
		Human population	Dogs	Cats	Reptiles	Small mammals
1	Country	(millions)	(millions)	(millions)	(millions)	(millions)
2	U.S.	278.1	59.3	75.1	7.7	13.1
3	China	1273.0	20.4	48.8	0.6	1.5
4	Russia	145.5	9.5	12.5	4.5	4.5
5	France	59.6	8.1	9.0	1.1	1.8
6	Italy	60.0	7.4	9.2	1.1	1.2
7	U.K.	59.6	6.5	8.0	1.7	3.7
8	Canada	31.6	3.8	4.8	0.19	1.1

a) State what is shown in
   i) column E   iii) cell E5
   ii) row 5   iv) cell B8

b) Which country has the most pet dogs?

c) Explain how you could find the total number of pets (in millions) for each country using a calculator. What formula could you use on a computer?

d) Explain how you could find the mean number of pets per person for each country using a calculator. What formula could you use on a computer?

e) Use a computer to create a graph that shows the number of pet reptiles in the different countries.

6. Six Grade 7 students recorded how many hours they watched television during one school week.

	A	B	C	D	E	F	G	H
1	Student's name	Monday	Tuesday	Wednesday	Thursday	Friday	Total	Mean
2	Rishi	3	1.5	1.5	4	5		
3	Justin	2	2.5	2	2.5	4		
4	Ben	1.5	1	1.5	2	2.5		
5	Christine	0	2.5	2.5	3	3.5		
6	Cameron	4	4	3.5	4	4.5		
7	Tyler	3	0	3	4.5	4.5		
8	Total							
9	Mean							

a) Copy and complete the spreadsheet.

b) What information is given in cell G5?

6. c) What formula did you use to calculate the value for cell D9?

d) Use your data to create a graph for the school newspaper. Explain why your choice of graph suits the data.

e) Change the data for Cameron (row 6) to the following: Monday 5 h, Tuesday 5 h, Wednesday 0 h, Thursday 5 h, Friday 0 h. Which daily mean has the greatest change?

7. This table shows sales at Johnny's Hot Dog Stand.

Price of item	Sales						
	M	T	W	T	F	S	S
hot dog $2.00	57	62	48	62	87	130	114
sausage $3.00	33	23	21	39	50	64	50
soft drink $1.50	40	58	55	75	92	132	108
chips $1.50	12	18	7	20	12	37	34

a) Use the data to complete the following spreadsheet. (The first row has been done for you.)

	A	B	C	D	E	F	G	H	I	J
	Day	Hot dog sales	Total revenue @ $2	Sausage on a bun sales	Total revenue @ $3	Soft drink sales	Total revenue @ $1.50	Chip sales	Total revenue @ $1.50	Total daily revenue
1										
2	Monday	57	$114.00	33	$99.00	40	$60.00	12	$18.00	$291.00

b) Johnny wants a day off. Which day would be best, based on these daily sales?

c) How much did Johnny collect in sales for the week?

d) Create a graph to display the week's hot dog sales. Which type of graph did you use to show this information? Why?

8. Describe how a spreadsheet can be used in your favourite sport or hobby. What categories of information would be included? Why?

C Extending

9. A school has a vending machine that sells water, orange juice, apple juice, and grapefruit juice. Over the years, the school has found that the ratio of water to orange juice to apple juice to grapefruit juice sold is $5:6:8:1$.

a) If a total of 100 drinks are sold each week, how many apple juices are sold?

b) Make a computer spreadsheet to help the school order the correct number of drinks. If a number is entered into any drink column (as shown below), your spreadsheet should automatically update to show how many other drinks to order, as well as the total number of drinks to order.

c) Use a formula to calculate the percent of each kind of drink sold.

	A	B	C	D	E	F
1		Water	Orange juice	Apple juice	Grapefruit juice	Total drinks
2	Week 1	50				
3	Week 2		66			
4	Week 3			40		
5	Week 4				14	

10. Use the data in question 7.

a) What is the ratio of the total number of hot dogs sold during the week to the total number of sausages sold? Use your ratio to decide how many hot dogs Johnny should order next week, if he orders 300 sausages.

b) What is the ratio of the number of soft drinks sold to the number of chips sold? If Johnny sells 300 soft drinks next week, how many chips will he probably sell?

Mid-Chapter Review

Frequently Asked Questions

Q: What are some of the ways you can collect data?

A: Primary data are collected directly. You can collect primary data by conducting a survey, a census, an experiment, a questionnaire, or an interview. Secondary data are collected by someone else. Secondary data can be accessed by using reference materials, such as encyclopedias, almanacs, atlases, and the Internet.

Q: What is meant by bias in data collection?

A: Bias occurs if a sample is not similar enough to the total population. For example, suppose that the police want to make sure that people are driving at safe speed limits in a school zone. Sitting in a police car outside the school will produce a biased sample. Drivers will slow down as soon as they see the police car. Using an unmarked car will give better results. It is difficult to avoid bias completely, but it is important to reduce bias whenever possible.

Q: How are a database and a spreadsheet alike? How are they different?

A: They are both collections of information. They can both be created by hand or on a computer by entering data.

A database consists of different sections called fields. Each field contains specific data. For example, a school database might include numeric fields, with marks and phone numbers. Names and subjects are entered in text fields. Comments are entered in a memo field. The data can be sorted by any field except the memo field.

A spreadsheet organizes data using columns and rows. In this spreadsheet, the rows have students' names, and the columns have the number of hours the students watched television. The number of hours for each student is entered in a cell. A spreadsheet allows formulas to be entered in the cells. Calculations are automatically performed by substituting data from other cells into these formulas.

	A	B	C	D	E	F	G	H
1	Student's name	Monday	Tuesday	Wednesday	Thursday	Friday	Total	Mean
2	Rishi	3	1.5	1.5	4	5		
3	Justin	2	2.5	2	2.5	4		
4	Ben	1.5	1	1.5	2	2.5		
5	Christine	0	2.5	2.5	3	3.5		
6	Cameron	4	4	3.5	4	4.5		
7	Tyler	3	0	3	4.5	4.5		
8	Total							
9	Mean							

Practice Questions

(3.1) **1.** Suppose that you want to find out if more Canadians prefer eating toast or cereal in the morning. List four ways that you could find secondary data.

(3.2) **2.** Suppose that you are in charge of organizing the music for a school dance. You survey five of your friends about the kinds of music they like best.

   a) Did you use a census or a sample?

   b) Will your decisions based on this survey be biased? Why or why not?

Use the following database to answer questions 3 to 6.

Order date	Order #	Last name	Phone #	Frame style	Colour	Length (cm)	Width (cm)	Notes
10/28/2004	3321	Smith	555-4542	238	Black	34	78	Call after 5 P.M.
10/28/2004	3322	Brown	555-8964	711	White	67	32	Attach certificate to back.
10/28/2004	3323	Jones	555-9075	338	Golden Oak	45	45	Customer needs it by Friday.
11/1/2004	3324	Wong	555-8754	338	Mahogany	65	45	Use non-glare glass.

Record: 1 ▶▶ of 4

(3.3) **3.** Find an example of each field in the database above.

   a) a numeric field   **c)** a date field

   b) a text field   **d)** a memo field

(3.3) **4.** If this database was sorted in decreasing order by "Frame style," which order would be first?

(3.3) **5.** Give examples of three fields that might be useful additions to this database.

(3.3) **6.** What is the order number for the square frame?

(3.4) **7.** Describe one feature a database has that a spreadsheet does not have.

Use the following spreadsheet about Best Friends Pet Shop to answer questions 8 to 11.

	A	B	C	D	E	F
1	Type of fish	Quantity	Price per fish	Total price before tax =B*C	Tax = 0.15*D	Total price including tax =D+E
2	neon tetra	10	$2.00	$20.00	$3.00	$23.00
3	guppy	20	$1.50	$30.00	$4.50	$34.50
4	angelfish	5	$4.00	$20.00	$3.00	$23.00
5	black molly	5	$5.00	$25.00	$3.75	$28.75
6	swordtail	4	$6.25	$25.00	$3.75	$28.75

8. a) What is the heading for column B in the spreadsheet?

   b) What is the heading for row 6?

   c) What information is given in cell B6?

   d) What is one formula that is used in the spreadsheet?   (3.4)

9. a) Using the spreadsheet data, create a graph to compare the prices for the different fish.

   b) What type of graph did you use? Why?   (3.4)

10. The pet store also sells platyfish at 4 for $3.00. Kyle wants to add 10 platyfish to his order. Show how line 7 of the spreadsheet should be filled in.   (3.4)

11. Add a new column, column G, to the spreadsheet to show the profit. If the neon tetras cost $0.50 each, develop a formula to calculate the profit in column G.   (3.4)

Frequency Tables and Stem-and-Leaf Plots

You will need
• a ruler
• centimetre grid paper

▶ **GOAL**

Organize data using frequency tables and stem-and-leaf plots.

Learn about the Math

The students in Tonya's gym class were practising their standing long jumps. They jumped the following distances, in centimetres:

187	205	221	186	185	212	222	215	198
200	205	207	193	186	172	208	223	175
206	215	227	228	230	218	188	173	196
202	221	214	220	229	189	193	212	212

The students want to sort the data into levels of achievement. One group of students decides to create a **frequency table**. Another group thinks that a **stem-and-leaf plot** will show the levels better. A third group wants to use a bar graph. Each group has to decide on appropriate **intervals** to use.

frequency table

a count of each item, organized by categories or intervals

? **How can you compare the data in different achievement levels?**

stem-and-leaf plot

an organization of numerical data into categories based on place values; the digits representing greater values are the stems, and the other digits are the leaves

Example 1: Organizing data in a frequency table

Organize the data for standing long jumps in a frequency table.

Group 1's Solution

Interval (cm)	Frequency
170–179	3
180–189	6
190–199	4
200–209	7
210–219	7
220–229	8
230–239	1

We noticed that the longest jump is 230 cm and the shortest jump is 172 cm.

If we use every possible whole number length in centimetres from 172 to 230, the frequency table will be too large.

We decided to organize the data in intervals of 10.

interval

the space between two values; for example, 0–9 represents the interval from 0 to 9, including 0 and 9

Example 2: Organizing data using a stem-and-leaf plot and a bar graph

Organize the data for standing long jumps using each type of display.

a) a stem-and-leaf plot

b) a bar graph

a) Group 2's Solution

Long Jump Distances (cm)	
Stem	**Leaf**
17	2 3 5
18	5 6 6 7 8 9
19	3 3 6 8
20	0 2 5 5 6 7 8
21	2 2 2 4 5 5 8
22	0 1 1 2 3 7 8 9
23	0

We used intervals of 10 to organize the data in a stem-and-leaf plot.

We used the hundreds and tens digits as the stems, and the ones digits as the leaves.

b) Group 3's Solution

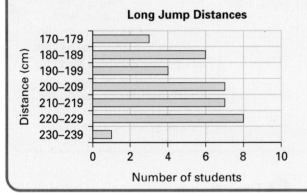

Long Jump Distances

Since we are graphing long jumps, we decided to put the bars horizontally.

We used intervals of 10 rather than including all the distances, because we think this will make the graph easier to read.

Reflecting

1. a) How is a stem-and-leaf plot like a frequency table?

  b) How is a stem-and-leaf plot like a bar graph?

2. When making a frequency table, stem-and-leaf plot, or bar graph, how do you choose appropriate intervals?

3. Which method is easiest for you to use to organize the data and visually compare the count in each category? Justify your choices.

4. Which method can be used to organize data without losing individual values?

Work with the Math

Example 3: Changing a stem-and-leaf plot into a frequency table

A company cafeteria has two lunch shifts. The second shift starts 1 h after the first shift. The cafeteria staff needs 10 min after the first shift to clean the tables and prepare for the second shift. To do this, at least 90% of the tables must be empty within 50 min.

The cafeteria staff records how much time the employees in the first shift take to finish lunch and leave the tables during one shift. Are 90% of the tables empty within 50 min, or do the lunch times have to change?

Amount of Time the Tables Are Occupied by First-Shift Employees (min)

37	45	28	45	52	38	30	25	46	43
27	39	47	44	29	30	35	45	44	56
60	36	33	25	25	43	27	30	45	42

Kwami's Solution

Time at Tables (min)	
Stem	**Leaf**
2	5 5 5 7 7 8 9
3	0 0 0 3 5 6 7 8 9
4	2 3 3 4 4 5 5 5 5 6 7
5	2 6
6	0

Interval (min)	Frequency
20–29	7
30–39	9
40–49	11
50–59	2
60–69	1

There are 30 tables, and 90% must be empty within 50 min. So, this means that 27 of the 30 tables must be empty within 50 min. Only 3 tables were occupied for 50 min or more. So, the lunch times can stay as they are.

A Checking

5. Choose appropriate intervals to organize each set of data in a frequency table.

   a) lengths of short stories (number of words): 120, 173, 287, 599, 183, 298, 376, 922

   b) times for candles to burn completely (min): 120, 125, 129, 128, 125, 122, 120, 123

   c) attendance at shows (number of people): 120, 4989, 2998, 2774, 1487, 159, 3992

   d) heights of plants (cm): 120, 387, 428, 127, 287, 125, 332, 487

B Practising

6. Would you use a stem-and-leaf plot, a frequency table, or either to organize each set of data? Explain your choice.

   a) the absences in each class in a school

   b) the heights of NBA basketball players

   c) an inventory of textbooks in a classroom

7. a) If you were collecting data on the colours of vehicles in a parking lot, why would a stem-and-leaf plot not be appropriate?

   b) Could you use a frequency table? Explain your answer.

8. Use this stem-and-leaf plot to answer the questions below.

Vehicle Speed on Highway (km/h)	
Stem	Leaf
8	0 5 5 7 7 7 9 9 9 9 9
9	2 3 3 4 4 6 6 6 8 8 9 9 9 9 9
10	0 1 1 1 1 4 6 8 8 8 9

a) How many vehicles had their speeds measured?

b) How many vehicles were travelling over 90 km/h?

c) If the speed limit on the highway is 100 km/h, how many vehicles were speeding?

d) What percent of vehicles were speeding?

9. If you collected the sets of data in question 6, which would be primary data and which would be secondary data? Explain your thinking.

10. Rosa is in a bowling league. These are her scores for the season.

132 118 122 106 94 94 112 118
104 120 108 104 96 122 130 116
104 118 106 124

a) Display Rosa's scores in a stem-and-leaf plot.

b) What is the range of scores? The range is the difference between the highest and lowest score.

c) The team's mean score was 110. In what percent of her games did Rosa score above 110 points?

d) Create a bar graph to show Rosa's scores. Label the horizontal axis "Game number" and the vertical axis "Points scored."

11. The quality control engineers at a potato chip factory want to ensure a consistent product. They count the number of chips in every 100th bag. Their results are shown below.

Number of Potato Chips per Bag

135 154 188 137 123 151 122 134
123 119 108 119 143 150 132 128
129 144 123 145 127 126 107 150
127 132 133 127 142 117 108 125
122 137 96 99

a) Organize these results in a stem-and-leaf plot.

b) Use your stem-and-leaf plot to create a bar graph that the company can display in the employee cafeteria.

12. The following stem-and-leaf plot shows the number of passengers who rode a train along a certain route during a one-month period. Record the data in a frequency table using intervals of 5.

Number of Monthly Passengers on Train	
Stem	Leaf
9	5 7 9
10	0 0 3 5 8 9
11	2 4 4 6 8 8 8
12	3 4 7 7 9 9 9 9
13	2 4 4 6 8
14	2 3 3 3 5
15	0 1

ⓒ Extending

13. The points scored for and against a basketball team are listed below. Create a two-sided stem-and-leaf plot to show the points scored for and against the team.

Points scored for team: 129, 108, 114, 125, 132, 107, 97, 127, 108, 124, 117, 94, 99, 108

Points scored against team: 113, 127, 132, 109, 101, 90, 88, 112, 109, 122, 119, 102, 110, 97

3.6 Mean, Median, and Mode

▶ **GOAL**

Describe data using the mean, median, and mode.

Learn about the Math

Asha, Peter, and Winnie are the captains of the school math teams. They wrote all their contest results in this table.

Team Captain	Contest								
	# 1	# 2	# 3	# 4	# 5	# 6	# 7	# 8	# 9
Asha	82	82	88	100	74	81	87	83	83
Peter	84	84	90	71	78	87	89	88	86
Winnie	85	85	85	81	81	85	82	85	83

? **Why is it difficult to determine which is the top math team?**

A. Use a calculator to calculate the **mean** for each team. Round your answer to one decimal place. Which is the top math team based on the mean?

B. Determine the **median** mark for each team. Which is the top math team based on the median mark?

C. Determine the **mode**. Which is the top math team based on the mode?

D. Using the mean, median, and mode, which do you think is the top math team? Why might someone else disagree?

Reflecting

1. The mean, median, and mode are all used to determine similar information. What do they all measure?

2. If you were asked to determine the most common hair colour in your class, why is only one of the mean, median, or mode possible to use?

mean

the sum of a set of numbers divided by the number of numbers in the set

median

the middle value in a set of ordered data; when there is an even number of numbers, the median is the mean of the two middle numbers

mode

the number that occurs most often in a set of data; there can be more than one mode or there might be no mode

Work with the Math

Example 1: Using a stem-and-leaf plot to calculate the median

Kwami, Indira, and Simon planted seeds as part of their science project. They waited two weeks and then measured the heights of the seedlings. How can they determine the average height?

Kwami, Indira, and Simon's Solution

Height of Plants (cm)	
Stem	**Leaf**
3	9
4	2 4 4 7 7 7
5	0 9 9
6	1 4
7	3 5 6 7
8	7 8 9
9	0 0 2 8 8 8
10	5 8
11	2 6
12	0

We constructed a stem-and-leaf plot to display the measurements.

We decided to use the median as the average height. The median is the number in the middle.

Since there are 30 numbers, the median is the number halfway between the 15th and 16th numbers.

The 15th number is 76.

The 16th number is 77.

$(76 + 77) \div 2 = 76.5$

The median height of the plants is 76.5 cm.

Example 2: Using a spreadsheet to determine the mean

For their science project, Mohammed and Marie wanted to know how the amount of water affects the growth of flower seeds. How can they use a spreadsheet program to analyze their data?

Mohammed and Marie's Solution

	A	B	C	D	E	F
	Amount of water	**Height of plant A (cm)**	**Height of plant B (cm)**	**Height of plant C (cm)**	**Height of plant D (cm)**	**Mean height (cm)**
1						
2	no water	2	2.5	1.5	2	=AVERAGE (B2:E2)
3	lightly watered	20	34	28	33	=AVERAGE (B3:E3)
4	moderate watering	74	84	66	82	=AVERAGE (B4:E4)
5	heavy watering	17	8	23	4	=AVERAGE (B5:E5)

There's no mode in our results.

The median doesn't seem useful because there are big gaps between some numbers.

We decided to use the mean. To find the mean, we entered a formula in each row of the spreadsheet.

A Checking

3. Calculate the mean, the median, and the mode for each set of data.

   a) 5, 8, 4, 8, 7, 8, 6, 10

   b) 4, 5, 7, 10, 7, 2, 5, 7

   c) 18, 22, 22, 17, 30, 18, 12

   d) 5, 5, 5, 0, 5, 10, 10

B Practising

4. Calculate the mean for each set of data.

   a) 23, 52, 40, 23, 56, 96

   b) 208, 112, 321, 207, 308, 171

   c) 6.2, 7.4, 6.74, 8.33, 8.8, 3.2

   d) 55%, 58%, 92%, 74%, 63%, 78%

   e) 42, 84, 99, 103, 33, 61

5. Calculate the median for each set of data.

   a) 4, 8, 2, 9, 3, 3, 0

   b) 32, 88, 13, 54, 84

   c) 80%, 69%, 72%, 86%, 91%, 42%

   d) 312, 221, 873, 992, 223, 224

   e) 5.40, 2.88, 1.71, 3.50, 9.02

6. Determine the mode for each set of data.

   a) 7, 8, 9, 9, 8, 6, 6, 9, 8, 4

   b) 18, 19, 19, 12, 17, 16, 18, 18, 12, 16

   c) 4.3, 7.1, 8.8, 7.1, 7.2, 7.6, 4.3, 7.1, 8.8, 7.0

   d) B, G, F, G, G, A, F, F, C, D

   e) 93%, 75%, 61%, 93%, 75%, 93%

7. Suppose that you are planning a winter vacation. You want to spend a lot of time outdoors, so you are checking the "average" temperature in different locations. Would you prefer to have the "average" temperature reported as a mean, a median, or a mode? Why?

8. Andrew is conducting a survey to determine the most common eye colour. Why does he need to use the mode, not the mean or the median, to analyze his results?

9. Chocolate bars are on sale for the prices shown in this stem-and-leaf plot.

Cost of a Chocolate Bar (in cents) at Several Different Stores	
Stem	**Leaf**
7	7
8	5 5 7 8 9
9	3 3 3
10	0 5

   a) Calculate the mean, median, and mode prices for the chocolate bars.

   b) Which measure (mean, median, or mode) do you think is the most appropriate for analyzing this set of data? Why?

10. The following table shows the number of picture books that Grade 7 students read to their Grade 1 reading buddies during the first three months of the year.

Number of Books Read So Far This Year		
Student	**Tally**	**Frequency**
Maryann	ＨＨＨ ＨＨＨ ＨＨＨ ＨＨＨ I	21
Rachel	ＨＨＨ ＨＨＨ ＨＨＨ III	18
Peter	ＨＨＨ ＨＨＨ ＨＨＨ	
Raj	ＨＨＨ ＨＨＨ ＨＨＨ IIII	
Terry	ＨＨＨ ＨＨＨ ＨＨＨ IIII	
Sam	ＨＨＨ ＨＨＨ III	
Petra	ＨＨＨ ＨＨＨ ＨＨＨ IIII	
Siobhan	ＨＨＨ ＨＨＨ ＨＨＨ II	
Alex	ＨＨＨ ＨＨＨ II	
Suki	ＨＨＨ ＨＨＨ ＨＨＨ IIII	

   a) Complete the frequency table.

   b) What is the mean, median, and mode number of books read by all the students?

11. The ages, in years, of the members of a children's choir are listed below.

9   10   11   7   14   11
10   8   7   7   11   11

a) Calculate the mean, median, and mode for the ages.

b) If two new 7-year-olds join the choir, how do the mean, median, and mode change?

12. If the mean of six numbers is 48, does one of the numbers have to be 48? Explain why or why not. Give an example with six numbers to show your answer.

13. A company employs 16 people. The chief executive earns $200 000 a year, the director earns $150 000, the two supervisors earn $100 000 each, the four managers earn $50 000 each, and the eight factory workers earn $35 000 each.

a) Calculate the mean, median, and mode salary for the company.

b) If the director wants to attract new factory workers, how should she report the "average" salary?

c) If one of the factory workers wants a raise, should he use the mean, the median, or the mode to justify his demands?

d) Which measure (mean, median, or mode) best describes the "average" salary at this company? Explain your answer.

14. We often use the word "average" when we are referring to something that is the most common or frequent. In each phrase below, does "average" refer to the mean, median, or mode? Justify your answer.

a) the average monthly temperature

b) just an average day in the life of a student

c) the average hairstyle in the 1980s

d) the average number of cars in the parking lot

● Extending

15. A teacher records the test marks of 25 students and calculates the class mean to be 72. Jean-Pierre finds out that his mark of 86 was recorded incorrectly as 36. What will the corrected class mean be?

16. The mean of five different numbers is 4. When the greatest number is removed from the set, the mean of the remaining numbers is 2. What number is removed?

17. Six numbers are all less than 10. The greatest number is 9, and the least number is 1. The mean of these numbers is 6, the median is 7, and the mode is 8. When the number 6 is removed, the mean of the remaining five numbers is 6, the median is 8, and the mode is 8. What three numbers are missing?

	Numbers: 9, 1, 6, ■, ■, ■	Numbers: 9, 1, 6̸, ■, ■, ■
Mean	6	6
Median	7	8
Mode	8	8

Math Game

TARGET MEAN

The goal of this game is to make four two-digit numbers that, when averaged, equal or are close to a target mean.

Number of players: 2 or more

Rules

1. Decide on a target mean between 11 and 66; for example, 35.

2. To make the two-digit numbers, players take turns rolling one die and deciding whether to put the digit rolled in the tens place or the ones place. After a player rolls the die on turn eight, the player will have four numbers.

3. Players calculate the mean, rounded to two decimal places.

4. The player whose mean is closest to the target mean wins 1 point.

5. The first player to reach 3 points wins.

Example: Target mean is 35:	Simon's card	Jody's card
Simon rolls a 2. He decides to put the 2 in the tens place for his first number, since 2 is fairly close to 3.	2 _ __ __ __	__ __ __ __
Jody rolls a 6. She puts it in the ones place.	2 _ __ __ __	_ 6 __ __ __
Simon rolls a 3. This must go in the ones place to complete the first two-digit number.	2 3 __ __ __	_ 6 __ __ __
Jody rolls a 1. She must put it in the tens place.	2 3 __ __ __	1 6 __ __ __
Simon rolls a 4. He puts it in the tens place to begin his second number.	2 3 4 _ __ __	1 6 __ __ __
The players continue taking turns until they fill in all four numbers.	2 3 4 2 3 5 6 6	1 6 2 4 5 1 4 6
The players calculate each mean, to two decimal places. Jody wins the first round, since her mean is closer to the target of 35. She gets 1 point.	Mean = 41.50	Mean = 34.25

THE DVORAK KEYBOARD

When the first typewriter was invented, typists often typed so fast that the mechanical keys jammed. To avoid jamming, the standard QWERTY keyboard was developed. Modern keyboards do not jam, but the QWERTY keyboard remains.

To improve efficiency, different arrangements of letters have been recommended, based on letter frequency. One of these arrangements, invented by Dr. August Dvorak, is called the Dvorak keyboard.

Letters in each row (Dvorak keyboard)	Combined frequency of letters in row (using secondary data from letter frequency table in lesson 3.1)	
	Count	Percent
Row above home row: pyfgcrl	232	$\frac{232}{1000}$ or $\frac{23}{100}$ = 23%
Home row: aoeuidhtns	691	$\frac{691}{1000}$ or $\frac{69}{100}$ = 69%
Row below home row: qjkxbmwvz	77	$\frac{77}{1000}$ or $\frac{8}{100}$ = 8%

1. a) Compare the arrangement of the letters on the Dvorak keyboard with the arrangement of the letters on the QWERTY keyboard, shown in lesson 3.1.

b) Compare the Dvorak table, shown above, with the QWERTY table, shown in lesson 3.1.

2. Can you do better than Dvorak? Design the perfect keyboard, based on the letter frequency table in lesson 3.1.

- Put the most frequently used letters on the 10 keys for the home row.
- Put the next frequently used letters on the 8 keys above the home row.
- Put the least frequently used letters on the keys in the row below the home row.

You may want to add symbols and punctuation marks.

3. Test your keyboard against both the QWERTY and Dvorak keyboards.

a) Create a tally chart for the letters in the paragraph at the top of this page. There are 345 letters in the paragraph.

b) Calculate how many of these letters can be typed using only the letters on the home row of your keyboard, the QWERTY keyboard, and the Dvorak keyboard.

3.7 Communicating about Graphs

▶ **GOAL**

Make inferences and convincing arguments that are based on analyzing data and on trends.

Communicate about the Math

Kevin is in charge of milk sales at his school. He has prepared a report for the principal to show the sales for the past five months, and to recommend the quantities of chocolate milk and white milk that should be ordered for the month of February.

Kevin's friend Jody reads his report and makes some comments.

? **How can Kevin improve his report?**

Kevin's Report

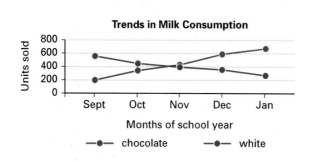

Trends in Milk Consumption

The double broken-line graph shows changes in the sales of white milk and chocolate milk over the past five months. The graph shows that more students used to buy white milk than chocolate milk, but now more students buy chocolate milk. I think they weren't allowed to buy chocolate milk at the start of the year, but they convinced their parents to let them buy it ...

Jody's Questions

How do you know that?

What do you mean by "start of the year"?

Why do you think that's the reason?

But how much of each kind of milk should be ordered for February? You need to make a recommendation.

A. Which of Jody's questions do you agree with? Why?

B. How could Kevin respond to Jody's questions and improve his report?

C. What other questions can you ask to help Kevin improve his report?

Reflecting

1. Which parts of the Communication Checklist did Kevin cover well? Explain.

2. Why might it be useful to communicate about a graph?

3. How can you use the Communication Checklist to write a better report?

Work with the Math

Example: Analyzing a circle graph

Kevin wants to show that sales of chocolate milk are higher at the end of a week, so that more chocolate milk should be ordered then. He creates the following graph. Interpret his graph.

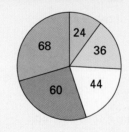

Chocolate Milk Sold in a Week
- ○ Monday
- ○ Tuesday
- ○ Wednesday
- ● Thursday
- ○ Friday

Jody's Solution

- The different numbers on the graph tell you the average number of chocolate milks sold each day.

- The sizes of the sections in the graph show that the amounts increase every day from Monday to Friday.

- Since more than half the circle is for Thursday and Friday, this means that more than half the chocolate milk is sold on the last two days in a week.

- On a Friday, more than $\frac{1}{4}$ of the week's chocolate milk is sold. So, if you order for a week that has a holiday on Friday, you need to reduce the usual order by $\frac{1}{4}$.

- On a Monday, only about $\frac{1}{10}$ of the week's chocolate milk is sold. So, if you order for a week that has a holiday on Monday, you need to reduce the usual order by only $\frac{1}{10}$.

A Checking

4. The school principal wants to know the quantities of chocolate milk and white milk that should be ordered each week, so Janet created a graph and interpreted it. Improve Janet's explanation, using the Communication Checklist to help you.

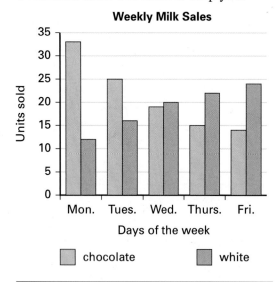

Weekly Milk Sales

chocolate   white

> If you look at the graph, you can see
>
> that more people drink chocolate milk
>
> at the beginning of the week. So, we
>
> should order more chocolate milk than
>
> white milk since there are more people
>
> drinking chocolate milk. I think we
>
> should order 100 white milks each week
>
> since there are five days in a week and
>
> the greatest sales number is 24. I think
>
> we should order 150 chocolate milks
>
> each week since the highest sales day
>
> is about 30 milks.

B Practising

Use the Communication Checklist to help you answer the following questions.

5. Robert surveyed all 24 family members at his family reunion about their favourite kinds of pie. Then he prepared the following graph.

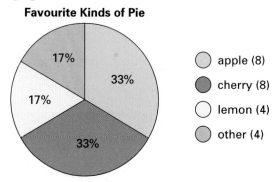

Favourite Kinds of Pie

apple (8)
cherry (8)
lemon (4)
other (4)

What advice could Robert give about the pies to serve at the next family reunion?

6. Karen and Ahmed work at a children's clothing store. They use a database to keep track of the kinds of pants that are sold each month. Here are their data:

Pants	January	February	March	April	May	June	Total
corduroy	235	215	203	185	170	120	1128
cargo	180	195	204	240	245	268	1332
denim	300	310	308	293	304	294	1809
dress	110	105	95	112	104	111	537
capri	28	36	52	78	102	185	481

Record: ◄◄ ◄ 1 ► ►► ►* of 5

a) Create a graph to show their data.

b) Use your graph to write a report that will help the store manager know what kinds of pants to carry in the store. Look at the trends. Include answers to the following questions:
- Which pants are increasing in popularity?
- Which pants are decreasing in popularity?
- Which pants are remaining steady?
- How can you use the data to order pants for the next six-month period?

Chapter Self-Test

1. Mr. Connelly, the school teacher-librarian, wants to purchase new books for the school library. He has the following ideas for collecting data:

   A. Observe which books are most commonly taken out, and buy more books by the same authors.

   B. Ask all the students who come into the library what new books they would like to have in the library.

   C. Visit a Web site that lists the award-winning books for the past three years.

   D. Create a questionnaire to give to all the students in the school, asking about their favourite kinds of books.

   a) Which ideas involve collecting primary data?

   b) Which ideas include a bias in favour of students' opinions?

   c) Which idea includes a bias against students' opinions?

2. Suppose that you want to find out which restaurant in your community makes the best hamburgers. You decide to conduct a survey of 25 people to gather their opinions. What biases would probably occur if you use each sample below?

   a) 25 students in your classroom

   b) people leaving one of the nearby restaurants

   c) people entering a grocery store

   d) players on the local football team

3. The videos in a school media centre are organized in the following database.

Catalogue #	Title	Length (min)	Subject	Grade level(s)	Current status
378	Stable Structures	30	science	7	in
559	Introducing Canada	45	history	7	out (Grant Ave. P.S.)
3177	Database Dilemma	40	math	7, 8	in
4569	Intro to Integers	25	math	7, 8	out (Coronation P.S.)
8892	Art Is for All	30	art	1-6	out (Pinewoods P.S.)
9772	Our Changing Earth	50	science, geography	7	in

   Record: 6 of 6

   a) Currently, the videos are sorted by the field "Catalogue number" in ascending order. Identify the video that will appear first if the videos are sorted by

   i) title, in descending order

   ii) time, in ascending order

   iii) subject, in ascending order

   b) From the database, give one example of

   i) a numeric field

   ii) a text field

   iii) an entry

   iv) a record

4. Schools often use spreadsheets to order supplies. Use the following spreadsheet to answer the questions below.

	A	B	C	D	E	F
1	Item number	Item	Unit price	Quantity	Total cost =C*D	Price with tax =E*1.15
2	3342	bond paper, case of 5M	$20.50	6		
3	3348	const. paper, black, 50	$3.35	8		
4	4511	paintbrush, acrylic, each	$0.49	12		
5	4899	paint, acrylic, blue, 2L	$15.23	3		
6	5788	pencils, w. eraser, 12	$1.25	3		

   a) What is the entry in cell A6?

   b) What formula should be in cell E2?

   c) What amount should appear in cell E4?

   d) Mr. Singh wants to place an order for four jugs of green acrylic paint (item no. 4898) at a cost of $15.23 per 2 L jug. Show how this would appear on line 7 of the spreadsheet.

5. Describe one feature that a spreadsheet has that a database does not have.

6. A biologist who is studying gulls measured and recorded these wingspans (in centimetres).

132	145	162	135	142
122	138	124	135	140
128	122	145	138	139
122	146	150	167	128
134	147	151	122	

a) What intervals should the biologist use to record the wingspans in a stem-and-leaf plot?

b) Record the wingspans in a frequency table.

c) Which interval had the greatest frequency?

d) Which value is the mode for the wingspans?

7. The daily high temperatures (in °C) for the month of June are listed below.

19	22	18	23	21
20	20	18	21	24
25	24	27	28	26
24	25	26	27	25
26	27	30	31	33
33	33	30	33	27

a) Calculate the median, mean, and mode for the temperatures.

b) A newscaster reports that the average temperature for the month of June was 33°C. Was the median, the mean, or the mode used to calculate the average? Why would this be misleading?

c) Which measure (mean, median, or mode) would best describe the average temperature for the month of June? Explain.

8. The following graph shows how much space in *The Tribune* is devoted to the various sections.

Space Devoted to Sections in *The Tribune*

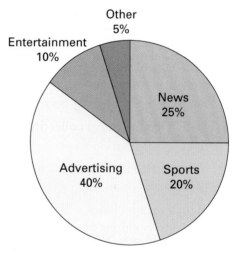

a) How could you use this graph to respond to readers who say that too much of the newspaper is filled with advertising? What would you tell them?

b) How could you use this graph to explain to advertisers that you value their business? What would you tell them?

Chapter Review

Frequently Asked Questions

Q: How are frequency tables, stem-and-leaf plots, and bar graphs alike? How are they different?

A: All three show numerical data organized in different intervals. A frequency table shows the total in each interval as a number. A bar graph shows the total visually. A stem-and-leaf plot shows both and also shows all the original data.

Number of Years Our School's Teachers Have Taught					
Frequency table		**Stem-and-leaf plot**		**Bar graph**	

Number of years taught	Frequency	Stem	Leaf
0–9	7	0	1 1 4 6 6 8 9
10–19	4	1	0 4 4 9
20–29	5	2	1 3 5 5 8
30+	1	3	0

Bar graph: Number of years taught (0–9, 10–19, 20–29, 30+) vs. Number of teachers (0, 2, 4, 6, 8)

Q: How are the mean, median, and mode the same? How are they different?

A: The mean, median, and mode are all single numbers that describe a typical piece of data.

For example, suppose that your marks on six tests are 61, 73, 82, 88, 88, and 88. The mean is calculated by adding all six marks and dividing by 6: $(61 + 73 + 82 + 88 + 88 + 88) \div 6 = 80$.

The result, 80, would be the mark on each test if the marks on the tests were equal.

The median is the middle value when the marks are written in order from least to greatest. For the test marks, the median is the mean of the two middle numbers 82 and 88: $(82 + 88) \div 2 = 85$.

The mode is the value that occurs most frequently. The mode for the test marks is 88, which occurs three times. The mode is the only average that you can use to describe non-numerical data, such as favourite colour or eye colour.

Practice Questions

(3.2) **1.** Suppose that you are collecting data to compare your classmates' interests in sports with the Canadian averages. Would you use a survey, a census, an experiment, a questionnaire, an interview, or research to gather each type of data below? Explain.

   a) the sports that your classmates play

   b) the number of hockey games that a typical classmate watches per year

   c) the favourite sport for your classmates

   d) the most common brand of running shoe that is worn by the boys in your class

(3.2) **2.** Suppose that you want to collect data on how frequently people in your community get sick. You decide to ask the first 100 people leaving a local doctor's office.

   a) Why might your results be biased?

   b) What bias might you expect from this sample?

(3.3) **3.** The following database shows the last seven players to play the game Spaceship Sirius on Katherine's computer.

Player's name	Date and time	Total play time (min)	Score this round	Highest score	Bonus rounds	Total points
Jabbar	11/12/13:00	22	3400	6500	3	200 870
Redoo	11/12/12:20	18	2300	8900	7	107 040
Fraglin	11/12/11:15	24	4500	4500	2	39 800
Lizzaboo	11/12/10:50	3	300	6800	4	450 000
Triman	11/12/9:30	33	6700	9860	9	180 550
Gonzer	11/12/9:02	28	450	3440	1	55 990
Chelspat	11/11/15:15	14	2200	6790	8	785 000

Record: ◄◄ ◄   1 ► ►► ►◄ of 7

   a) How is the database sorted now?

   b) Whose name will appear first if the database is sorted in descending (highest to lowest) order using the field "Highest score"?

   c) Whose name will appear first if the database is sorted in ascending (lowest to highest) order using the field "Total play time (min)"?

4. Some information from the database in question 3 was copied into this spreadsheet. (3.4)

	A	B	C	D	E	F	G
1	Player's name	Total play time (min)	Score this round	Highest score	Total points	Points per minute this round	Rank based on highest score of all rounds
2	Jabbar	22	3400	6500	200 870		
3	Redoo	18	2300	8900	107 040		
4	Fraglin	24	4500	4500	39 800		
5	Lizzaboo	3	300	6800	450 000		
6	Triman	33	6700	9860	108 550		
7	Gonzer	28	450	3440	55 990		
8	Chelspat	14	2200	6790	785 000		

   a) What information is in cell A7?

   b) In which cell is the value 55 990?

   c) What value should appear in cell G5?

   d) What formula could you use to calculate "Points per minute this round"?

5. Omar wants to know what people thought of the school play. He collected the following opinions. (Key: E = excellent, G = good, S = satisfactory, P = poor) (3.5)

  E   E   S   P   G   G   G   G   S
  E   P   S   S   G   S   E   E   E
  P   S   P   E   G   G   G   E   E
  P   S   G   G   E   G   E   P   S

   a) Organize these opinions using a frequency table.

   b) What did most people think of the play?

6. Sandra asks 20 people entering a music store how old they are. Here are her data: (3.5)

  17 25 33 38 24 8 45 27
  27 15 26 37 8 14 38 4
  42 17 25 31

   a) Organize the data that Sandra collected using a stem-and-leaf plot.

   b) Describe three ways that the music store could use the data.

   c) Use the data to create a graph that shows the most common age of the people entering the music store.

(3.5) **7.** What intervals would you use to create a stem-and-leaf plot for each set of data?

    a) 25, 87, 92, 29, 33, 98, 19, 33, 45

    b) 446, 440, 440, 442, 444, 442, 440, 443, 440

(3.6) **8.** Calculate the mean, median, and mode for each set of data.

    a) 4, 8, 8, 9, 3, 4, 4

    b) 125, 83, 115, 94, 109, 115, 89, 104

(3.6) **9.** David's French test scores are 87%, 88%, 82%, 83%, 88%, 86%, and 88%. His latest test score is 100%. Which measure (mean, median, or mode) will be most affected by his latest score? Explain.

(3.6) **10.** Shirley's golf scores are listed below.

    118   112   116   120   112   117

    96   90   90   92   81   83

    92   92   92   90

    a) Organize these scores using a stem-and-leaf plot.

    b) Calculate the mean, mode, and median for Shirley's golf scores.

    c) Which measure (mean, median, or mode) is the most appropriate representation of Shirley's golf ability? Why?

11. Anne recorded the colours of 100 cars in the parking lot of a supermarket. These are her results. (3.7)

Colour of car	Number of cars of each colour
white	20
silver	32
black	18
red	12
blue	18

    a) What type of graph would you use to display Anne's data? Why?

    b) Construct this graph.

12. The following graph shows the accumulated rainfall over a 6 h period. Use the graph to answer the following questions. (3.7)

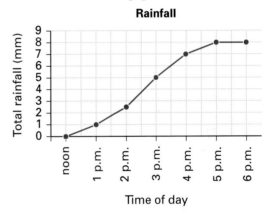

    a) How much rain fell in total?

    b) When was the rainfall heaviest?

    c) When did the least amount of rain fall?

    d) How much rain fell between 3:00 p.m. and 4:00 p.m.?

    e) What was the mean rainfall for the 6 h period?

    f) Approximately how much rain had fallen by 2:30 p.m.?

Chapter Task

Planning a Playlist

Task Checklist

- ☑ Does your playlist reflect the preferences of each grade?
- ☑ Is your graph appropriate for the data?
- ☑ Did you clearly explain your recommendations?
- ☑ Will your computer program work if the data change from year to year?

Simon and James are on the committee for the school dance. They want to determine music preferences so that they can report to the student council and inform the disc jockey who has been hired for the dance.

The boys conduct a survey of all the students who will be attending the dance. Their results are shown to the right.

> (Key: C = country, R = rap, P = pop/rock,
> A = alternative/punk, S = ska/reggae,
> D = disco/retro, H = hiphop)

? **How can you organize and interpret the data to help plan a playlist and justify it to the student council?**

A. Organize and summarize the data the boys collected.

B. Use the data to determine the percent of students who preferred each kind of music.

C. Make a 25 song playlist to give to the disc jockey. Why is your playlist appropriate?

D. Use a graph to represent the data for the student council. What points do you want to make? How does your graph help you make your points?

E. How can you use a computer program to update the data for future dances?

Grade	Music preference
6	R, C, C, P, D, R, R, A, A, A, A, H, D, S, S, C, C, D, R, P, P, H, D, P, S, C, S, D, H, P, R, C, C, C, D, C, C
7	D, D, R, A, A, A, H, S, S, C, P, P, R, R, S, S, C, R, P, P, H, P, S, R, A, A, C, R, S, H, R, H, S, S, A, A, P, D, P, P, P, D, P
8	C, C, A, A, A, A, A, P, S, S, H, P, P, P, R, R, D, D, H, S, S, A, A, D, C, C, S, P, D, H, H, A, A, P, P, C, C, R, S, S, D, P, H, D, P, A

116 Chapter 3

NEL

Cumulative Review
Chapters 1–3

Cross-Strand Multiple Choice

(1.1) **1.** The LCM of 16 and 40 is 80. Which is the next common multiple?

    A. 320   **B.** 200   **C.** 120   **D.** 160

(1.3) **2.** Which is the GCF of 36 and 42?

    A. 3   **B.** 12   **C.** 2   **D.** 6

(1.5) **3.** Which is the exponential form of $3 \times 3 \times 3 \times 3 \times 3 \times 3$?

    A. 6^3 **B.** 3^5 **C.** 3^6 **D.** 729

(1.7) **4.** Which is the value of the expression $(4 + 2^2 \times 4) - 6 \div 3$?

    A. 30   **B.** 34   **C.** 18   **D.** 10

(2.2) **5.** Which is the missing term in the proportion $35 : 28 = \blacksquare : 20$?

    A. 36   **B.** 5   **C.** 24   **D.** 25

(2.3) **6.** Cory earns $32 for working 4 h. How much will he earn for working 9 h, if he is paid at the same hourly rate?

    A. $1152   **B.** $36   **C.** $14   **D.** $72

(2.6) **7.** Linda sold 200 T-shirts at her stall last weekend. The graph shows how much of each size were sold.

T-shirt Sizes Sold

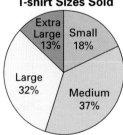

How many medium-size T-shirts did Linda sell?

    A. 37   **B.** 36   **C.** 64   **D.** 74

8. Which is in order from least to greatest? (2.5)

    A. $20\%, 0.3, \frac{3}{5}, \frac{3}{4}, 90\%$

    B. $0.3, \frac{3}{5}, 20\%, \frac{3}{4}, 90\%$

    C. $\frac{3}{5}, 0.3, 20\%, \frac{3}{4}, 90\%$

    D. $20\%, \frac{3}{5}, 30\%, 90\%, \frac{3}{4}$

9. When Mei stopped 80% of the shots on goal in the last game, she stopped 16 shots. How many goals were scored? (2.6)

    A. 8   **B.** 16   **C.** 12   **D.** 4

10. Which is the quotient for $3.64 \div 0.04$? (2.8)

    A. 91   **B.** 9.1   **C.** 0.91   **D.** 0.091

11. Attendance at the circus was counted for each performance during the last two weeks.

| Number of People at Circus ||
Stem	Leaf
18	3 5 6 7 7 7 9
19	0 1 6 8
20	2 4 4
21	0 1 6 6 8
22	4

Which number was not an attendance at one of these circus performances? (3.5)

    A. 190   **B.** 201   **C.** 196   **D.** 224

12. Which is the median for this set of plant heights? (3.6)

    8 cm, 14 cm, 9 cm, 14 cm, 13 cm, 11 cm

    A. 14 cm   **C.** 12 cm

    B. 11.5 cm   **D.** 11 cm

Cross-Strand Investigation

Westside School's soccer team is named The Westside Wild. The team plays 16 games each year. Copy and complete the table below. Then use the table to answer the questions that follow.

Note: If possible, use a spreadsheet program to complete this investigation. If this is not possible, use a calculator.

The Westside Wild's Record, 1997–2004

Year	Wins	Losses	Ties	Fraction form of wins to total games played	Decimal form of wins to total games played (to nearest hundredth)	Percent of games won (to nearest percent)
1997	7	9	0	$\frac{7}{16}$	0.44	44%
1998			0	$\frac{2}{16}$		
1999		5		$\frac{10}{16}$		
2000	9	7	0			
2001			1			56%
2002	6	10	0			
2003	11		2			
2004	1	15				

13. a) Determine the mean, median, and mode for the number of wins from 1997 to 2004. Which measure gives the best picture of how successful The Westside Wild was from 1997 to 2004?

b) Create a graph to display The Westside Wild's wins. Explain your choice of graph.

c) Would a stem-and-leaf plot be a good way to display the winning percents? Explain.

d) If the 2005 season is expanded by 25%, how many games will The Westside Wild play?

e) Each team at a soccer tournament has 11 players (10 plus a goalie) on the field at a time. Show how to use multiples to calculate the number of players on the field in four games.

CHAPTER 4

Patterns and Relationships

▶ GOALS

You will be able to

- recognize and extend patterns, and predict how they will continue
- use appropriate language to discuss number patterns
- use tables of values and scatter plots to show relationships in number patterns

Getting Started

Finding Calendar Patterns

? **What number patterns can you find in the calendar?**

A. Copy the calendar page. Draw a box around any 2-by-2 square. Add the pairs of numbers along the diagonals. What do you notice?

B. Now choose different 2-by-2 squares. Add the pairs of numbers along the diagonals of these squares. What pattern do you find?

C. Look at the squares in steps A and B.

   a) What pattern do you find when you add the numbers in the columns in each square?

   b) What pattern do you find when you add the numbers in the rows in each square?

D. Pick another 2-by-2 square on the calendar. Are the patterns similar?

E. Explain why the patterns you found in steps A, B, and C work the way they do.

F. Repeat steps A through E for 3-by-3 squares on the calendar page.

Do You Remember?

1. Write the next three numbers, words, or letters in each pattern. Then describe the rule for each pattern.

   a) 1, 4, 7, 10, …

   b) three, six, nine, twelve, …

   c) b, e, h, k, …

   d) 11, 22, 33, 44, …

   e) 11, 101, 1001, 10 001, …

   f) 1, 7, 14, 22, …

2. **a)** Draw the next four arrows in the following pattern.

   E   N S   W   E   N S   W   E   N S   W

   b) Describe what happens at each stage.

3. Write the next three numbers in each number pattern. Describe the pattern.

   a) 60, 52, 44, 36, …

   b) 2, 4, 3, 6, 5, 10, 9, 16, …

   c) 7, 5, 8, 6, 9, 7, 10, …

4. A computer program gives this output for the input numbers 1 to 10. Copy and complete the **table of values**.

Input	Output
1	7
2	13
3	19
4	
5	
6	
7	
8	
9	
10	

5. **a)** What is the 23rd number in the pattern 2, 4, 6, 8, 2, 4, 6, 8, 2, 4, 6, 8, …?

   b) Explain how you determined your answer in part (a).

6. The first three figures in a pattern are shown.

   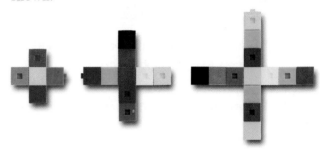

   a) Draw the next two figures in the pattern.

   b) Describe the pattern.

   c) Predict how many cubes will be needed to build the 6th figure in the pattern.

7. Kyle plants a bean seed for his Science Fair project. He records the growth of the seed in a table of values.

Number of days	Height of bean plant (mm)
1	2
2	6
3	10
4	14
5	18

   a) Use Kyle's data to make a **scatter plot**.

   b) Use your scatter plot to predict the height of the bean plant on day 6.

Exploring Number Patterns

You will need
• triangle dot paper
• coloured pencils
• a calculator

▶ **GOAL**

Identify and describe number patterns.

Explore the Math

In 1653, Blaise Pascal used this triangular arrangement of numbers to solve a problem.

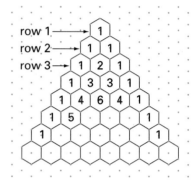

? **What patterns can you find in Pascal's triangle?**

A. Copy Pascal's triangle. Find a pattern that relates the numbers in a row to the numbers in the row above. Use the pattern to fill in the missing numbers.

B. Describe a pattern in the sums of the horizontal rows.

C. Describe patterns in the diagonals.

D. Use your calculator to calculate 11^1, 11^2, 11^3, and 11^4. Record the results.

E. Describe a pattern in Pascal's triangle that relates to the powers of 11.

F. How does your pattern from step E work for the 8th row in Pascal's triangle?

G. Highlight the odd and even numbers in Pascal's triangle, using two different colours. What patterns do you see?

H. Describe how to find the "hockey stick" pattern in Pascal's triangle. Use the pictures to help you.

I. Look for other patterns in Pascal's triangle, and give them names.

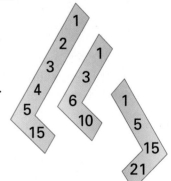

Reflecting

1. Suppose that you were asked about the numbers in row 10 of Pascal's triangle.

a) Which numbers in row 10 could you fill in without knowing the numbers in the previous row? Explain how you would do this.

b) Which numbers could you not fill in unless you knew the numbers in row 9? Explain why not.

2. Describe the patterns you found in Pascal's triangle.

This game consists of rows of counters. The players take turns removing any number of counters from a single row.

Number of players: 2

> **You will need**
> • 15 counters

Rules

1. Start with 15 counters.

2. Arrange the counters in any number of rows. Each row may contain any number of counters.

3. Each player takes a turn removing some or all of the counters from one row only. At least one counter must be removed on every turn.

4. Continue until all the counters have been removed. The player who takes the last counter from the last remaining row is the winner.

Play this game a few times with different numbers of rows until you begin to see the winning patterns.

Variation 1: Traditional Triangular Nim

Set up your rows with one counter in the first row, two counters in the second row, three in the third row, four in the fourth row, and five in the fifth row. Players take any number of counters from one row on a turn. The player who takes the last counter loses the game.

Variation 2: Square Nim

Set up as many counters as you want in a rectangular pattern of columns and rows. Players take a "square" of counters on a turn; that is, they must take one counter (a 1-by-1 square), four counters (a 2-by-2 square), nine counters (a 3-by-3 square), and so on. Before you begin, decide whether the player who takes the last counter will win or lose the game.

Applying Pattern Rules

▶ **GOAL**

Recognize patterns, and use rules to extend and create the patterns.

Learn about the Math

Anthony tells Rosa about a story called "Pigs Is Pigs," by Ellis Parker Butler. In this story, two pet guinea pigs (a male and a female) are in a pen at a railway station. They keep having babies! Soon the station is overrun with guinea pigs.

? **How many pairs of guinea pigs will there be after one year?**

Anthony and Rosa decided to model the way that the number of guinea pigs would grow. They wrote the following rules for their model:

• The guinea pigs must be two months old before they can have babies. Then they can have a new litter every month.
• There will be only two babies (a male and a female) in each litter.
• There is enough food and water, and the animals are healthy and keep living.

A. Copy Rosa's diagram to show how the number of pairs of guinea pigs would increase. Extend the diagram for one more month.

Month	Guinea pigs	Number of pairs
January		1
February		1
March		2
April		3
May		5
June		8

B. Copy this table that Anthony made.

C. What is the rule for the pattern in the table?

D. Write the missing numbers in the **sequence**.

E. How many pairs of guinea pigs will there be at the end of one year?

Month	Number of pairs of guinea pigs
January	1
February	1
March	2
April	3
May	5
June	8
July	13
August	
September	
October	
November	
December	

sequence

a set of numbers or things arranged in order, one after the other, according to some rule; for example, the sequence of numbers 1, 1, 2, 3, 5, …

Reflecting

1. Suppose that you only knew the first three **terms** of the sequence in step D. Explain why you would not have enough information to fill in the remaining terms.

2. Suppose that you wanted to determine the number of pairs of guinea pigs at the end of two years. Why would you have to do many calculations using the pattern rule for this sequence?

3. Can you find the 12th term in this sequence without knowing the 11th term? Explain your answer.

term

each number or item in a sequence; for example, in the sequence 1, 3, 5, 7, …, the third term is 5

Work with the Math

Example 1: Recognizing a multiplication pattern and describing the pattern rule

Describe a pattern rule for the sequence 2, 4, 8, 16, 32, … . Then write the next four terms.

Kaitlyn's Solution

2, 4, 8, 16, 32, 64, 128, 256, 512

×2 ×2 ×2 …

Starting with 2, each number is doubled or is multiplied by 2.

You could also say that each number in the sequence is a power of 2. The pattern could be written as $2^1, 2^2, 2^3, …$, where the base is 2 and the exponent increases by 1 each time.

The rule for a sequence is "Start with 0 and add 1 more each time: +1, +2, +3, and so on." Write the next eight numbers in the sequence.

Omar's Solution

0, 1, 3, 6, 10, 15, 21, 28, 36

+1 +2 +3 ...

I just kept adding 1 more than I added the last time.

Ⓐ Checking

4. These sequences have the same kind of addition rule as the guinea pig sequence. Write the next three terms in each sequence.

a) 2, 2, 4, 6, 10, ...

b) 3, 3, 6, 9, 15, ...

5. Describe the pattern rule for each sequence. Write the next three numbers.

a) 8, 16, 24, 32, ...

b) 4, 8, 12, 16, 20, ...

c) 5, 8, 11, 14, ...

d) 10, 100, 1000, ...

Ⓑ Practising

6. a) Describe the rule for this sequence. 100, 99, 97, 94, 90, ...

b) Write the next three numbers.

7. Use the addition rule for the guinea pig sequence to determine the next three terms in each sequence.

a) 4, 4, 8, 12, 20, 32, ...

b) 5, 5, 10, 15, 25, ...

c) 10, 10, 20, 30, 50, ...

d) 13, 13, 26, 39, 65, ...

8. Describe the pattern rule for each sequence. Write the next three numbers.

a) 3, 6, 12, 24, ...

b) 1, 3, 9, 27, ...

c) $1, \dfrac{1}{2}, \dfrac{1}{4}, \dfrac{1}{8}, \ldots$

d) 576, 288, 144, 72, ...

9. The pattern rule for a sequence is "Start at 10, double the number before this one, and subtract 1." Write the first five numbers in the sequence.

10. The pattern rule for a sequence is "Start at 2, triple each number, and add 1." Write the first five numbers in the sequence.

11. The rule for a pattern is "Add 2, subtract 1, add 3, subtract 1, add 4, subtract 1, and so on." For example, if you start at 10, the next six numbers are 12, 11, 14, 13, 17, and 16.

+2 −1 +3 −1

10 12 11 14 13 17 16

a) Start at 50, and write the next five numbers in the sequence.

b) Start at 29, and write the next four numbers in the sequence.

12. Use this diagram to determine the sum of the first 10 odd numbers. Explain how you found your answer.

13. Create a rule that will result in an interesting number pattern. List the first seven numbers in your sequence.

ⓒ Extending

14. Using the same rule as in question 11, write the next three numbers in a sequence that starts with

a) 65

b) 309

15. a) Describe the rule for this pattern.
  0, 1, 2, 3, 6, 11, 20, 37, 68, …

b) Write the next three numbers.

Mental Imagery   FOLDING SQUARES

- Colour eight squares on 2 cm grid paper as shown.
  Colour the back of each square the same colour as the front.

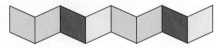

- Cut out the eight squares to form a long strip. Fold on the lines between the squares to form an accordion.

- Follow the steps below to fold the accordion to form a two-square red pattern.

  A. Fold the *left* blue and green behind the *middle* red and yellow.
  B. Fold the *right* red and yellow in front to cover the *middle* blue and green.
  C. Fold the reds under the yellows to show two yellows at the front.
  D. Complete the last folding to form the red–red pattern.

  1. Fold your accordion along the lines between the squares to form the following.

  a)

  b)

  c)

  d)

Using a Table of Values to Represent a Sequence

You will need
- counters
- a calculator
- linking cubes

▶ **GOAL**

Use tables of values to represent number sequences.

Learn about the Math

Kaitlyn and Tynessa used counters to show a sequence of square numbers. They want to know whether they have enough counters for the 6th figure in the sequence.

? **How can you determine the number of counters needed without building the figure?**

Communication Tip

- The figure number is called the **term number** because it tells the position of the term in the sequence. The term numbers in Kaitlyn and Tynessa's sequence start at 1.
- The number of counters needed for each term number is called the **term value**.

Example 1: Using a table of values to represent and analyze a sequence

Make a table of values. Determine the number of counters needed to model the 6th square number in the sequence.

Tynessa's Solution

Term number (figure number)	Picture	Term value (number of counters)
1	○	1
2		4
3		9
4		16
5		25
6		36

+ 3
+ 5
+ 7
+ 9
+ 11

I noticed that when you go down the "Term value" column, you add the next odd number. I can write this pattern rule as "Start with 1 and add the next greatest odd number."

The value for the 6th square is 25 + 11 = 36.

This pattern rule uses the previous term in the sequence.

Term number	Rule	Term value
1	$1 \times 1 = 1^2$	1
2	$2 \times 2 = 2^2$	4
3	$3 \times 3 = 3^2$	9
4	$4 \times 4 = 4^2$	16
5	$5 \times 5 = 5^2$	25
6	$6 \times 6 = 6^2$	36

I also noticed that the term value is the square of the term number. I can write this rule:

term value = term number^2

The value for the 6th square is 6^2 or 36.

This rule uses the position of the term in the sequence.

Reflecting

1. How are Tynessa's tables of values alike? How are they different?

2. The pattern rule in the first table uses an "adding on" strategy. What strategy was used for the rule in the second table?

3. Which table would you use to calculate the 20th term in the sequence? Explain why.

Example 2: Using a table of values to solve a problem

The rule to build a toothpick pattern is "Start with a square, and add 3 toothpicks each time to make another square." How many toothpicks do you need to build the 5th figure and the 20th figure in the pattern?

Colin's Solution

Term number (figure number)	Picture	Term value (number of toothpicks)
1		4
2		7
3		10
4		13
5		16

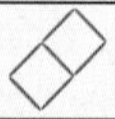

+ 3
+ 3
+ 3
+ 3

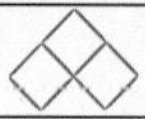

My picture shows 16 toothpicks in the 5th figure. The rule for the number pattern is "Start with 4 and add 3 each time."

You can calculate the number of toothpicks for the 20th figure by starting with 4 and adding on 3 a total of 19 times.

$4 + (3 \times 19) = 61$

Term number	Rule	Term value
1	× 3 + 1	4
2	× 3 + 1	7
3	× 3 + 1	10
4	× 3 + 1	13
5	× 3 + 1	16

Another way to solve the problem is to figure out how the term number and the term value are related.

The value of each term is 1 more than three times the term number. This rule is:

term value = (3 × term number) + 1

For the 20th term, $(3 \times 20) + 1 = 61$.

You get the same answer with both methods. There are 61 toothpicks in the 20th term.

Ⓐ Checking

4. a) Complete the table of values for the pattern shown.

b) Write a rule that tells how the value of each term can be calculated from the previous term in the sequence.

c) Write a rule that tells how the value of each term can be calculated from its term number.

d) Predict the value of the 8th term in the sequence.

Term number (figure number)	Picture	Term value (number of stars)
1		3
2		5
3		7
4		
5		

5. Peter and Heidi are looking at this table of values. Peter says that the pattern rule is "Start with 3 and add 3."

Term number	Term value
1	3
2	6
3	9
4	12
5	15

Heidi says that the pattern rule is "Multiply the term number by 3." Who is right? Explain.

B Practising

6. Copy and complete the table of values.

Term number (figure number)	Picture	Term value (number of tiles)
1		6
2		11
3		16
4		
5		

7. Make a table of values for each sequence. Include pictures. Extend the table of values to show the next three terms in the sequence.

a) 6, 12, 18, 24, …

b) 5, 9, 13, 17, …

8. a) Use a table of values to predict how many cubes you would need to build the 6th figure in this sequence.

b) Explain how the pattern rule works.

9. a) Use linking cubes to build these figures. Then build the next two figures in the sequence.

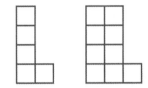

b) Make a table of values to record the number of cubes you used to build each figure in part (a).

c) Write a rule that describes the pattern.

10. Asha has only 61 toothpicks. If she uses as many of these toothpicks as she can, what is the largest figure that she can build in the pattern?

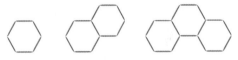

11. Matthew and Suki are trying "The Cubic Challenge." They must build the 6th figure in the pattern shown below. Matthew says they will need a total of 216 cubes. Suki says they will need 108 cubes. Who is right? Explain.

C Extending

12. This "staircase" sequence shows a growing pattern. Predict the total number of cubes you will need to build the 50th staircase. Show your thinking.

Frequently Asked Questions

Q: What is a sequence?

A: A sequence is a list of items that are in a logical order or follow a pattern. For example, the terms in the sequence 9, 18, 27, 36, 45, ... are the multiples of 9.

Also, pictures can show a sequence. For example, this picture shows the sequence 1, 4, 9, 16, ...

Q: How do you use a table of values to analyze a pattern or sequence?

A: A table of values shows what happens at each stage in a pattern. The term numbers are listed in the first column. The term values are listed in the last column. Sometimes there is a middle column with pictures or numbers to illustrate the pattern.

Term number	Picture	Term value
1		3
2		6
3		9
4		12
5		15

+ 3
+ 3
+ 3
+ 3

There are different ways of determining a rule to calculate the value of a term in a sequence.

- One way uses the previous terms in the sequence. Look at the numbers in the last column of the table of values above. There is an addition pattern of + 3 for each new value, beginning at 3.

- Another way uses the term's position in the sequence. Look for a relationship between the term number and the term value. In the sequence above, the term value is calculated by multiplying the term number by 3.

Practice Questions

(4.1) **1.** Use mathematical language to describe three different patterns in the following triangle.

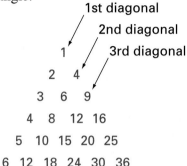

1st diagonal
2nd diagonal
3rd diagonal

```
        1
       2  4
      3  6  9
    4  8  12  16
   5  10 15 20 25
  6 12 18 24 30 36
```

(4.2) **2.** The pattern rule for a sequence is "Start at 8, double each number, and add 2." Write the first five numbers in the sequence.

(4.3) **3.** Determine a pattern rule to predict the number of squares you would need to build the 6th figure in this pattern. Explain your thinking.

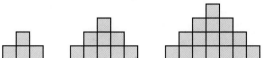

(4.3) **4.** Determine the number of toothpicks in the 5th figure in this sequence.

(4.3) **5. a)** Describe the patterns in this table of values.

b) Copy and complete the table.

Term number	Term value
1	3
2	7
3	11
4	
5	
6	

6. Use a table of values to determine the number of counters in the 9th term in this sequence. (4.3)

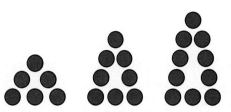

7. a) Draw the 4th figure in this pattern. (4.3)

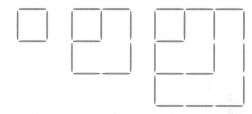

b) Use a table of values to determine the number of toothpicks in the 7th figure.

8. a) Use a table of values to represent the sequence 2, 7, 12, 17, … . (4.3)

b) Write a rule that tells how to calculate the value of a term using the terms before it.

c) Write a rule that tells how to calculate the value of a term using the term number.

d) Calculate the value of the 10th term in the sequence.

9. To raise funds for the music department, a class decides to sell frozen cookie dough. The class makes a profit of $2 on each container. The class sells 5 containers on the first day, 8 on the second day, and 11 on the third day. This pattern continues for 10 days. What is the difference between the profit for the 1st day and the profit for the 10th day? (4.3)

4.4 Solve Problems Using a Table of Values

▶ **GOAL**

Use a table of values to solve patterning problems.

You will need
• a calculator

Learn about the Math

Sarah is setting up a display at her family's convenience store. She wants to stack cases of soft drinks in a triangle arrangement with 10 rows. There will be 1 case at the top, 2 cases in the 2nd row, 3 cases in the 3rd row, and so on. Sarah wants to know the number of cases she needs to get from the storeroom to set up her display.

? How many cases does Sarah need to make 10 rows?

1 Understand the Problem

Sarah knows that each row has one more case than the row above it. She also knows how many cases are in the top three rows. She has to figure out how many cases are in 10 rows.

2 Make a Plan

Sarah decides to sketch the top three rows of the display. Then she will use her sketch to make a table of values and look for a number pattern.

3 Carry Out the Plan

Row number	Picture	Cases in row	Total cases used
1		1	1
2	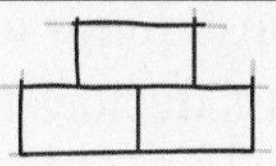	2	3
3	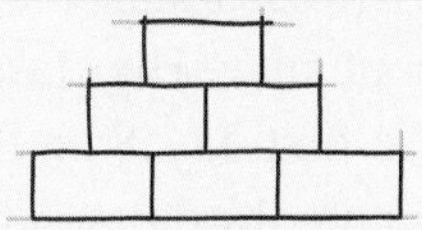	3	6

Sarah notices that the number of cases in each row matches the row number. This means the 10th row will have 10 cases in it, the 9th row will have 9 cases in it, and so on.

Sarah also notices another pattern. When you add the number of cases in the top row to the number of cases in the bottom row, you get a total of 11 cases. The same thing happens when you add the number of cases in the second row and the second last row.

Using this pattern, Sarah calculates the total number of cases:

$$\text{Total number of cases} = (10 + 1) + (9 + 2) + (8 + 3) + (7 + 4) + (6 + 5)$$
$$= 11 + 11 + 11 + 11 + 11$$
$$= 55$$

4 **Look Back**

To make sure that she is right, Sarah checks by extending her table of values.

Row number	Cases in row	Total cases used
1	1	1
2	2	3
3	3	6
4	4	10
5	5	15
6	6	21
7	7	28
8	8	36
9	9	45
10	10	55

Reflecting

1. How did using a table of values help Sarah solve her problem by looking at a similar, but simpler, problem?

2. How many cases of soft drinks would Sarah need for the 11th row of her display?

3. How can you use a table of values to calculate the total number of cases Sarah would need for a display with 15 rows?

4. a) What other strategies could Sarah have used to solve her problem?

   b) How are these other strategies similar to using a table of values? How are they different?

Example: Using a table of values to solve a patterning problem

Rana is the coordinator of a regional volleyball league. Eight teams must play each team once. How many games must Rana schedule for the volleyball season?

Rana's Solution

1 Understand the Problem

I know that the teams play in twos and that each of the eight teams has to play each of the other seven teams once. I want to figure out the least number of games that I need to schedule.

2 Make a Plan

I will sketch what would happen with fewer numbers of teams. Then I can use my sketches to make a table of values. This will help me see if there is a pattern.

3 Carry Out the Plan

1 team   2 teams   3 teams   4 teams

Number of teams	Number of games required	
1	0	+ 1
2	1	+ 2
3	3	+ 3
4	6	+ 4
5	10	+ 5
6	15	+ 6
7	21	+ 7
8	28	

I see a possible pattern rule. For each new row in the table of values, I have to add one more game than I added the time before. So I'll need to schedule 28 games for the eight teams.

4 Look Back

I can sketch the picture for 8 teams and use different colours to help me keep track of the number of games.

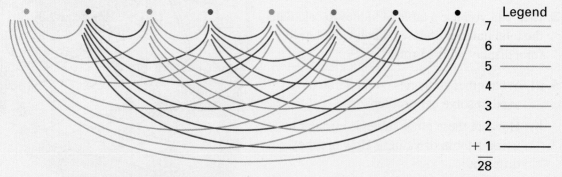

Legend
7 ———
6 ———
5 ———
4 ———
3 ———
2 ———
+ 1 ———
28

The numbers for the colours follow a pattern, so I think my solution is correct.

A Checking

5. Use a table of values to find the number of toothpicks you would need to build the 7th figure. Explain your thinking.

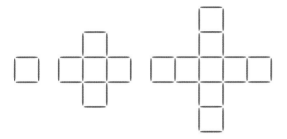

B Practising

6. How many volleyball games would Rana have had to schedule if there were 10 teams in the league and each team played each other twice?

7. During a canned food drive, the number of students who donate food doubles each day. Three students donate food on day 1. On which day will 96 students donate food? Show your thinking.

8. There are 14 people at a meeting, and they all shake hands with each other. Use a table of values to find the total number of handshakes.

9. A fire has destroyed 200 trees in a nearby forest. Your school's Environmental Club has volunteered to do the replanting. Each member of the club is given four seedlings. Two members plant their seedlings on the first day. They are followed by four members on the second day and six members on the third day. If this pattern continues, on which day will the replanting be complete?

10. Heather wants to make 250 origami peace cranes for a school display. She makes 12 cranes the first day, 15 cranes the second day, 18 cranes the third day, and so on. If she continues this pattern for 10 days, will she reach her goal of 250 peace cranes?

11. The school office staff has organized a telephone tree system to use in case of emergency. The first caller phones three parents. These parents, in turn, each call three more parents, and so on.

   a) How many parents will receive a call on the third round of calls?

   b) How many parents in total will receive calls by the end of the third round of calls?

   c) If there are 363 parents to be called, how many rounds of calls will be needed?

12. The students challenge the teachers to a basketball shootout, with the money from ticket sales going to charity. The goal is to raise $175. On the first day, 16 tickets are sold. On the following days, the number of tickets sold increases by three tickets per day.

   a) How many tickets in total are sold by the end of the 5th day?

   b) Each ticket costs $2. On which day of ticket sales will the students reach their goal?

13. Create a problem you can solve by making a table of values. Show how to solve your problem.

Using a Scatter Plot to Represent a Sequence

You will need
- grid paper
- a ruler
- coloured pencils
- toothpicks
- a calculator

▶ **GOAL**

Use scatter plots to represent number sequences.

Learn about the Math

Omar is designing a path for the school garden.
- The path will have 30 equilateral triangle paving stones, which line up to form a straight path.
- On each side of the path there will be border pieces. Each border piece is half as long as one side of a triangle stone.

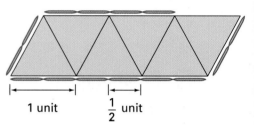

1 unit $\frac{1}{2}$ unit

? **How many border pieces will Omar need for the path?**

A. The table of values below shows how the perimeter of the path increases as triangle stones are added. Copy and complete the table of values.

B. Draw the scatter plot using the data in your table of values. Use values from "Number of triangles" and the corresponding "Perimeter of path (units)" as the **coordinates** for each point.

coordinates

an ordered pair, used to describe a location on a grid labelled with an x-axis and a y-axis; for example, the coordinates $(2, 3)$ describe this location:

Number of triangles	Perimeter of path (units)
1	3
2	4
3	5
4	
5	
6	
7	
8	

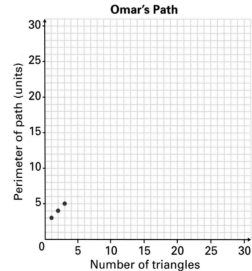

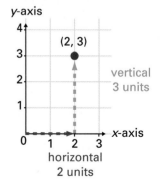

C. Use the pattern of the points on the scatter plot to predict the perimeter of the path using 30 triangle stones.

D. Add a third column on the right side of your table of values. Label it "Number of border pieces." Record the data.

Number of triangles	Perimeter of path (units)	Number of border pieces
1	3	
2	4	
3	5	

E. Draw a new scatter plot. Use "Number of triangles" and the corresponding "Number of border pieces" as the coordinates for each point.

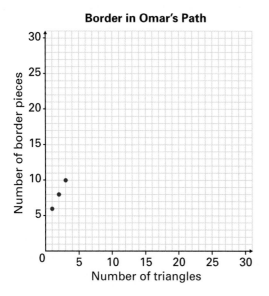

Border in Omar's Path

F. Use the scatter plot you drew in step E to determine how many border pieces are needed for 30 triangle stones.

Reflecting

1. How are your two scatter plots the same? How are they different?

2. Explain how you can use a scatter plot to determine the number of border pieces needed for any number of triangle stones.

3. How is using a scatter plot like using a table of values to predict the perimeter and the number of border pieces?

4. Why is using a scatter plot not as accurate as using a pattern rule from a table of values to make predictions about a sequence?

Example: Using a table of values to make a scatter plot for a sequence

This flowerbed will be 10 triangles long. Each triangle will contain one type of flower. Use a scatter plot to predict the total number of border pieces needed to make the flowerbed.

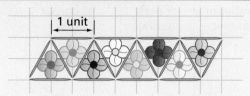

1 unit

Kaitlyn's Solution

I made a table of values for the first five terms in the sequence.

Number of triangles	Number of border pieces
1	3
2	5
3	7
4	9
5	11

I used the coordinates to draw a scatter plot.

The points all go in a straight line.

This makes me think that the values follow a pattern rule. The values increase by the same amount for each term.

If I follow the line past (5, 11), I can see that I will need 21 border pieces for 10 triangles.

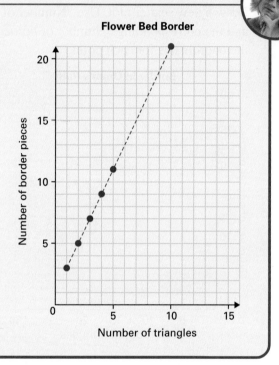

Flower Bed Border

Number of border pieces (vertical axis)
Number of triangles (horizontal axis)

A Checking

5. Mohammed is designing fences for a summer job. One kind of rail fence has three rails and one post in each section. Use a scatter plot to predict how many posts and rails Mohammed will need for a fence that is 12 sections long. Explain what you did.

6. a) Make a scatter plot to represent this toothpick pattern.

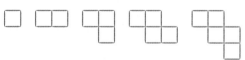

b) How many toothpicks would you need to make the 9th figure in the sequence?

c) Which figure in the sequence could you make with 22 toothpicks?

B Practising

7. Create a scatter plot for the table of
values. Use your scatter plot to find the
missing values in the table.

Term number	Term value
4	8
6	10
8	12
?	14
?	16
14	?

8. Mohammed is building
another rail fence that has
sections like this.

Use a scatter plot to determine the number
of posts and rails Mohammed will need to
build each fence.

a) 8 sections long   **b)** 14 sections long

9. a) How many squares of each colour
would you need to build each of the
first four figures? Make a table of
values to record your answers.

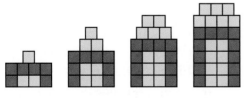

b) Draw a scatter plot to show how many
squares of each colour you would need
to make the 10th figure. Use different
colours and a legend to show the
different colours of squares.

10. Create a toothpick pattern of your own,
and draw the first four figures in your
sequence. Then make a table of values or
draw a scatter plot for the first 10 figures.

11. There are two triangular patterns in the
marching band shown above. In the inner
triangle, there is one person at the tip
(in front of the conductor), then three people,
and so on. Use a table of values or a scatter
plot to answer parts (b), (c), and (d).

a) Describe the pattern of the inner
triangle.

b) What is the total number of band
members in the inner triangle?

c) How many more players would be
needed to increase the total number of
rows in the inner triangle to 15?

d) How many more band members would
be needed to have 17 players in the
longest row?

e) Write your own math question about
this photograph. Explain how to solve it.

C Extending

12. Omar has $350 to spend on the garden he
is designing. The triangle stones cost $20
each, and the border pieces cost $5 each.
How long a path can Omar afford to make?

The sequence you used in lesson 4.2 for the guinea pigs (1, 1, 2, 3, 5, 8, 13, …) is called the *Fibonacci sequence*. It is named after an Italian mathematician who lived about 800 years ago. The Fibonacci sequence often occurs in nature.

1. a) How many spirals go in each direction on the pine cone?

b) How are the numbers related?

iii)

b) How are the numbers related to the Fibonacci sequence?

2. a) How many petals does each flower have?

i)

ii)

3. A number that occurs in the Fibonacci sequence is called a *Fibonacci number*. For example, 3, 5, 8, 13 are four sequential Fibonacci numbers.

a) Take any four sequential Fibonacci numbers. Multiply the first and last numbers. Multiply the two middle numbers. What is the difference between the two products?

b) Repeat part (a) for at least two other sets of four sequential Fibonacci numbers.

c) Write a rule that describes the relationship you observed in parts (a) and (b).

Chapter Self-Test

1. Write the next three terms in each sequence. Describe the pattern rule.

  a) 12, 24, 36, 48, ...

  b) 5, 12, 26, 54, ...

  c) 8, 8, 16, 24, 40, ...

  d) 2, 4, 7, 11, 16, 22, …

2. a) Make a table of values for the first seven terms of this pentagonal pattern.

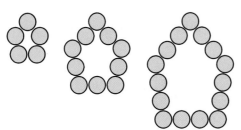

  b) State a rule that describes how to calculate the value of a term if you know its position number.

3. A computer program gives this output for the input numbers.

Input	Output
1	2
2	8
3	14
4	20
5	
6	
7	
8	
9	
10	

  a) Copy and complete the table of values.

  b) Explain how a scatter plot could be used to find the output for an input of 30.

  c) Describe a rule that the computer follows to produce the output number.

4. a) Use the scatter plot for this sequence to complete the table of values.

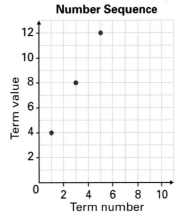

Number Sequence

Term number	Term value
1	
	6
3	8
4	
	12
6	
7	

  b) Explain why there are at least two ways to describe the pattern for this sequence.

5. Zach and Indu are arranging tables and chairs in the school gym for a multicultural feast. Use a scatter plot or a table of values to determine the number of guests that can be seated at a row of 14 tables.

6. There are nine members in a chess club. Each member wants to play two games against each of the other members. How many games will be played in total?

7. The student council held a car wash to raise $200 for the Children's Hospital. In the first hour, they washed three cars. Each hour after that, the number of cars they washed increased by two.

  a) How many cars did they wash by the end of 6 h?

  b) If they charged $5 per car, did they reach their goal?

Chapter Review

Frequently Asked Questions

Q: How can you use a table of values as a problem-solving strategy?

A: A table of values organizes information to make figuring out a pattern rule easier. The pattern rule can then be used to predict and verify the solution to the problem. A table of values may help you treat a problem as similar, but simpler, problems. The solutions to these simpler problems often have a pattern that you can apply to solve the original problem.

Term number	Term value
1	2
2	4
3	6
4	8

Q: How does a scatter plot represent the pattern in a table of values for a sequence?

A: A scatter plot shows the relationship between two quantities in a table of values. The term number is plotted on the *x*-axis. The term value is plotted on the *y*-axis.

If the coordinates follow a straight line or a curve, the term number and term value are likely related. For example, this scatter plot represents a pattern in which the value of each term is twice the term number.

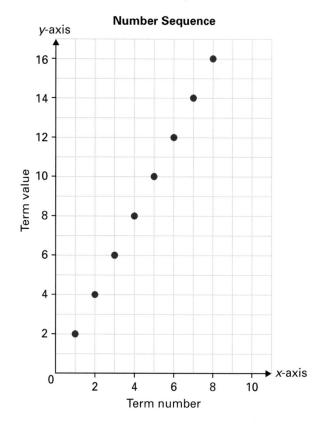

Q: What are the advantages and disadvantages of using a scatter plot to represent a pattern?

A: Advantages: A scatter plot gives you a picture of the pattern. You can use a scatter plot to predict values instead of calculating them.

Disadvantage: A scatter plot may not be as accurate as a pattern rule for predicting values, especially if the scale is too large or small.

Practice Questions

(4.2) **1.** Write the next three terms in each sequence. Describe the pattern rule.

   a) 2, 5, 8, 11, …

   b) 100, 98, 94, 88, …

   c) 6, 6, 12, 18, 30, …

(4.3) **2. a)** Draw the next two figures in this sequence.

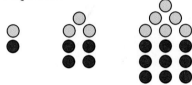

   b) Copy and complete the table of values.

Figure number	Number of green counters	Number of red counters	Total number of counters
1			
2			
3			
4			
5			

   c) Describe any patterns you see.

   d) How many counters in total would you need to make the 6th figure in the sequence? How many counters would you need to make the 10th figure?

(4.4) **3.** How many line segments would it take to join 15 dots to every other dot?

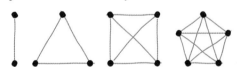

(4.4) **4.** Mike reads 10 pages of a novel on night 1. Each night after that, he reads three pages more than the night before. At this rate, Mike thinks he can finish reading his 219 page novel in 10 days. His brother says it will take 22 days. Who is right? Explain.

5. a) Use the scatter plot for the sequence to find the missing numbers in the table of values.   (4.5)

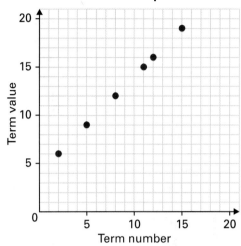

Term number	Term value
1	
	6
3	
4	
	9
6	
7	
	12

   b) Explain how you could use a scatter plot to find the value of the 20th term.

6. a) Make a table of values for the numbers of squares of each colour used to build each of the first four figures.   (4.5)

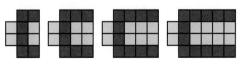

   b) Draw a scatter plot to show how many squares of each colour you would need to build the 10th figure. Use different colours and a legend to show the different colours of squares.

Chapter Task

Design a Beaded Necklace

Rana is designing a beaded necklace. Her necklace will use green, white, and purple beads, laid out in three rows.

- Rana makes a basic shape of white beads to separate the shapes made from green and purple beads.
- She makes shapes from green and purple beads that grow from an upside-down T. Each new shape grows from the previous shape by adding a new bead of the colour already in each row.
- When Rana makes the T shape that has 10 beads in the top row, she then reduces the number of beads in each row by one, until she gets back to the original green and purple T shape.
- She repeats the pattern until her necklace is complete.

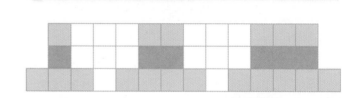

? How many beads does Rana need to make a necklace?

A. Suppose that Rana has 200 white beads. How many green beads and purple beads will she need to make a necklace with the pattern described above?

B. Rana's necklace will be closed with a clasp at the back. She wants to finish each end of the necklace with the white bead shape. How many white beads will she need?

C. Design a bead pattern of your own using three colours. It should follow rules similar to those used by Rana. How many beads of each colour will be required? Use a table of values, a pattern rule, or a scatter plot to justify your answers.

Task Checklist

- ☑ Did you make a plan?
- ☑ Did you include diagrams, a table of values, and scatter plots?
- ☑ Did you show all your steps?
- ☑ Did you use appropriate math vocabulary?

Math in Action

Games Designer

If you can identify and create number patterns, you're qualified to play *Ready or Not*™. This board game was invented by three Canadians—Rick Newell, Art Mathies, and Jim Kyriacou—to help their children and grandchildren develop number skills.

Ready or Not™ is similar to *Scrabble*®. Instead of creating words, however, players create number patterns that reveal various relationships in a given set of numbers.

Problems, Applications, and Decision Making

Look at each pattern shown in the photograph above.

1. Describe each pattern.

2. Players create patterns by building on existing patterns. Create a new pattern by adding three tiles, all in a row, to the existing pattern.

Points are calculated by multiplying the sum of the numbers on the tiles played by the number of tiles played.

For example, $(2 + 4 + 6 + 8) \times 4 = 80$.

You receive bonus points if the sum of the numbers on the tiles is evenly divisible by the number of tiles played.

For example, suppose that you put down 5, 6, 7, 8, and 9.

$5 + 6 + 7 + 8 + 9 = 35$ and $35 \div 5 = 7$

There is no remainder, so you receive a bonus.

3. Look at the board shown at the top of the previous page. Which patterns on the board receive a bonus? Explain.

4. You can use three to seven tiles on a turn. Create a pattern that will receive a bonus.

The next three questions show the types of patterns you can create in the game. Work with a partner to answer these questions.

5. Describe each pattern.

a) 17, 18, 19, 20   c) 3, 6, 9, 6, 3

b) 16, 20, 24, 28   d) 125, 25, 5

6. Predict the next two numbers in each pattern.

a) 16, 20, 24, 28, …   c) 6, 36, 216, …

b) 27, 30, 33, …   d) 19, 16, 13, 10, 7, …

7. During a game, the players create the patterns shown on the board to the right.

a) Describe each new pattern.

b) Do any of the new patterns receive bonus points? Which ones? Explain.

Advanced Applications

Game designers Rick Newell and Art Mathies say that a good game should have certain qualities. "It should be motivational, challenging, fun, and easy to understand. It should have a basic concept, as well as rules and scoring."

8. Create your own game to help students develop the mathematics patterns you learned about in this chapter. Rick and Art based their ideas on *Scrabble*®. You might consider a game such as *Trivial Pursuit*® or *Monopoly*® as a starting point.

CHAPTER 5

2-D Measurement

▶ GOALS

You will be able to

- develop and apply formulas for the areas of parallelograms, triangles, and trapezoids
- calculate the areas and perimeters of irregular shapes
- solve problems involving the perimeter and area of 2-D shapes

Getting Started

You will need
- centimetre grid paper
- coloured pencils
- scissors
- a calculator
- pattern blocks

Designing a Flag

For a math project, Yuki and Ryan designed a flag on centimetre grid paper. They want to describe how much coloured paper they will need to make the areas of the rectangles, parallelograms, and triangles on their flag.

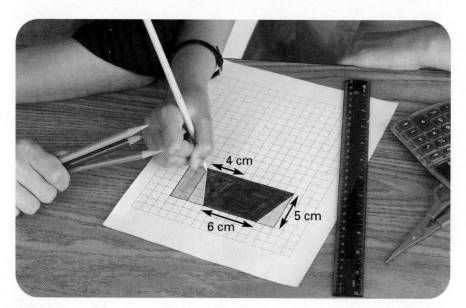

? **What is the area of each coloured region on the flag?**

A. Draw the flag on centimetre grid paper. Then cut out the pieces.

B. Calculate the area of the green rectangle.

C. Compare the area of the green rectangle with the total area of the yellow triangles.

D. What is the area of each yellow triangle? How do you know?

E. What part of the red parallelogram has the same area as a yellow triangle? Cut out this part of the red parallelogram, and move it so that the red pieces form a rectangle. Calculate the area of the red rectangle. What is the area of the red parallelogram?

F. Use the two yellow triangles and the blue triangle to make a rectangle. Calculate the area of this rectangle.

G. What is the area of the blue triangle? How do you know?

H. Write each ratio.
   a) yellow : green   **b)** yellow : blue   **c)** yellow : red

Do You Remember?

1. Use $<$, $>$, or $=$ to make each statement true.

   a) 1 km ▨ 1000 cm   **c)** 40 mm ▨ 4 m

   b) 300 cm ▨ 3 m   **d)** 89 m ▨ 890 cm

2. Choose the unit you would most likely use to measure each item.

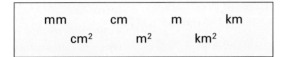

   a) the length of a soccer field

   b) the area of a checkerboard

   c) the area of carpet in a room

   d) the distance from Toronto to Ottawa

   e) the thickness of a dollar coin

3. For a rectangle, Area = length × width. Calculate the missing values and the perimeter of each rectangle.

  a)

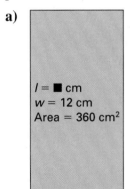

$l = $ ■ cm
$w = 12$ cm
Area $= 360$ cm²

  b)

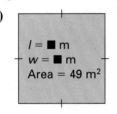

$l = $ ■ m
$w = $ ■ m
Area $= 49$ m²

  c)

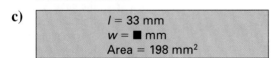

$l = 33$ mm
$w = $ ■ mm
Area $= 198$ mm²

  d)

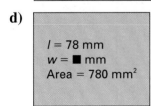

$l = 78$ mm
$w = $ ■ mm
Area $= 780$ mm²

4. Calculate the perimeter and the area of the orange region in each shape.

  a)

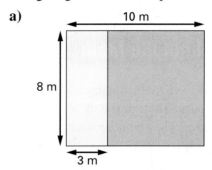

  b)

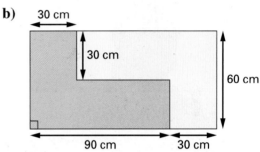

5. Heather and her friends are painting the walls of a storage room in the basement. Each wall is 9 m long and 2 m high. There are two doors, each 1 m wide and 2 m high. The doors are not being painted.

   a) One litre of paint covers 10 m². How many litres of paint will Heather need?

   b) The price of the paint is $3.75/L. How much will the paint cost?

6. Use a calculator to multiply. Round each answer to the nearest tenth.

   a) 7.0×9.36

   b) 25.1×17.5

   c) 3.14×6.8

   d) 6.5×4.09

7. Describe each type of polygon in a set of pattern blocks.

Area of a Parallelogram

▶ **GOAL**

Develop and apply the formula for the area of a parallelogram.

You will need
• centimetre square dot paper
• a calculator
• a geoboard and elastic bands

Learn about the Math

Sandra and Ravi are designing a logo for a new amusement park called Adventurepark. They start with a 5-by-4 rectangle and make it into a parallelogram. They wonder what the **formula** for the area of a parallelogram is.

formula

a rule represented by symbols, numbers, or letters, often in the form of an equation; for example, Area of a rectangle = base × height

? **What formula can you use to calculate the area of the parallelogram?**

A. Draw Sandra and Ravi's parallelogram on centimetre square dot paper. (You can use any side as the base.) Draw a vertical line from the top to the bottom to make a right triangle. This line is the height. Label the height and the base.

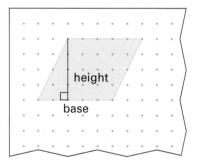

Communication Tip

• The little square in the diagram means that the height and the base form a right angle (90°). The height is always perpendicular to the base. Any side can be a base.

• Units of area always have a small, raised "2" written after them, as follows: 12 m². This indicates that two dimensions, length and width, are involved.

B. Cut out the triangle from the logo. Move the triangle to the right side of the parallelogram to form a rectangle. What is the area of the rectangle? What is the area of the parallelogram?

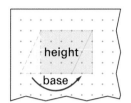

Reflecting

1. How can cutting and then moving the right triangle help you develop the formula for a parallelogram?

2. When calculating the area of a parallelogram, why don't you multiply the two side lengths?

3. What is the formula for the area of a parallelogram with base *b* and height *h*?

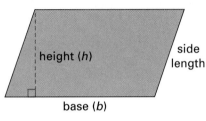

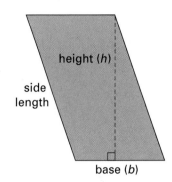

4. Why would you use a formula to calculate the area of a parallelogram instead of counting squares?

Work with the Math

Example: Calculating the area of a parallelogram

What is the area of parallelogram *WXYZ*?

Solution A

$A = b \times h$ The formula for the area of a parallelogram is $A = $ base $\times$ height.
$= 5.0 \text{ cm} \times 2.4 \text{ cm}$ If side *ZY* is the base, then the height is 2.4 cm.
$= 12.0 \text{ cm}^2$ Use these values in the formula and multiply.
              The area of the parallelogram is 12 cm².

Solution B

$A = b \times h$ Look at the shape a different way. (Turn it.)
$= 3.0 \text{ cm} \times 4.0 \text{ cm}$ If side *YX* is the base, then the height is 4.0 cm.
$= 12.0 \text{ cm}^2$ The area of the parallelogram is 12 cm².

A Checking

5. Copy and complete the table, based on the parallelograms shown.

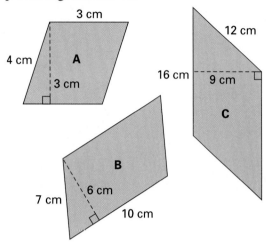

Parallelogram	Base (cm)	Height (cm)
A		
B		
C		

6. Create three parallelograms on a geoboard.

   a) Use an elastic band to show the height of each parallelogram. Record your results on centimetre square dot paper.

   b) Count the squares to describe the area of each parallelogram.

   c) Use a formula to calculate the area of each parallelogram.

7. Calculate the area of parallelogram *WXYZ* in two different ways. Show your work.

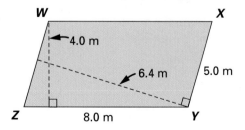

B Practising

8. Calculate the area of each parallelogram to the nearest tenth.

   a)

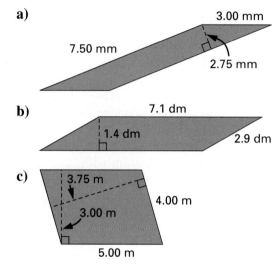

   b)

   c)

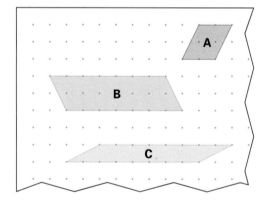

9. a) Draw the following parallelograms on centimetre square dot paper. For each parallelogram, label a base and the corresponding height.

   b) Calculate the area of each parallelogram.

10. Draw an example of a parallelogram in which both the base and the height are the sides. Explain your thinking.

11. Draw possible base and height combinations for three different parallelograms, each with an area of 36 cm^2.

12. Copy and complete the table.

	Base	Height	Area of parallelogram
a)	4 m	▩ m	28 m²
b)	20 cm	11 cm	▩ cm²
c)	▩ cm	9 cm	63 cm²
d)	1.7 dm	2.6 dm	▩ dm²
e)	0.6 m	▩ m	4.2 m²
f)	27.5 mm	32.6 mm	▩ mm²

13. Calculate for each parallelogram.

a) the area of the auditorium

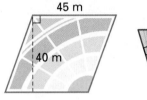
45 m
40 m

c) the area and perimeter of the table top

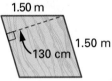
1.50 m
1.50 m
130 cm

b) the area and perimeter of the wall tile

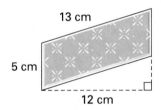
13 cm
5 cm
12 cm

14. Calculate each floor area in this apartment. Each room is a parallelogram.

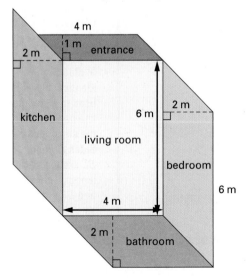
4 m
1 m
2 m entrance
kitchen
6 m
2 m
living room
bedroom
6 m
4 m
2 m bathroom

15. Draw a parallelogram, and label it A. Now draw three more parallelograms, as described below.

  a) a parallelogram that is half the area of A

  b) a parallelogram that is twice the area of A

  c) a parallelogram that is three-quarters the area of A

16. Adventurepark will need a parking lot for staff vehicles. The parking spaces will be angled in a row. Each parking space will be a parallelogram with a base of 5.0 m and a height of 2.8 m. The cost to pave each parking space is $21.50.

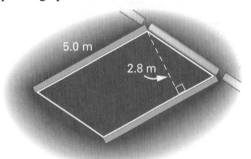
5.0 m
2.8 m

Copy and complete the following table. Show your calculations.

Number of parking spaces	Area (m²)	Total cost ($)
1		
5		
10		
50		

You will need
• a ruler
• centimetre grid paper
• a calculator

5.2 Area of a Triangle

▶ **GOAL**

Develop and apply the formula for the area of a triangle.

Learn about the Math

Chang and Sandra are calculating the cost of painting the floor of an outdoor stage. The measurements are shown in the diagram below. The price of the paint is $2.25 per square metre.

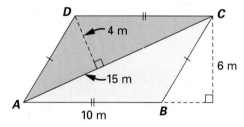

? **What is the cost of painting the yellow half of the stage?**

Sandra says, "We know how to calculate the area of a parallelogram. Can we use the area of the parallelogram to calculate the area of the yellow triangle?"

Chang says, "The area of the parallelogram is $b \times h = 10$ m $\times$ 6 m, or 60 m². If I cut the parallelogram along line segment AC, I can match the two triangles. The area of the yellow triangle is half the area of the parallelogram. So, the area of $\triangle ABC = 60$ m² $\div$ 2, or 30 m²."

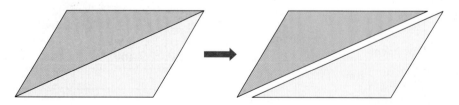

Sandra says, "The cost of painting 1 m² is $2.25. So, the cost of painting 30 m² is 30 × $2.25, or $67.50."

Example 1: Calculating the area of a triangle by forming a parallelogram

How can you calculate the area of $\triangle ABC$ by forming a parallelogram?

Chang's Solution

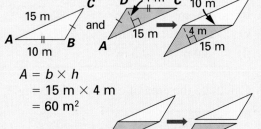

I matched sides AB and DC by sliding the blue triangle underneath the yellow one. This formed a parallelogram. The base is 15 m. The height is 4 m.

$$A = b \times h$$
$$= 15 \text{ m} \times 4 \text{ m}$$
$$= 60 \text{ m}^2$$

I multiplied to find the area of this parallelogram.

The two triangles split the parallelogram in half. So, the area of one triangle is half the area of the parallelogram.

$$A = 60 \text{ m}^2 \div 2$$
$$= 30 \text{ m}^2$$

The area of $\triangle ABC$ is 30 m².

Sandra's Solution

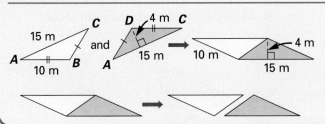

I matched sides CB and DA to form a different parallelogram than Chang's. The base is 15 m, and the height is 4 m.

My calculations would be the same as Chang's, so the area of $\triangle ABC$ is 30 m².

Reflecting

1. Why can every triangle be thought of as half a parallelogram?

2. What is the formula for the area of a triangle in terms of the base (b) and height (h) of the related parallelogram?

Work with the Math

Example 2: Calculating the area of a triangle

Calculate the area of $\triangle ABC$ to the nearest tenth.

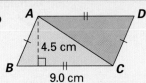

Solution

$$A = (b \times h) \div 2$$
$$= (9.0 \text{ cm} \times 4.5 \text{ cm}) \div 2$$
$$= 40.5 \text{ cm}^2 \div 2$$
$$= 20.25 \text{ cm}^2$$

To find the area of a triangle, multiply the base (b) by the height (h), and divide by 2.

$\triangle ABC$ has a base of 9.0 cm and a height of 4.5 cm.

The area of $\triangle ABC$, rounded to the nearest tenth, is 20.3 cm².

A Checking

3. Show how each triangle is half a parallelogram. Then calculate the area of the triangle.

a)

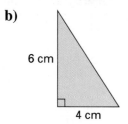

6 cm
8 cm

b)

6 cm
4 cm

c)

8 cm
2 cm

4. On centimetre grid paper, draw two triangles with each area.

a) 12 cm²   **c)** 10 cm²

b) 18 cm²   **d)** 8 cm²

5. Draw the following triangles on centimetre grid paper, and estimate the area of each triangle. Based on your estimates, predict whether any of the triangles have the same area. Check by measuring and calculating.

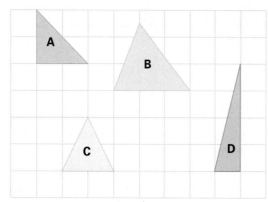

B Practising

6. Copy and complete the table.

	Base	Height	Area of triangle
a)	4 cm	9 cm	▮ cm²
b)	12 cm	45 cm	▮ cm²
c)	3.5 mm	6.0 mm	▮ mm²
d)	6.0 m	7.5 m	▮ m²

7. Calculate the area of each purple triangle.

a)   **b)**

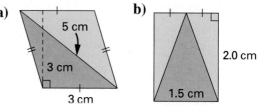

5 cm
3 cm
3 cm

2.0 cm
1.5 cm

8. Measure a base and a height for each triangle. Then determine each area.

a)   **c)**

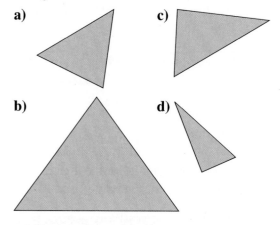

b)   **d)**

9. △ABC has three different pairs of bases and corresponding heights, as shown.

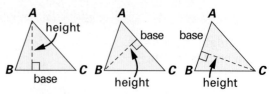

A
height
B base C

A
base
B height C

A
base
B height C

a) Does every triangle have three different pairs of bases and corresponding heights? Explain.

b) When you use a formula to calculate the area of a triangle, does it matter which height-base pair you use? Explain.

10. Calculate the area of the triangle.

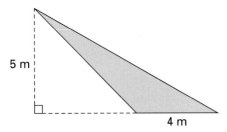

5 m

4 m

11. Copy and complete the table.

	Base	Height	Area of triangle
a)	6 cm	▨ cm	72 cm²
b)	▨ m	6.0 m	10.2 m²
c)	40 mm	9 cm	▨ cm²
d)	250 mm	▨ cm	625 cm²

12. Calculate the area of each shape by measuring the sides.

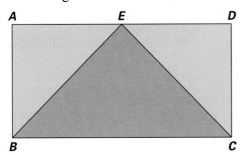

A   E   D

B   C

a) △BEC   **c)** △ABE

b) △EDC   **d)** rectangle ABCD

13. a) Calculate the area of the yellow fabric in the flag.

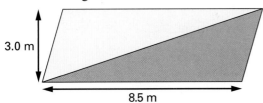

3.0 m

8.5 m

b) The price of the yellow fabric is $8.40/m². Calculate the cost of the yellow fabric for one flag.

c) Calculate the area of yellow fabric that is needed for 30 flags.

ⓒ Extending

14. The perimeter of △ABC is 58 cm.

a) How long is AB?

b) Calculate the area of △ABC.

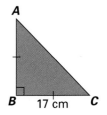

A

B   17 cm   C

15. The perimeter of △DEF is 16 cm.

a) How long is EF?

b) Calculate the area of △DEF.

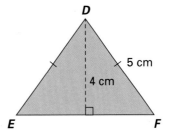

D

5 cm

4 cm

E   F

16. Determine each missing value.

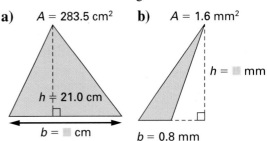

a)   A = 283.5 cm²   **b)**   A = 1.6 mm²

h = 21.0 cm

b = ▨ cm

h = ▨ mm

b = 0.8 mm

17. Design a kite by combining four triangles. Calculate the area of each triangle and the total area of the kite.

2-D Measurement **159**

5.3 Calculating the Area of a Triangle

You will need
- a geoboard and elastic bands
- centimetre square dot paper

▶ **GOAL**

Explore the area of a triangle on a geoboard.

Explore the Math

Matthew and Susan are designing a triangular sign. The sign must have a base of 3 m. It must slide into a frame that is made of two parallel metal pieces, as shown.

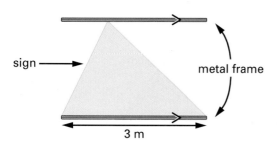

Matthew and Susan want to explore what the area of the sign will be.

? How do the areas of possible triangular signs compare?

A. Select two parallel line segments on your geoboard to represent the metal frame.

B. Create a triangle that has a base of 3 units and will fit exactly between the top and bottom of the frame. Calculate the area of your triangle.

C. Move the top vertex to other positions on the top of the frame. Calculate the area of each new triangle you create.

D. Repeat steps A to C using the same base but a different height.

E. Repeat steps A to C using the same height but a different base.

F. Describe your observations. Draw diagrams on square dot paper to support your observations.

> **Communication Tip**
>
> In diagrams, arrows are used to show that line segments are parallel.

Reflecting

1. How is the area of a triangle related to its base and its height?

2. Which of your triangles had the easiest area to calculate? Why?

3. Why might you have expected the results you reported in steps B to E?

You can calculate the area of some shapes on a geoboard by counting squares or using a formula. You can also use *Pick's theorem*. Unlike the other two methods, Pick's theorem works for any enclosed shape on a geoboard.

Using Pick's Theorem

To calculate the area of an enclosed shape on a geoboard, follow these steps:

Step 1: Count the number of pegs (p) that the elastic band touches. These are the *perimeter* pegs.

Step 2: Divide by 2.

Step 3: Add the number of pegs (i) inside the shape. These are the *interior* pegs.

Step 4: Subtract 1.

For example, $\triangle ABC$ has 3 perimeter pegs and 3 interior pegs.

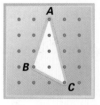

Area of $\triangle ABC = p \div 2 + i - 1$

$\qquad\qquad = 3 \div 2 + 3 - 1$

$\qquad\qquad = 3.5$ square units

1. Check the area of $\triangle ABC$.

   a) Calculate the area of the rectangle around $\triangle ABC$.

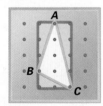

   b) Use the formula for the area of a triangle to calculate the area of each blue triangle. Add the areas of the blue triangles.

   c) Subtract the total area of the blue triangles from the area of the rectangle. Does your answer agree with the answer using Pick's theorem?

2. Use Pick's theorem to calculate the area of each shape. Check by counting squares or using a formula.

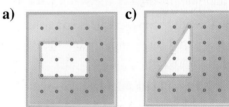

   a)   c)

   b)   d)

3. Make an irregular polygon on a geoboard. Use Pick's theorem to calculate the area of your polygon.

Area of a Trapezoid

You will need
• centimetre grid
  paper
• a calculator

▶ **GOAL**

Develop and apply the formula for the area of a trapezoid.

Learn about the Math

A new water park has the shape of a **trapezoid**.
The parallel sides are 60 m apart. In the winter,
the water park will be covered with canvas.
Yuki and Sandra want to calculate the area
of the water park to find out how much canvas
is needed.

? **What is the area of the water park?**

A. Draw a diagram of the trapezoid on centimetre grid paper. Make two
copies.

B. On one copy, divide your trapezoid into two or three simpler shapes
(rectangles and/or triangles).

C. Use a formula you already know to calculate the area of each simpler
shape.

D. Calculate the area of the water park using the areas in step C.

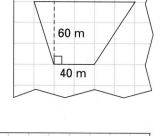

E. On the other copy, create a congruent trapezoid next to
the original trapezoid, so that the two trapezoids form
a parallelogram.

F. Use a formula to calculate the area of the parallelogram.

G. Calculate the area of the trapezoid using the area of the
parallelogram.

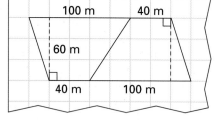

Reflecting

1. Can a trapezoid always be divided into a rectangle and two triangles?
Draw examples on grid paper to support your answer.

2. Why can a trapezoid always be divided into two triangles that have the
same height? What formula would this give for the area of a trapezoid?

3. Why can a trapezoid always be thought of as half a parallelogram?
What formula would this give for the area of a trapezoid?

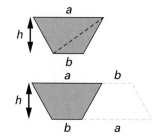

Work with the Math

Example 1: Calculating the area of a trapezoid by dividing it into triangles

Calculate the area of the wading pool.

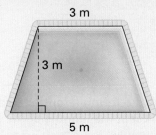

3 m

3 m

5 m

Yuki's Solution

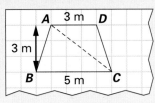

I drew a trapezoid and divided it into 2 triangles. In △ABC, the base is side BC, which is 5 m. The height is 3 m.

Area of △ABC = (b × h) ÷ 2
           = (5 m × 3 m) ÷ 2
           = 7.5 m²

Area of △ADC = (b × h) ÷ 2
           = (3 m × 3 m) ÷ 2
           = 4.5 m²

In △ADC, the base is side DA, which is 3 m. The height is still 3 m.

7.5 m² + 4.5 m² = 12 m²

The area of the trapezoid is the sum of the areas of the triangles.

The area of the wading pool is 12 m².

Example 2: Calculating the area of a trapezoid by making a parallelogram

Calculate the area of the wading pool above.

Sandra's Solution

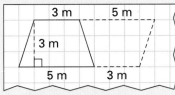

I drew the trapezoid. Then I copied it to form a parallelogram.

The base of this parallelogram is 5 m + 3 m = 8 m. Its height is 3 m.

Area of parallelogram = b × h
                     = 8 m × 3 m
                     = 24 m²

Area of trapezoid = 24 m² ÷ 2
                 = 12 m²

The trapezoid is half the area of the parallelogram, so its area must be half of 24 m².

The area of the wading pool is 12 m².

Calculate the area of the wading pool.

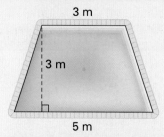

3 m

3 m

5 m

Solution

$A = (a + b) \times h \div 2$

$= (3\ m + 5\ m) \times 3\ m \div 2$

$= 8\ m \times 3\ m \div 2$

$= 24\ m^2 \div 2$

$= 12\ m^2$

This is the formula for the area of a trapezoid, where a and b represent the lengths of the parallel sides. You can choose these values: $a = 3$ m, $b = 5$ m, and $h = 3$ m. (You could have chosen $a = 5$ m and $b = 3$ m instead.)

Substitute these values into the formula.

The area of the wading pool is 12 m².

Ⓐ Checking

4. Draw the following trapezoid on grid paper. Calculate the area of the trapezoid by dividing the shape into simpler shapes.

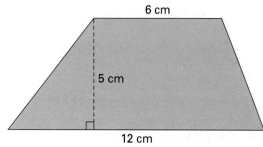

6 cm

5 cm

12 cm

5. Trace trapezoid *ABCD*, and show how two trapezoids form a parallelogram. Calculate the area of trapezoid *ABCD*.

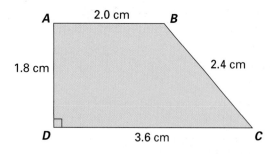

A 2.0 cm B

1.8 cm

2.4 cm

D 3.6 cm C

6. Calculate the area of this trapezoid using a formula.

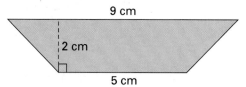

9 cm

2 cm

5 cm

Ⓑ Practising

7. a) Which side length of the trapezoid below is not required to calculate its area?

b) Calculate the area and the perimeter of the trapezoid to the nearest tenth.

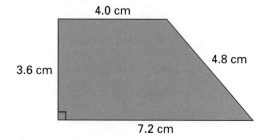

4.0 cm

4.8 cm

3.6 cm

7.2 cm

8. On centimetre grid paper, draw a trapezoid with an area of 13 cm².

9. Calculate the area of each trapezoid.

a)

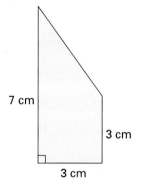

7 cm

3 cm

3 cm

b)

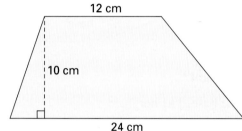

12 cm

10 cm

24 cm

10. A garden has the dimensions shown below. Which calculation could you use to find the area of the garden? Explain your thinking.

a) Area = (3 m + 1 m) × 4 m ÷ 2

b) Area = 1 m + (4 m × 3 m) ÷ 2

c) Area = (4 m + 1 m) × 3 m ÷ 2

d) Area = (1 m + 3 m) × 4 m ÷ 2

e) Area = (4 m + 3 m) × 1 m ÷ 2

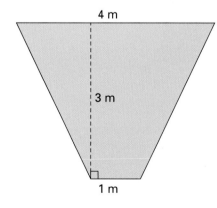

4 m

3 m

1 m

11. A trapezoid has an area of 10.5 cm² and parallel sides that measure 5.0 cm and 2.0 cm. What is the height of the trapezoid? Explain what you did.

12. a) A trapezoid has parallel sides that are 28 cm and 40 cm long, and 18 cm apart. Calculate the area of the trapezoid.

b) Another trapezoid has parallel sides that are 2 m and 5 m long, and 3 cm apart. Calculate its area.

13. A trapezoid has an area of 20 cm² and a height of 5 cm. What is the sum of its top and bottom sides?

14. A kindergarten classroom has 10 trapezoid-shaped tables. Some Grade 7 students have volunteered to paint game boards to cover the top surfaces of the tables. Each table has the measurements shown. What is the total area the students will paint?

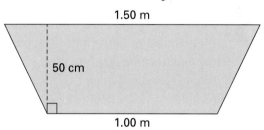

1.50 m

50 cm

1.00 m

❂ Extending

15. Use what you know about trapezoids to calculate the area of this regular hexagon. Explain what you did.

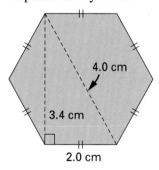

4.0 cm

3.4 cm

2.0 cm

16. The length of the shortest side of a red pattern block is 25 mm. What is the area of the trapezoid?

Mid-Chapter Review

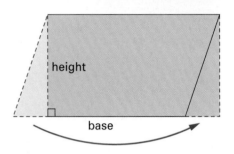

Frequently Asked Questions

Q: What is the formula for the area of a parallelogram?

A: The formula for the area of any parallelogram is
$A = b \times h$, which means the *base* times the *height*.

This is because the area of a parallelogram is the
same as the area of a rectangle with the same height
(length) and base (width).

**Q: When calculating the area of a parallelogram, which
measurements do you use for the base and the height?**

A: You can use any side of a parallelogram as the base. The height
depends on which side you use as the base. The height can be
drawn anywhere between the parallel sides, but it must be
perpendicular to the base. The parallelogram on the left can be
turned. Its area does not change, but what we call the base and
the height do change.

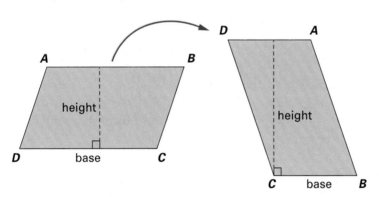

Q: What is the formula for the area of a triangle?

A: The formula for the area of any triangle is
$A = (b \times h) \div 2$, which means the *base* times
the *height*, divided by 2. The area of a triangle
is half the area of a parallelogram with the
same base and height.

The area of the green triangle is half the area
of the blue and green parallelogram.

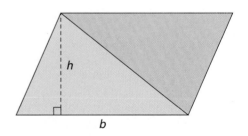

Q: **When using the formula for the area of a triangle, which side do you use as the base?**

A: You can use any side of a triangle as the base. The height depends on which side you use as the base. The height intersects the base at a 90° angle and reaches to the opposite vertex. Sometimes the height is outside the triangle.

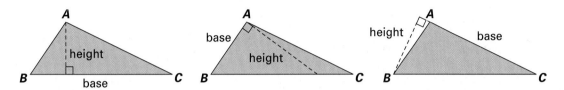

Q: **How can you calculate the area of a triangle shown on a geoboard?**

A1: You can use a formula. For this triangle, the base is 4 units and the height is 3 units. Area = 4 × 3 ÷ 2, or 6 square units.

A2: You can subtract the area of the other shapes from the area of the rectangle around the triangle. For this triangle, the whole area is 4 × 4 = 16 square units. The areas of the other shapes are

- small rectangle, 4 square units
- small triangle on the left, 2 × 3 ÷ 2 = 3 square units
- small triangle on the right, 2 × 3 ÷ 2 = 3 square units

Area of yellow triangle is 16 − (4 + 3 + 3) = 6 square units.

Q: **What is the formula for the area of a trapezoid?**

A: The formula for the area of any trapezoid is $A = (a + b) \times h \div 2$, which means the sum of bases a and b times the *h*eight, divided by 2.

The area of the blue trapezoid is half the area of the blue and purple parallelogram.

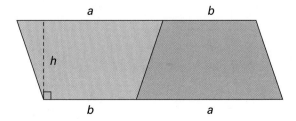

Practice Questions

(5.1) **1.** Braydon is creating an obstacle course in the shape of a parallelogram for his dog. The parallelogram will cover an area of 48 m^2. Its base and height must be whole numbers of metres. Draw and label the perimeter that gives the longest obstacle course.

(5.2) **2.** Copy and complete the table.

	Shape	Height	Base	Area
a)	triangle	■ cm	3 cm	21 cm^2
b)	parallelogram	5.0 km	1.4 km	■ km^2
c)	triangle	3.20 cm	■ cm	7.84 cm^2
d)	parallelogram	■ mm	15.2 mm	30.4 mm^2

(5.4) **3.** Calculate the area of each shape.

a)

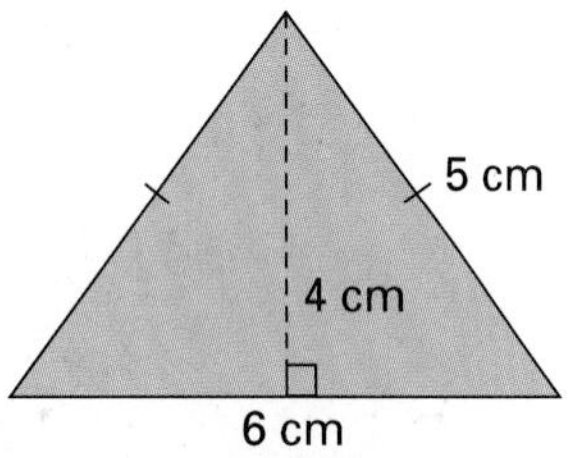
5 cm
4 cm
6 cm

b)

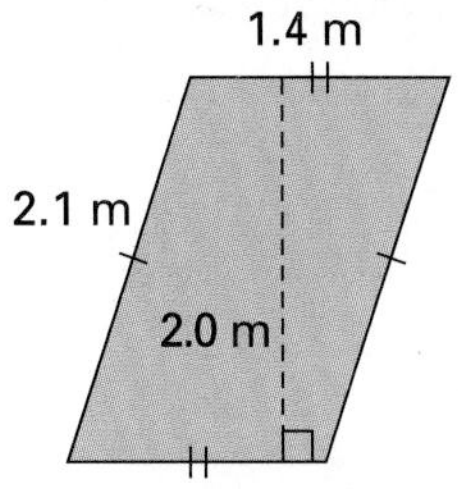
1.4 m
2.1 m
2.0 m

c)

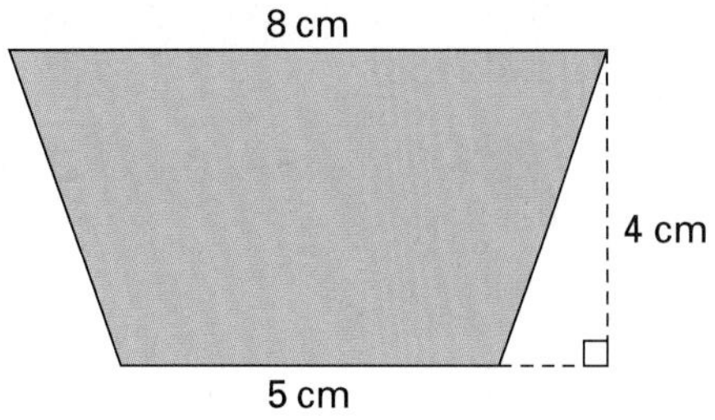
8 cm
4 cm
5 cm

4. Use any method to calculate the area of each shape. (5.4)

a)

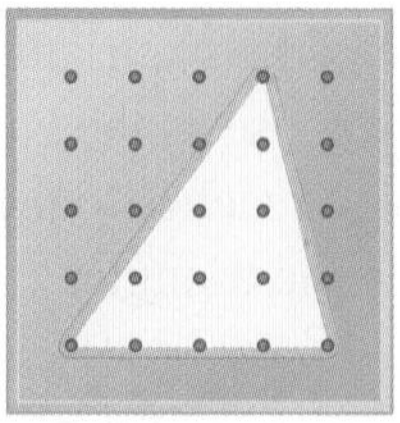

d)

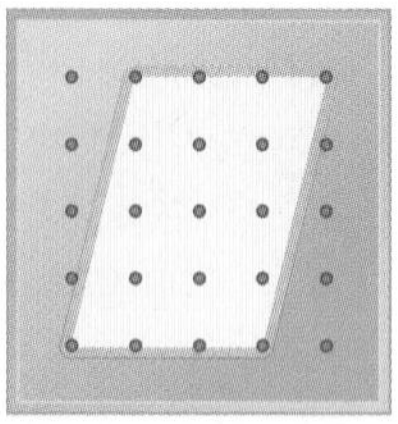

b)

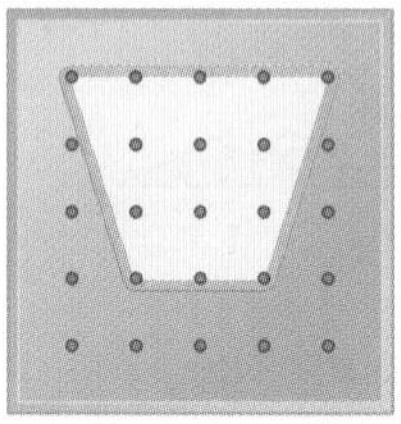

e)

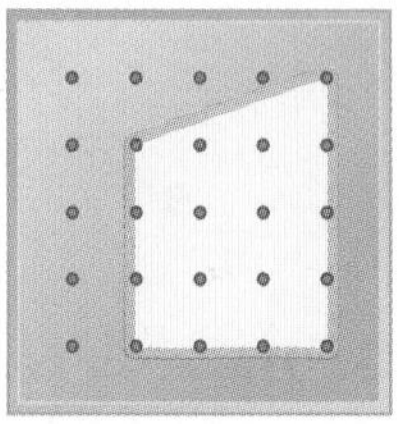

c)

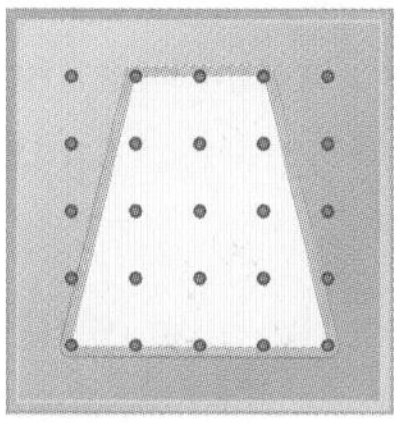

f)

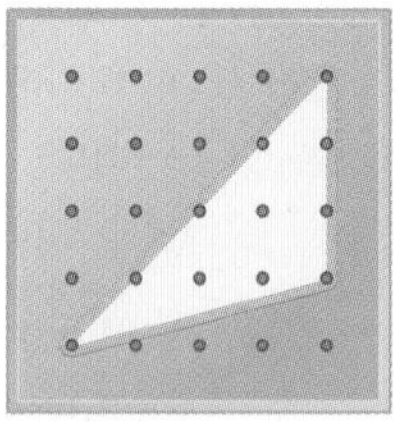

5. Draw each shape on centimetre grid paper.

a) a triangle with an area of 27 cm^2 (5.4)

b) a parallelogram with an area of 64 mm^2

c) a trapezoid with an area of 16 cm^2

d) a trapezoid with an area of 21 cm^2

6. The following diagram shows Meagan's lawn, which she plans to re-sod. The price of sod is \$12 per square metre. How much will it cost Meagan to re-sod her lawn? (5.4)

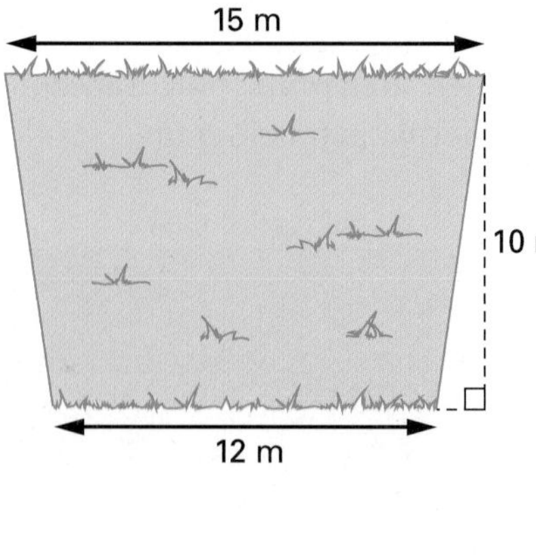
15 m
10 m
12 m

In this game, you will find all the shapes that each have a perimeter of about 7 units (rounded to the nearest whole number).

On a geoboard, the horizontal or vertical distance between two pegs is 1 unit.

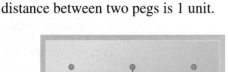

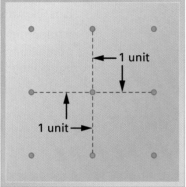

The diagonal distance between two pegs is *about* 1.5 units.

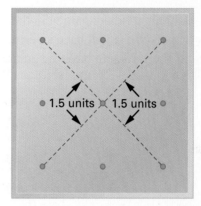

Here is one shape that has a perimeter of *about* 7 units on a geoboard.

Number of players: 2 players or 2 teams

Rules

1. Players (or teams) take turns. One player creates a shape on a geoboard with an elastic band. The other player verifies that the perimeter of the shape is about 7 units, and records the shape by drawing it on the dot paper.

2. When the players have found all of the eight other shapes, the game is over. Everyone wins!

Exploring the Area and Perimeter of a Trapezoid

You will need
- a 24 cm piece of string
- tape
- a ruler

▶ **GOAL**

Explore the relationship between the area and the perimeter of a trapezoid.

Explore the Math

Brooke is planning a flower garden in the shape of an **isosceles trapezoid**. It will have a perimeter of 24 m. Brooke wants the garden to have the greatest possible area.

isosceles trapezoid

a trapezoid where the non-parallel sides have equal lengths

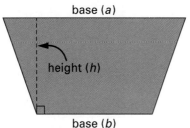

base (*a*)

height (*h*)

base (*b*)

❓ **What dimensions will give the garden the greatest area?**

A. Use tape to help you arrange a 24 cm piece of string into different isosceles trapezoids. Let 1 cm represent 1 m.

B. Copy the table and add more rows.

Perimeter (cm)	Sketch of possible trapezoid	Side length (cm)	Side length (cm)	Base *a* (cm)	Base *b* (cm)	Height *h* (cm)	Area (cm²)
24	11 cm / 9 cm / 2 cm	2	2	9	11	1.5	15

C. Continue the table for more possible trapezoids. Determine which dimensions give the greatest area.

D. Suppose that the perimeter of Brooke's garden will be 32 m. Create a new table to show possible trapezoids. Determine which dimensions give the greatest area.

Reflecting

1. What happens to the area of the garden as its height increases?

2. What happens to the area of the garden as its sides and its height become closer to the same measurement?

3. Which trapezoid has the greatest area?

Mental Math

USING A STAIRCASE TO CONVERT LENGTHS

You can use a metric staircase to convert lengths from one unit to another.

To convert 1.59 km to metres, the number of metres must be *greater than* 1.59.

You go down three steps and multiply 1.59 by 10 three times.

1.59 km = 15.9 hm = 159 dam = 1590 m
       × 10   × 10   × 10

To convert 1590 m to kilometres, the number of kilometres must be *less than* 1590.

You go up three steps and divide 1590 by 10 three times.

1590 m = 159 dam = 15.9 hm = 1.59 km
       ÷ 10   ÷ 10   ÷ 10

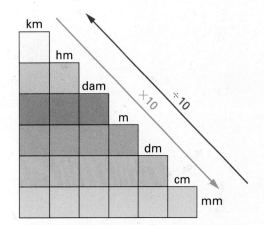

1. How can you use mental math to multiply numbers by 10?

2. How can you use mental math to divide numbers by 10?

3. Convert each length.

  a) 2.75 km to metres
  b) 8.6 m to millimetres
  c) 14 dam to centimetres
  d) 1.788 km to decametres

  e) 19 245 mm to metres
  f) 455.5 m to kilometres
  g) 0.5 mm to centimetres
  h) 0.15 cm to decimetres

Calculating the Area of a Complex Shape

▶ **GOAL**

Calculate the area of an irregular 2-D shape by dividing it into simpler shapes.

Learn about the Math

Chang is designing an extreme sports park for Adventurepark. He is using pattern blocks to try different designs.

? **How can you calculate the area of the extreme sports park?**

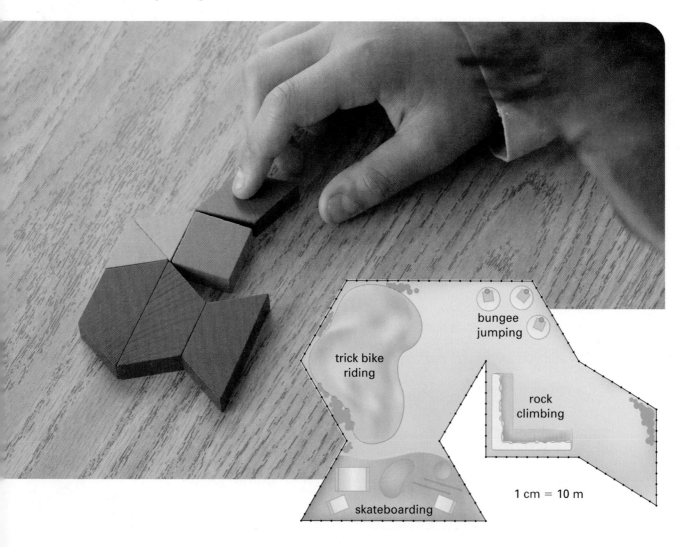

What is the area of the extreme sports park?

Chang's Solution

I used three trapezoids, two triangles, one parallelogram, and one square to make the complex shape. I know how to calculate the area of each of these simpler shapes.

I measured the base and the height of each pattern block, and labelled the measurements on sketches. Then I used a formula to calculate each area. I added the areas to find the total area of the complex shape.

Simpler polygon	Area of simpler polygon	Total area
2.5 cm / 2.0 cm / 5.0 cm	$A = (a + b) \times h \div 2$ $= (5.0 \text{ cm} + 2.5 \text{ cm}) \times 2.0 \text{ cm} \div 2$ $= 7.5 \text{ cm} \times 2.0 \text{ cm} \div 2$ $= 7.5 \text{ cm}^2$	There are three trapezoids with the same area. $3 \times 7.5 \text{ cm}^2 = 22.5 \text{ cm}^2$
2.0 cm / 2.5 cm	$A = b \times h \div 2$ $= 2.5 \text{ cm} \times 2.0 \text{ cm} \div 2$ $= 2.5 \text{ cm}^2$	There are two triangles with the same area. $2 \times 2.5 \text{ cm}^2 = 5.0 \text{ cm}^2$
2.0 cm / 2.5 cm	$A = b \times h$ $= 2.5 \text{ cm} \times 2.0 \text{ cm}$ $= 5.0 \text{ cm}^2$	5.0 cm^2
2.5 cm / 2.5 cm	$A = l \times w$ $= 2.5 \text{ cm} \times 2.5 \text{ cm}$ $= 6.25 \text{ cm}^2$	6.25 cm^2
To find the total area of the complex shape, add the smaller areas. 1 cm = 10 m, so 1 cm^2 = 10 m $\times$ 10 m, or 100 m^2		38.75 cm^2 The area of the sports park is about 3880 m^2.

Reflecting

1. Explain how starting with simpler polygons helps you calculate the area of a complex polygon. Use Chang's example to support your answer.

2. Chang calculated the area of one trapezoid and the area of one triangle. Explain why he did not need to calculate the areas of the other two trapezoids and the other triangle.

3. Divide the complex polygon into a different set of simpler shapes. What is the sum of their areas?

Example 2: Calculating the area of an irregular shape

Calculate the total area of the orange shapes in this rectangle.

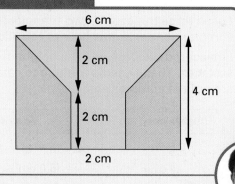

Ravi's Solution

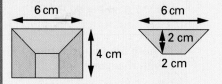

Area of rectangle	Area of trapezoid	Area of square	I divided the green shape into a trapezoid and a square.
$= l \times w$	$= (a + b) \times h \div 2$	$= l \times w$	I used formulas to calculate the areas of the shapes.
$= 6 \text{ cm} \times 4 \text{ cm}$	$= (6 \text{ cm} + 2 \text{ cm}) \times 2 \text{ cm} \div 2$	$= 2 \text{ cm} \times 2 \text{ cm}$	The area of the rectangle is 24 cm².
$= 24 \text{ cm}^2$	$= 8 \text{ cm} \times 2 \text{ cm} \div 2$	$= 4 \text{ cm}^2$	The area of the trapezoid is 8 cm².
	$= 8 \text{ cm}^2$		The area of the square is 4 cm².

Area of rectangle minus area of trapezoid and square	I subtracted the area of the trapezoid and the area of the square from the area of the rectangle.
$= 24 \text{ cm}^2 - (8 \text{ cm}^2 + 4 \text{ cm}^2)$	
$= 24 \text{ cm}^2 - 12 \text{ cm}^2$	
$= 12 \text{ cm}^2$	The total area of the orange shapes is 12 cm².

Ⓐ Checking

4. Calculate the area of the green shapes in each diagram. Show your work.

a)

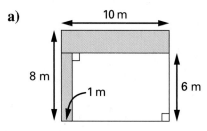

b)

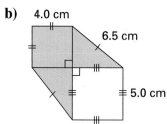

B Practising

5. Use pattern blocks to create an irregular polygon. Trace your polygon on paper, and calculate its total area.

6. Calculate the area of the purple part of each diagram.

a)

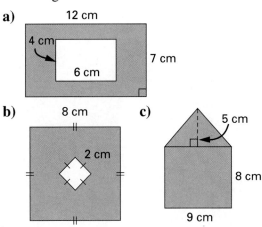

b)   8 cm   c)

7. Heidi is designing a two-lane running track for an athletic club. The track will go around a rectangle that will have an area of 3600 m². The length and width of the rectangle must be whole numbers of metres.

   a) Draw five possible tracks.

   b) Which track is the longest?

   c) Which track is the shortest?

8. The following diagram shows a picnic area. Each square represents 1 m². The green areas are grass. The grey areas need to be paved. If a paving company charges $12 to pave 1 m², how much will the company charge to pave all the grey areas?

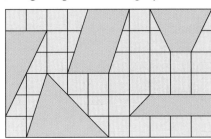

9. What is the area of the triangle on this geoboard?

10. Calculate the area of each polygon.

a)

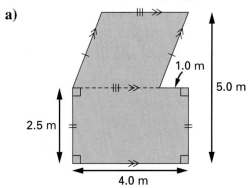

b)

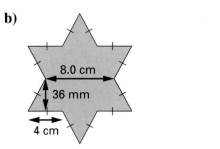

C Extending

11. Richard built a cement barrier that is 100 blocks long and 1 block high. He plans to paint the front, top, and back of each block, as well as any other visible sides. He does not plan to paint the sides that touch one another or the sides that are on the bottom. What is the total area of the surfaces that Richard will paint? (*Hint:* Start with four blocks, then consider ten blocks, and so on.)

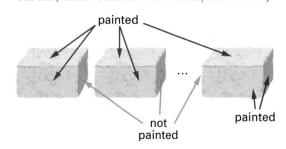

painted

not painted   painted

5.7 Communicating about Measurement

▶ **GOAL**

Describe perimeter and area using appropriate mathematical language.

Communicate about the Math

Sarah says, "I was asked to describe a real-life example of area and perimeter. I remember that my family bought outdoor carpet for part of our deck and fastened it with moulding around the edges. The carpeted area is 10 m long and 2 m wide. I can describe how to calculate the cost of the carpet and the moulding."

Sarah asked Rosa to proofread and comment on her description.

? **How can Sarah improve her description?**

Sarah's Description	**Rosa's Questions**
First I measured the perimeter.	How did you measure the perimeter?
I got 24 m. Next I calculated the area. I used length × width. The area equalled 20. To calculate the	Why did you need to calculate the area?
	Why did you use length and width to calculate the area?
amount of moulding, I multiplied 24 × $0.50. To calculate the amount of carpet, I multiplied 20 × $1.25. The total cost is $37.	How did you know which price to use for each measurement?

A. Did Sarah show that she understands how to calculate area and perimeter? Explain.

B. Which of Rosa's questions do you think is most important? Why?

C. What other questions would you ask about Sarah's description?

D. Rewrite Sarah's description. Include answers to Rosa's questions and your questions.

Reflecting

1. Which points in the Communication Checklist did Sarah cover well? Explain your answer.

2. Why is it important to communicate how you solve a problem?

Work with the Math

Example: Writing a complete answer

Ryan and Sandra built a window inside a frame for the school play, as shown. They painted the trim for the window blue and the frame purple.

a) What length of trim is painted?

b) What area of frame is painted? Explain what you did.

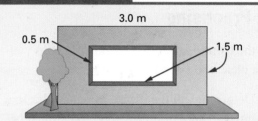

a) Ryan's Solution

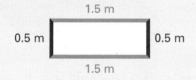

To calculate the length of trim, I added the measurements for all four sides of the window. I doubled 0.5 m to get the two ends of the window. I doubled 1.5 m to get the top and bottom of the window.

$2 \times (0.5 \text{ m}) + 2 \times (1.5 \text{ m}) = 4.0 \text{ m}$

4.0 m of trim is painted.

b) Sandra's Solution

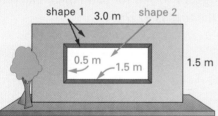

I labelled the window and the frame shape 1. I labelled the window shape 2.

Area of shape 1
$3.0 \text{ m} \times 1.5 \text{ m} = 4.5 \text{ m}^2$

The area of shape 1 is length times width, which is 4.5 m².

Area of shape 2
$0.5 \text{ m} \times 1.5 \text{ m} = 0.75 \text{ m}^2$

The area of shape 2 is 0.75 m².

Area of frame
$4.5 \text{ m}^2 - 0.75 \text{ m}^2 = 3.75 \text{ m}^2$

Then I subtracted the area of the window (shape 2) from the area of the window and the frame (shape 1).

To the nearest tenth, 3.8 m² of frame is painted.

I rounded to the nearest tenth because the measurements are given in tenths.

A Checking

3. A Grade 5 student asks you to explain how to calculate the perimeter and area of a square with sides that are 2 cm long. Write your explanation.

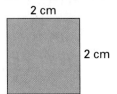
2 cm
2 cm

B Practising

Use the Communication Checklist to help you write clear and complete explanations.

4. Determine all the different rectangles that have an area of 36 m². The sides must be whole numbers of metres. Use words, numbers, and pictures to explain what you did.

5. Calculate the area of the top face of a yellow pattern block (hexagon). Explain what you did.

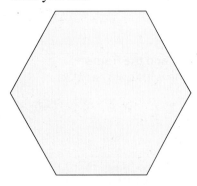

6. Calculate the area of triangles A, B, and C. Explain why triangles A, B, and C have the same area.

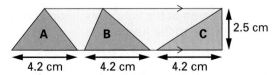
A   B   C   2.5 cm
4.2 cm   4.2 cm   4.2 cm

7. Calculate the area of the yellow shape in each diagram. Explain what you did.

a)
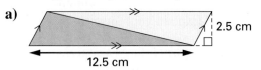
2.5 cm
12.5 cm

b)
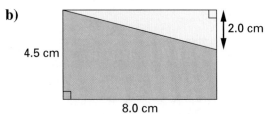
2.0 cm
4.5 cm
8.0 cm

c)
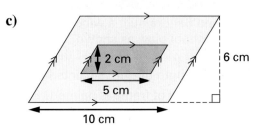
2 cm
5 cm
6 cm
10 cm

d)
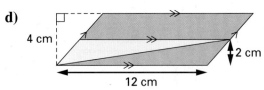
4 cm
2 cm
12 cm

8. Jean-Pierre drew the following diagram to show the patio that will be built beside the Adventurepark restaurant. Jean-Pierre has a budget of $1500 for patio tiles. The price of the tiles is $22/m².

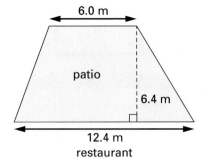
6.0 m
patio
6.4 m
12.4 m
restaurant

a) What is the area of the patio? Explain what you did.

b) How much will the patio tiles cost?

c) Will Jean-Pierre be over budget? Explain how you know.

1. Calculate each perimeter and area.

a)

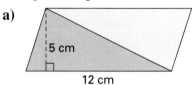

7.5 cm
3.0 cm
3.75 cm

b)

5.0 m
4.2 m
4.5 m

2. Calculate the area of each trapezoid.

a)

5.0 cm
2.0 cm
2.5 cm

b)

8 cm
4 cm
13 cm

3. Calculate the area of the orange triangle in each parallelogram.

a)

5 cm
12 cm

b)

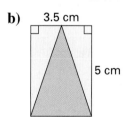

3.5 cm
5 cm

c)

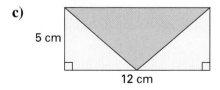

5 cm
12 cm

4. Calculate the area of the green shape in each rectangle.

a)

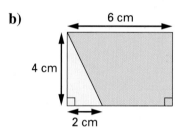

2 cm   4 cm
4 cm
4 cm   2 cm

b)

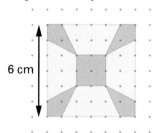

6 cm
4 cm
2 cm

5. What area of the large square is orange? Explain what you did.

6 cm

6. Asha is redecorating the room shown below.

a) What area of carpet does she need to cover the floor? Explain your solution.

b) What is the perimeter of the room?

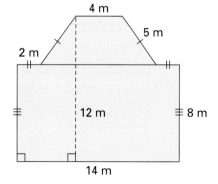

4 m
5 m
2 m
12 m
8 m
14 m

Frequently Asked Questions

Q: **How can you calculate the area of the blue arrow inside the rectangle?**

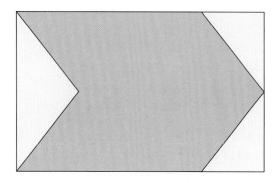

A1: Divide the arrow into simpler polygons. Use formulas to calculate the areas of the simpler polygons. Then add the areas.

Often there is more than one way to divide a complex shape. For example, the arrow can be divided into two small triangles, one rectangle, and a larger triangle.

The arrow can also be divided into two parallelograms.

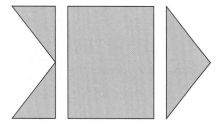

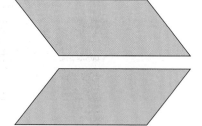

A2: Calculate the area of a rectangle that encloses the arrow. Then subtract the areas of the shapes around the arrow from the area of the rectangle.

Here you would subtract the areas of the three yellow triangles from the area of the rectangle.

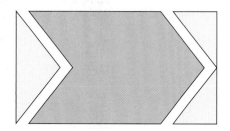

Practice Questions

(5.2) **1. a)** Draw three parallelograms that each have an area of 36 cm².

b) Draw three triangles that each have an area of 24 cm².

(5.5) **2.** If two trapezoids have the same area, do they always have the same perimeter? Explain your thinking with words and diagrams.

(5.6) **3.** Measure the dimensions of each shape, and calculate the area.

a)

b)

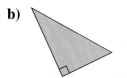

c)

d)

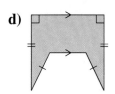

(5.6) **4.** Calculate the area of the grass around the pond in the following diagram.

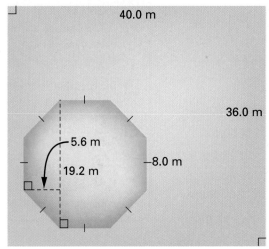

5. What is the area of the triangle on this geoboard? (5.6)

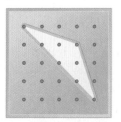

6. You need to choose a design for the Adventurepark sign. The cost of a sign is \$33.50/m². (5.6)

a) How much does each design cost?

b) Which design would you choose? Explain why.

Design 1

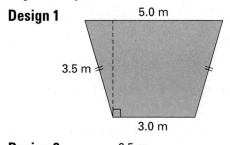

Design 2

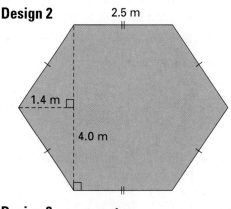

Design 3

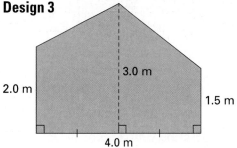

Adventurepark Design

You have been selected to design the grounds and attractions of Adventurepark. In your design, you need to include

- amusement rides
- a wave pool
- a water park
- an extreme sports park
- a restaurant
- walking paths
- landscaping

? **How will you organize your design of Adventurepark?**

A. Your design must include at least two of each of the following shapes: triangles, parallelograms, trapezoids, and complex polygons.

B. On a piece of grid paper, draw a scale diagram to represent the space needed for the grounds and the different attractions.

C. Create a table to show the area and perimeter of each shape.

D. Describe your design, and explain how you calculated the area and perimeter of each shape.

> **Task Checklist**
>
> ☑ Did you include and label at least two of each shape?
>
> ☑ Did you label the grounds and the attractions?
>
> ☑ Did you show all your calculations?
>
> ☑ Did you explain how you calculated the area and perimeter of each shape?

CHAPTER 6

Addition and Subtraction of Integers

▶ GOALS

You will be able to

- use models to represent integers
- add and subtract integers
- compare and order integers

Interpreting Data

Tynessa's science class did an experiment to see how long a piece of ice takes to melt and turn into boiling water. Tynessa's group put some ice in water, heated it, and recorded the temperature every 30 s.

They displayed their data in a table and as a broken-line graph.

Time (s)	Temperature (°C)
0	0
30	0
60	0
90	0
120	20
150	40
180	60
210	80
240	100
270	100
300	100
330	100
360	100
390	100

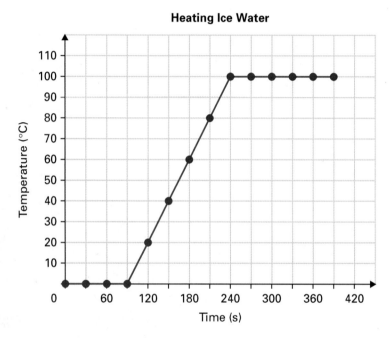

? ## What conclusions can you make about the group's experiment?

A. How long did the temperature of the water take to reach 70°C?

B. Estimate the temperature of the water after 105 s.

C. When was the temperature increasing?

D. When was the temperature not changing? What was happening then?

E. Why do you think different scales are used on the two axes?

F. Why do you think there are no negative numbers on either axis?

G. What other information can you tell about the group's experiment from the data?

Do You Remember?

1. Use < or > to make each statement true.

a) 8 ☐ 0   **c)** 24 ☐ 6

b) 12 ☐ 14   **d)** 99 ☐ 113

2. Which is true, $a < b$ or $a > b$?

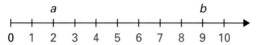

3. Add.

a) $12 + 19$ **c)** $8 + 105$

b) $32 + 45$ **d)** $23 + 38 + 93$

4. Subtract.

a) $37 - 21$ **c)** $126 - 58$

b) $65 - 65$ **d)** $107 - 36 - 14$

5. Use the thermometer to answer each question.

a) Today's temperature is 4°C. Locate 4°C on the thermometer.

b) Tomorrow's temperature is predicted to be 5°C lower than today's temperature. Locate the predicted temperature on the thermometer.

c) What is tomorrow's predicted temperature?

d) Why is $-6°C < -4°C$?

e) How many degrees colder is a temperature of $-2°C$ than a temperature of $+1°C$?

6. A positive sign (+) describes numbers above zero. A negative sign (–) describes numbers below zero. Use + or – to write the best number for each situation. Explain why you chose the sign you did.

a) Tara has $25 in her bank account.

b) Takuya owes a library fine of $2.

c) The George Massey Tunnel under the Fraser River in Richmond, British Columbia, is 20 m below sea level.

d) On March 8, the temperature in Grand Prairie, Alberta, was 38°C below zero.

e) On July 20, the temperature in Athens, Greece, was 41°C above zero.

7. In a countdown to a rocket launch, the time 5 s before takeoff is called "T minus five seconds," and the time 5 s after takeoff is called "T plus five seconds." Match each number to the correct countdown term.

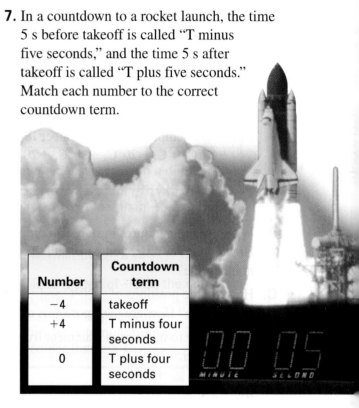

Number	Countdown term
−4	takeoff
+4	T minus four seconds
0	T plus four seconds

8. a) Write definitions for "positive" and "negative" in your own words.

b) Compare your definitions with the definitions in a dictionary. How are they the same? How are they different?

c) Keep the definitions near you as you progress through the chapter to see how your understanding of the words "positive" and "negative" changes.

Adding Integers Using the Zero Principle

You will need
- red and blue counters

▶ **GOAL**

Use the zero principle, with and without models, to add integers.

Learn about the Math

In a coin tossing experiment, Paul gained 1 point (+1) when he tossed Heads. He lost 1 point (−1) when he tossed Tails.

The following table shows Paul's results of 11 tosses.

Toss number	1	2	3	4	5	6	7	8	9	10	11
Result (+1) or (−1)	−1	−1	+1	+1	+1	−1	+1	−1	+1	−1	−1

opposite integers

two integers the same distance away from zero; for example, +6 and −6 are opposite integers

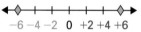

? **How can you add integers to calculate Paul's score after 11 tosses?**

You can use a blue counter ● to represent (−1) and a red counter ● to represent (+1). The integers (+1) and (−1) are **opposite integers**.

Adding (−1) and (+1) gives a net result of zero.

This is the **zero principle**.

zero principle

two opposite integers, when added, give a sum of zero; for example, (−1) + (+1) = 0

Example 1: Modelling the sum with counters

Use counters to calculate Paul's score after 11 tosses.

Paul's Solution

I modelled my first 11 tosses using counters. I used blue counters to represent (−1) and red counters to represent (+1).

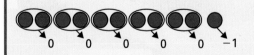

I changed the order to get pairs of blue and red counters.
In each pair, (+1) paired with (−1) is 0.
One blue counter was left over.
The answer is (−1).

Example 2: Using +1s and −1s

Use positive 1s and negative 1s to calculate Paul's score after 11 tosses.

Fawn's Solution

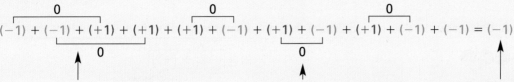

| I found pairs of positive 1s and negative 1s. | I added the +1 and the −1 in each pair to get 0. | There was a single −1 left over. |

Example 3: Combining integers greater than +1 and less than −1

Use integers greater than +1 and less than −1 to calculate Paul's score after 11 tosses.

Miguel's Solution

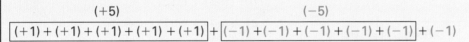

I wrote all the positive 1s first.   Then I wrote all the negative 1s.

$(+5) + (−5) + (−1)$

$= 0 + (−1)$ The zero principle says that $(−5) + (+5) = 0$.

$= (−1)$ $(−1)$ was left over.

Reflecting

1. a) How are Paul's and Fawn's solutions alike?

b) How are they different?

2. Miguel used the idea that the sum of any two opposite integers is always zero. Verify Miguel's solution using counters.

Work with the Math

Example 4: Using the zero principle

Add $(-4) + (+2)$.

Solution A: Using counters

⬤⬤⬤⬤⬤⬤⬤

(circled pairs) 0   0

The answer is (-2).

Solution B: Using numbers greater than +1 and less than −1

$(-4) + (+2) = (-2) + \boxed{(-2) + (+2)}$

$= (-2)$ → 0

The answer is (-2).

Ⓐ Checking

3. Add each expression using counters and numbers.

	Expression	Using counters	Using numbers
a)	$(-3) + (+2)$		
b)	$(-4) + (+6)$		
c)	$(+5) + (-6)$		
d)	$(-5) + (+7)$		
e)	$(+2) + (-8)$		
f)	$(-1) + (-9)$		

4. Use mental math to determine each sum.

a) $(+3) + (-3) = $ ▨

b) $(-7) + (+7) = $ ▨

Ⓑ Practising

You may use counters to answer the following questions.

5. Complete.

a) $(-3) + (-2) = $ ▨

b) $(+2) + (-2) = $ ▨

c) $(-4) + (+1) = $ ▨

d) $(-7) + (+6) = $ ▨

e) $(-5) + (-2) = $ ▨

f) $(-5) + (+2) = $ ▨

6. Explain why $(-25) + (+25) = 0$.

7. The following patterns are based on adding integers. Continue each pattern. Then write a rule to describe each pattern.

a) $0, -1, -2, -3, -4,$ ▨, ▨, ▨

b) $-3, -2, -1, 0,$ ▨, ▨, ▨

8. Fill in each ▨ with +1 or −1 to make each statement true.

a) $(+1) + $ ▨ $ + $ ▨ $ = (-1)$

b) $(-1) + $ ▨ $ + $ ▨ $ = (+1)$

c) $(+1) + $ ▨ $ + $ ▨ $ + $ ▨ $ + $ ▨ $ = (-1)$

d) $(+1) + $ ▨ $ + $ ▨ $ + $ ▨ $ + (+1) = (-1)$

9. Complete.

a) $(-3) + (+3) + (+5) = $ ▨

b) $(-1) + (-2) + (-1) = $ ▨

c) $(+2) + (+1) + $ ▨ $ = (-1)$

10. Use $=$, $<$, or $>$ to make each statement true.

a) $(-1) + (-2)$ ▦ (-4)

b) $(+2) + (-5)$ ▦ (-3)

c) $(-3) + (+6)$ ▦ $(+2)$

d) $(+5) + (-7)$ ▦ (-2)

e) $(-2) + (-4)$ ▦ (-5)

f) $(-2) + (+1)$ ▦ 0

11. Using $+1$ and -1 only, create an addition question that has each sum. Use at least four numbers for the question. Check your work using counters.

a) $+3$ **b)** -2 **c)** 0 **d)** -1

12. a) Calculate the sum. You can use counters or numbers.
$(+1) + (+1) + (-1) + (+1) + (-1) +$
$(-1) + (+1) + (+1) + (-1)$

b) Which method did you choose? Why?

13. Fill in each ▦ with an integer to make the equation true. Show three different solutions.

$$▦ + ▦ + ▦ = (-5)$$

14. Explain why you cannot complete this equation using only $+1$s or -1s.

$$(+1) + ▦ + ▦ + ▦ = (+1)$$

15. In a Magic Square, all rows, columns, and diagonals have the same sum. No number appears more than once.

a) This Magic Square uses integers from -6 to $+2$. Verify that the rows, columns, and diagonals all have the same sum. The sum of the third column is shown.

+1	−6	−1
−4	−2	0
−3	+2	−5

$(-1) + 0 + (-5) = (-6)$

b) This Magic Square uses integers from -1 to $+7$. Complete it. Check that all the sums are the same.

+2		
+7	+3	−1

c) This Magic Square uses integers from -4 to $+4$. Complete it. Check that all the sums are the same.

+3	−4	
−2		
	+4	

d) Create a Magic Square that uses integers from -10 to -2.

⦿ Extending

16. State whether each statement is true or false. Explain your reasoning.

a) The sum of two positive integers is positive.

b) The sum of two negative integers is negative.

c) The sum of a negative integer and a positive integer is always positive.

17. Continue each pattern. Write a rule to describe the pattern.

a) $0, +1, -1, +2, -2, +3, -3,$ ▦, ▦, ▦

b) $-1, 0, -1, -1, -2, -3, -5, -8,$ ▦, ▦, ▦

18. Without using a calculator, determine the sum of all the integers from -50 to $+50$. Describe your strategy.

19. a) Calculate the average daily high temperature for the four days.

Four-day weather forecast	High temperature (°C)	Low temperature (°C)
Wednesday	+10	0
Thursday	+5	−6
Friday	+9	−7
Saturday	+8	−3

b) Calculate the average daily low temperature for the four days.

c) What is the range between the highest high and the lowest low?

Adding Integers That Are Far from Zero

▶ **GOAL**
Add integers with and without models.

Learn about the Math

Suppose you are asked to calculate $(+35) + (-40)$ and you don't have enough counters.

? **How can you add integers that are far from zero?**

You can use a number line to model adding integers.

"We'll need 35 red counters and 40 blue counters. That's too many!"

Example 1: Using a number line

Add $(+35)$ and (-40) using a number line model.

Bonnie's Solution

I think of $(+35)$ as an increase in temperature from 0. I used a red arrow going from 0 to $(+35)$ to represent $(+35)$.

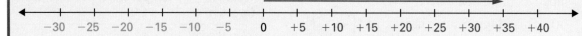

(-40) reminds me of a decrease in temperature. I represented (-40) with a blue arrow that started at $(+35)$ and went left 40 spaces.

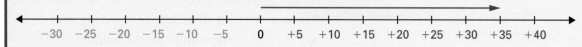

I ended up at (-5).
This means that $(+35) + (-40) = (-5)$.

Add (+35) and (−40) without using a model.

Fawn's Solution

I didn't have enough counters, so I imagined using 35 red counters to represent (+35) and 40 blue counters to represent (−40).

35 counters   40 counters

$(\bullet \; ... \; \bullet) + (\bullet \; ... \; \bullet\bullet\bullet\bullet\bullet\bullet\bullet)$

I know that (+1) + (−1) = 0 for each red/blue pair. There are 35 red/blue pairs.

$(\bullet\bullet \; ... \; \bullet\bullet) + (\bullet\bullet\bullet\bullet\bullet)$
   0   0

$= (\bullet\bullet\bullet\bullet\bullet)$

There would be 5 blue counters left over.
This means that (+35) + (−40) = (−5).

Reflecting

1. Fawn used the zero principle to find the sum with counters. Explain how Bonnie also used the zero principle with her number line.

2. a) How would the counter model and number line model have to change to represent (−35) + (+40)?

   b) How would the models have to change to represent (−35) + (−40)?

3. a) How can you predict whether the sum of two integers will be positive or negative, without adding the integers?

   b) How can you find the sum of two integers without using a model?

4. Imagine using counters to add (+35) + (−40). There will be more blue counters than red counters. How can knowing there will be more blue counters help you find the sum without using counters?

Example 3: Adding integers with different signs

Add $(-24) + (+39)$.

Solution A: Using counters

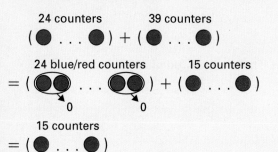

The sum is $(+15)$.

Solution B: Using a number line

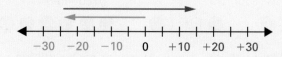

The sum is $(+15)$.

Solution C: Using logical reasoning

There are 15 more positives than negatives.
The sum is $(+15)$.

Example 4: Adding negative integers

Add $(-24) + (-39)$.

Solution A: Using counters

24 counters   39 counters

$(\bullet \ \ldots \ \bullet) + (\bullet \ \ldots \ \bullet)$

The sum is (-63).

Solution B: Using a number line

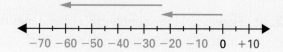

The sum is (-63).

Solution C: Using logical reasoning

You are adding two negatives, so the sign of the answer will be negative.
Add $24 + 39$ to get the actual number. The sum is (-63).

Ⓐ Checking

5. Use a number line to model the sum $(-27) + (+34)$.

a) Where does the first arrow start?

b) Where does the first arrow end?

c) Where does the second arrow start?

d) Where does the second arrow end?

e) What is the sum?

6. Think of how to use counters to model $(+39) + (-26)$.

a) How many counters of each type would you need?

b) How can you use the zero principle to find net results of 0?

c) How many counters will be left over?

d) What is the sum?

7. Calculate each sum without using a model.

a) $(-50) + (-20)$ c) $(-20) + (+50)$

b) $(-50) + (+20)$ d) $(-20) + (-50)$

B Practising

8. Add. Describe a counter model for each sum.

a) $(+5) + (+3)$ d) $(-10) + (-15)$

b) $(-5) + (-3)$ e) $(-15) + (+10)$

c) $(-4) + (-4)$ f) $(+11) + (-3)$

9. Add. Draw a number line model for each sum.

a) $(+5) + (-10)$ d) $(-10) + (+10)$

b) $(+10) + (-5)$ e) $(-30) + (-35)$

c) $0 + (-15)$ f) $(+35) + (-50)$

10. Find each sum.

a) $(-2) + (-5)$ d) $(+40) + (-70)$

b) $(-4) + (+5)$ e) $(-20) + (-50)$

c) $(-60) + (+20)$ f) $(+100) + (-80)$

11. How much greater is the second sum than the first sum? Show your work.

a) $(-25) + (+38)$ and $(-15) + (+38)$

b) $(+125) + (-52)$ and $(+125) + (-32)$

12. Is each statement always true? If you think that a statement is always true, explain why. If you think that a statement is not always true, provide an example of when it is false.

a) The sum of two negative integers is always negative.

b) The sum of a positive integer and a negative integer is always negative.

c) If the sum of two integers is zero, the integers must be opposites.

13. In this Magic Square, every row, column, and diagonal adds to 0. Copy and complete the square.

−1		+3
		−4

14. A hockey player's $+/-$ score is determined by the goals scored while the player is on the ice. A goal for the player's team counts as $+1$. A goal against the player's team counts as -1. Order the players from highest to lowest $+/-$ score.

Player	Goals for	Goals against
Heidi	110	94
Rana	103	89
Meagan	99	108
Sonya	105	97
Indu	101	102

15. Copy and complete the following table.

	Starting temperature (°C)	Temperature change (°C)	Final temperature (°C)
a)	−5	+1	
b)	−10	−6	
c)	0		−8
d)		−5	0
e)	+7		−2
f)		−10	+8

C Extending

16. This four-by-four Magic Square uses all the integers from −7 to +8. Copy and complete the square.

−7			
	−2		+1
		+3	
+5			

17. Make a four-by-four Magic Square in which each row, column, and diagonal adds to −2.

18. Ravi added $(+50) + (-30)$ and got $(+20)$. He noticed that this was like subtracting $50 - 30$. Describe when adding a negative number to a positive number works like subtracting.

Integer Addition Strategies

▶ **GOAL**

Learn integer addition strategies.

Learn about the Math

Miguel and his friends play soccer. The coach tracks the goals scored while each team member is playing.
• Each goal scored for the team counts as +1.
• Each goal scored against the team counts as −1.
• These goals are added through the season to produce a +/− score for each player.

Here are the current +/− scores for Miguel and his friends.

Player	Miguel	Paul	Lloyd	Sam	Matthew
+/− score	+18	−23	−31	+21	+13

? **What is the overall +/− score for the group of soccer players?**

You can **regroup** the integers to have the positive and the negative integers together.

regroup

change the order of terms in an arithmetic expression to form groups

---

Example 1: Separating positive and negative sums

Add $(+18) + (−23) + (−31) + (+21) + (+13)$.

Miguel's Solution

$(+18) + (−23) + (−31) + (+21) + (+13)$

$= \boxed{(+18) + (+21) + (+13)} + \boxed{(−23) + (−31)}$

$= (+52) + (−54)$

$= \boxed{(+52) + (−52)} + (−2)$

$= 0 + (−2)$

$= (−2)$

I can group all the positives together and all the negatives together.

The sum of the positives is (+52). The sum of the negatives is (−54).

$(−54) = (−52) + (−2)$

$(+52) + (−52) = 0$

The answer is (−2).

Example 2: Regrouping integers to produce more manageable sums

Add $(+18) + (-23) + (-31) + (+21) + (+13)$.

Paul's Solution

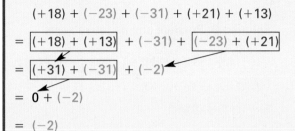

$$(+18) + (-23) + (-31) + (+21) + (+13)$$
$$= \boxed{(+18) + (+13)} + (-31) + \boxed{(-23) + (+21)}$$
$$= \boxed{(+31) + (-31)} + (-2)$$
$$= 0 + (-2)$$
$$= (-2)$$

I noticed that (-31) balances $(+18) + (+13) = (+31)$. I also noticed that $(+21)$ and (-23) can be added to get (-2).

$(+31)$ and (-31) are opposites, so they add to 0.

The answer is (-2).

Example 3: Using a calculator with a sign change key

Calculate $(+18) + (-23) + (-31) + (+21) + (+13)$.

Romona's Solution

18 ➕ 23 ➕⁄➖

I entered 18. Then I pressed ➕ to show that the next number will be added to it.

Next I entered 23 and pressed the ➕⁄➖ key to change the sign. This changed $+23$ to -23.

➕ 31 ➕⁄➖ ➕ 21 ➕ 13 ➁

I used ➕⁄➖ and ➕ to enter the rest of the numbers. I pressed ➁ to complete the calculation.

The answer is (-2).

Reflecting

1. Miguel used separate positive and negative sums to get the final overall score. Explain why his strategy will work with any integer sum.

2. Describe how mental math skills helped Paul work with more manageable sums compared to Miguel's strategy. Will Paul's strategy work for all integer sums?

3. Compare the three strategies used by Miguel, Paul, and Romona. How are they alike? How are they different?

Example 4: Adding positive and negative integers

Add $(-114) + (+35) + (+11) + (-15) + (-20) + (+14)$.

Solution A: Regrouping

$(-114) + \boxed{(+35)} + (+11) + \boxed{(-15) + (-20)} + (+14)$

$= (-114) + (+11) + (+14) + \boxed{(+35) + (-15) + (-20)}$

$= (-114) + (+11) + (+14) + 0$

$= (-114) + (+11) + (+14)$

$= (+11) + \boxed{(-114) + (+14)}$

$= (+11) + (-100)$

$= (-89)$

Solution B: Positive and negative sums

$(-114) + (+35) + (+11) + \boxed{(-15) + (-20)} + (+14)$

$= (-114) + (+35) + (-35) + (+11) + (+14)$

$= (-149) + (+60)$

$= (-89)$

The answer is (-89).

Example 5: Using a calculator to add positive and negative integers

Calculate $(+379) + (-106) + (-241)$.

Solution

1. Enter 379 and then press $\boxed{+}$.

2. Enter 106 and then press $\boxed{+/-}$.

3. Press $\boxed{+}$. Enter 241. Press $\boxed{+/-}$.

4. Press $\boxed{=}$. The answer is $(+32)$.

A Checking

4. Calculate, using the method indicated.

  a) $(-40) + (+55) + (+5) + (-40) + (-10)$
  Use regrouping to make more manageable sums.

  b) $(-13) + (+8) + (-12) + (+10) + (+9)$
  Use separate positive and negative sums.

  c) $(+225) + (-311) + (+110) + (-97)$
  Use a calculator.

B Practising

5. Verify that these expressions all have the same sum.

  a) $(-5) + (-2) + (-3) + (+5)$

  b) $(-5) + (+5) + (-2) + (-3)$

  c) $(-2) + (-3) + (+5) + (-5)$

  d) $(+5) + (-2) + (-3) + (-5)$

6. Consider the following expression:

$$(+4) + (-3) + (+1) + (+6) + (-2)$$

a) Add the integers in the order in which they appear.

b) Group together all the positive integers, and add them. Group together all the negative integers, and add them. Now add these two sums.

c) Why do you get the same answer when you regroup and when you add the terms in the order in which they appear?

7. Calculate each sum.

a) $(-12) + (+2) + (-5)$

b) $(+23) + (-14) + (-7)$

c) $(-18) + (+5) + (+18)$

d) $(+7) + (-3) + (-13) + (+6)$

e) $(-21) + (-30) + (+50) + (+10)$

f) $(-10) + (+48) + (-38) + (-9)$

8. Fill in each ▨ with a different two-digit integer to make the equation true. Find two different solutions.

$$▨ + ▨ + ▨ + ▨ + ▨ = (+4)$$

9. Add. Use the strategy of your choice.

a) $(+8) + (-4) + (+3) + (-5) + (-6) + (+4) + (+1) + (+5)$

b) $(-10) + (-15) + (+15) + (+20)$

c) $(+45) + (-35) + (+15) + (-25) + (+20) + (-5)$

d) $(-278) + (+415)$

e) $(+426) + (-242) + (-318)$

10. A hockey player had a +/− score of +11. The following table shows how it changed over seven games. What is the player's +/− score now?

Game	1	2	3	4	5	6	7
+/− score	−1	+4	−3	−2	+5	0	+1

11. Samantha bought some stock for $21 a share. She decided to sell the stock if it rose to $35 a share or dropped to $11 a share. Based on the price changes below, did Samantha sell the stock? Explain.

Week	1	2	3	4	5	6
Price change ($)	+10	+3	−12	−6	−15	+8

12. At 10:00, the position of a submarine is shown. Its changes in depth were recorded every hour in the table. What is the depth of the submarine at 15:00?

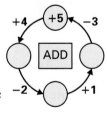

300 m

Time	11:00	12:00	13:00	14:00	15:00
Change in depth (m)	−53	−31	+18	−64	+85

C Extending

13. a) Copy the diagram below. Start with +5 in the top circle. Fill in the other circles by following the arrows and adding the indicated integers.

b) Why is the final sum +5 after you finish the last addition?

c) Copy the diagram again. Replace the numbers on the arrows with four different two-digit integers, so that you still end up with +5.

Mid-Chapter Review

Frequently Asked Questions

Q: How do you compare integers?

A: Look at a number line.

The integer on the left is less than the integer to its right. The $<$ and $>$ symbols are used to compare their positions on the number line.

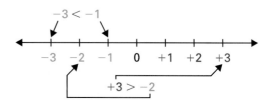

Q: What is the zero principle?

A: The zero principle states that you can add $(+1)$ and (-1), in either order, and get the same result: 0. In other words, if you have one thing and lose it, you have nothing left.

The zero principle can be used to show that any two opposite integers add to 0.

Q: How do you add two integers, such as $(-5) + (+3)$?

A1: Use a counter model. Based on the zero principle, you can add each pair of $+1$ and -1 counters to get 0. This model shows that $(-5) + (+3) = (-2)$.

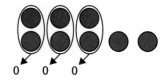

A2: Use a number line model. Based on the zero principle, you can add the overlap of positive and negative arrows to get 0. This model shows that $(-5) + (+3) = (-2)$.

The overlap of the arrows becomes 0.

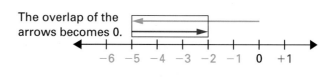

Q: How do you add $(-38) + (+51) + (-83) + (+64)$?

A1: Find the sum of all the positive numbers and the sum of all the negative numbers. Decide if there are more negatives or more positives. Then calculate how many more.

$$(-38) + (+51) + (-83) + (+64)$$
$$= \boxed{(-38) + (-83)} + \boxed{(+51) + (+64)}$$
$$= (-121) + (+115)$$
$$= (-6)$$

A2: Regroup and add parts.

$$\boxed{(-38) + (+51)} + \boxed{(-83) + (+64)}$$
$$= (+13) + (-19)$$
$$= (-6)$$

A3: Use a calculator with a sign change key.

Use the following key presses:
38 [+/−] [+] 51 [+] 83 [+/−] [+] 64 [=]
The answer is (-6).

Practice Questions

(6.1) **1.** Describe how to use positive and negative integers in each situation.

	Situation	Words to use
a)	business	profit, loss
b)	temperature	above freezing, below freezing
c)	sports	Choose your own words.
d)	car travel	accelerate, decelerate

(6.1) **2.** What number matches each description? Use the number line to help you.

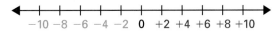

-10 -8 -6 -4 -2 0 +2 +4 +6 +8 +10

a) 6 more than −8 **c)** 5 less than −4

b) 3 more than −9 **d)** 9 less than +7

(6.1) **3.** Use < or > to make each statement true.

a) (-5) ▨ (-10) **c)** (-20) ▨ 0

b) (-14) ▨ (-10) **d)** $(+5)$ ▨ $(+10)$

(6.1) **4.** Add. Draw a number line for each sum.

a) $(-5) + (-2)$

b) $(+2) + (-5)$

c) $(-1) + (+3) + (-4)$

d) $(+30) + (-20) + (-10)$

5. Add. Draw a counter model for each sum. (6.3)

a) $(+5) + (-2)$ **c)** $(-4) + (+5)$

b) $(-2) + (-5)$ **d)** $(-1) + (-3) + (+4)$

6. Complete each equation. (6.3)

a) $(-8) +$ ▨ $= (-5)$

b) $(+2) +$ ▨ $= 0$

c) $(-5) +$ ▨ $+ (+7) = (+12)$

7. Think about adding two integers. (6.3)

a) What must be true about the integers for the sum to be positive?

b) What must be true about the integers for the sum to be negative?

8. Anthony hiked uphill from a valley that was 45 m below sea level. After an hour, he was 100 m higher. How high was he above sea level? (6.4)

9. Calculate each sum. (6.4)

a) $(-150) + (+50)$ **b)** $(-110) + (-20)$

10. Show how to add these integers. Keep the arithmetic as simple as possible. (6.5)

$(-34) + (+17) + (-20) + (-15) + (-2) + (+18)$

11. Calculate each sum. Use a calculator to check your answer. (6.5)

a) $(+11) + (-26) + (-15)$

b) $(-33) + (-20) + (+12)$

12. Georgina invested $1125 in a mutual fund in January. The monthly increases or decreases in the value of the fund are given below. Calculate the value of Georgina's investment at the end of June. (6.5)

Month	Jan.	Feb.	Mar.	Apr.	May	June
Value change ($)	−150	−55	+137	+91	−2	+8

Addition and Subtraction of Integers **205**

INTEGRO

When using a standard deck of cards, aces count as 1, numbered cards count as their face values, and jokers count as 0. Red cards are positive, and black cards are negative.

You will need
- integer cards numbered −10 to +10 (two of each) OR a standard deck of cards including jokers with face cards removed

Version 1

Number of players: 2 or 4

Rules

1. If there are four players, remove the jokers. Shuffle the cards. Deal the cards equally to all the players.

2. In a round, each player places one card face up on the table.

3. The first player to call out the sum of the cards wins all the cards in the turn. These cards go into the player's bank pile.

4. If there is a tie, the tied players play additional rounds until one of them wins.

5. When a player runs out of cards, the player shuffles his or her bank pile and continues playing. If the player's bank is empty, the player is out of the game.

6. The game ends when one player has won all the cards.

Version 2

Number of players: 2

Rules

1. Shuffle the cards. Deal half the deck to each player.

2. In a round, each player places three cards face down and a fourth card face up on the table.

3. The player whose fourth card is greater wins all the cards played. These cards go into the player's bank pile.

4. If the fourth cards produce a tie, each player adds the cards she or he played on the turn. The player with the greater sum wins. If there is another tie, the dealt cards are kept on the table and the players play again. The winner of this round takes all the cards on the table.

5. When a player runs out of cards, the player shuffles his or her bank pile and continues playing.

6. The game ends when one player has won all the cards.

Earth makes one complete 360° turn from west to east every 24 h. Therefore, Earth turns 360° ÷ 24 = 15° every hour.

Since there are 24 h in a day, Earth is divided into 24 time zones. Places that are located within the same time zone have the same time. This map shows the time zones of the world. It is the same time in Tuvalu and Kiribati, but they are one calendar day apart.

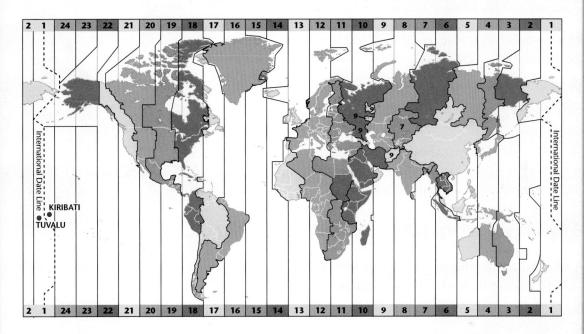

1. Find a map of Canada that shows time zones. How many time zones are in Canada?

2. If it is noon where you live, what is the time in each of the other provinces and territories?

3. Survey some teachers and students at your school. In what other time zones do they have relatives or friends?

4. It is 09:00 in Ottawa. Use the clocks to calculate how far behind or ahead of Ottawa each of the other cities in Canada is. Express your answers using positive and negative integers.

VANCOUVER

CALGARY

WINNIPEG

MONTREAL

HALIFAX

6.6 Using Counters to Subtract Integers

▶ **GOAL**
Subtract integers using a model.

Learn about the Math

Romona and Fawn were trying to calculate the difference $(-5) - (+2)$.

Fawn tried to use counters.

She got stuck. "I have no positive counters in the first group. I can't subtract the two positive numbers in the second group from anything."

? How can you model subtraction using counters?

A. Romona suggested using blue counters. They tried $(-5) - (-2)$. Use counters to model this subtraction. What is the result?

B. Explain why $(-5) - (-2)$ is easier to model with counters than $(-5) - (+2)$.

C. Romona and Fawn went back to the original question: $(-5) - (+2)$. Fawn then remembered the zero principle, which states that $(+1) + (-1) = 0$. She said, "If we put a blue counter and a red counter together as a pair, the value is 0. If we use enough of these pairs, we will be able to do the subtraction."

What is the smallest number of blue/red pairs the girls can add in order to take away two red counters?

D. What is the difference $(-5) - (+2)$?

Reflecting

1. Describe how using counters to model $(-5) - (-2)$ is different from using counters to model $(-5) - (+2)$.

2. Zeros did not need to be added to model $(-5) - (-2)$, but they were needed to model $(-5) - (+2)$.

a) Write a subtraction question with integers that requires adding zeros. Why is adding zeros required?

b) Write a subtraction question with integers that does not require adding zeros. Why is adding zeros not required?

3. Think about using counters to model a difference of two integers.

a) Describe when adding zeros is required.

b) Describe when adding zeros is not required.

4. When zeros are added during an integer subtraction, they are always added to the first term. Why?

Work with the Math

Example 1: Subtracting integers with the same sign

Calculate $(-2) - (-5)$ using a counter model.

Fawn's Solution

$(\bullet\bullet) - (\bullet\bullet\bullet\bullet\bullet)$

$= (\bullet\bullet \boxed{\bullet\bullet} \boxed{\bullet\bullet} \boxed{\bullet\bullet}) - (\bullet\bullet\bullet\bullet\bullet)$

$= (\bullet\bullet \boxed{\bullet\bullet} \boxed{\bullet\bullet} \boxed{\bullet\bullet}) - (\bullet\bullet\bullet\bullet\bullet)$

$= (\bullet\bullet\bullet)$

I added three zeros to the counters in the first parentheses, so there are enough (-1) counters to subtract.

Three positive counters are left. The answer is $(+3)$.

Example 2: Subtracting integers with opposite signs

Calculate $(+20) - (-10)$ using a counter model.

Romona's Solution

20 counters · · · 10 counters

$(\bullet \ldots \bullet) - (\bullet \ldots \bullet)$

I didn't have enough counters to model this difference, so I had to imagine what it would look like.

20 counters · · · 10 zeros · · · 10 counters

$= (\bullet \ldots \bullet \boxed{\bullet\bullet} \ldots \boxed{\bullet\bullet}) - (\bullet \ldots \bullet)$

$= (\bullet \ldots \bullet \boxed{\bullet\bullet} \ldots \boxed{\bullet\bullet}) - (\bullet \ldots \bullet)$

30 counters

$= (\bullet \ldots \bullet)$

There were no negatives in the first group. I had to add 10 zeros to subtract the 10 negatives.

After subtracting the 10 negatives from each group, there were 30 positives left.

The answer is $(+30)$.

Addition and Subtraction of Integers **209**

A Checking

5. Write the subtraction question represented by each model.

a) (●●●●) − (●●)

b) (●●●) − (●●)

c) (●●●) − (●●)

d) (●●●) − (●●)

e) (●●) − (●●●)

6. a) Identify the differences in question 5 that require zeros to be added to the first term before the subtraction can be modelled.

b) Calculate the differences in question 5.

B Practising

7. Use a counter model to calculate $(-4) - (-1)$. Why do you not need to add zeros to the model for this subtraction?

8. Use a counter model to calculate $(-1) - (-4)$. Why do you need to add zeros for this subtraction?

9. Use a counter model to show that $(-3) - (+4) = (-3) + (-4)$.

10. Use counters to explain why $(+6) - (-4)$ and $(-4) - (+6)$ are not equal.

11. Draw a counter model for each difference. Calculate each difference.

a) $(+4) - (+3)$ e) $(+3) - (-4)$

b) $(+3) - (+4)$ f) $(-4) - (+3)$

c) $(-4) - (-3)$ g) $(+4) - (-3)$

d) $(-3) - (-4)$ h) $(-3) - (+4)$

12. Draw a counter model for each difference. Calculate each difference.

a) $(+6) - (+2)$ e) $(+2) - (-6)$

b) $(+2) - (+6)$ f) $(-6) - (+2)$

c) $(-6) - (-2)$ g) $(+6) - (-2)$

d) $(-2) - (-6)$ h) $(-2) - (+6)$

13. Which subtraction statements have the same result? Use counters to explain why.

a) $(+9) - (+5)$ d) $(+2) - (-3)$

b) $(-4) - (-9)$ e) $(-4) - (+1)$

c) $(+7) - (+2)$ f) $(-3) - (-2)$

14. Complete each pattern.

a) $(+5) - (+4) = (+1)$
$(+5) - (+3) = \blacksquare$
$(+5) - (+2) = \blacksquare$
$(+5) - (+1) = \blacksquare$

b) $(-5) - (+4) = (-9)$
$(-5) - (+3) = \blacksquare$
$(-5) - (+2) = \blacksquare$
$(-5) - (+1) = \blacksquare$

c) $(-5) - (-9) = (+4)$
$(-5) - (-8) = \blacksquare$
$(-5) - (-7) = \blacksquare$
$(-5) - (-6) = \blacksquare$

d) $(-1) - (-2) = (+1)$
$(-1) - (-3) = \blacksquare$
$(-1) - (-4) = \blacksquare$
$(-1) - (-5) = \blacksquare$

15. Rosa used counters to represent $(-4) - (+6)$, as shown below.

Step 1   Step 2   Step 3

a) Was Rosa correct?

b) Draw your own counters to show how you would solve this question.

c) Describe each step.

16. Which differences require zeros to be added to the first term to complete the subtraction?

a) $(+100) - (+10)$ e) $(+10) - (-100)$

b) $(+10) - (+100)$ f) $(-100) - (+10)$

c) $(-100) - (-10)$ g) $(+100) - (-10)$

d) $(-10) - (-100)$ h) $(-10) - (+100)$

17. Copy and complete the following table. Use counters to model each difference. Part (a) is done for you.

	Day 1 (°C)	Day 2 (°C)	Difference
a)	−3	−1	$(-1) - (-3) = (+2)$
b)	+5	−4	
c)	−10	+6	
d)	−10	0	
e)	+10	+7	
f)	−8	−1	
g)	0	−10	
h)	−21	+20	

18. A golf tournament lasted two days. The players' scores for each day are shown in the following table. Miguel and Bonnie are figuring out the missing score. Copy and complete the table.

Golfer	Day 1	Day 2	Change (Day 2 − Day 1)
Ming	−5	−1	
Kaitlyn	+5	+10	
Omar	−10	+6	
Anthony	−10		−8
Braydon		−5	+10
Tynessa		+7	+2
Rana	−7		−2

19. Is each statement true or false? If a statement is sometimes true or never true, give an example to support your answer.

a) The difference between two negative numbers is always negative.

b) The difference between two positive numbers is always positive.

c) When you subtract a negative number from a positive number, the difference is always positive.

d) When you subtract a positive number from a negative number, the difference is always positive.

e) The difference between two integers always has the sign of the greater number.

20. −3 is the opposite of +3.

a) Why does $(-2) - (+3) = (-2) + (-3)$?

b) Is subtracting an integer always the same as adding its opposite? Use counters to explain your answer.

ⓒ Extending

21. Copy and complete the following table.

	a	b	$a - b$
a)	−215	+20	
b)	−150		−100
c)		+200	+150

22. Calculate.

a) $(+4) + (+2) - (+3)$

b) $(-4) + (-3) - (-2)$

c) $(+3) - (-8) + (-10)$

23. a) Write three different integers that make the following equation true.

$$\blacksquare + \blacksquare - \blacksquare = (-10)$$

b) Find two other solutions.

Using Number Lines to Subtract Integers

You will need
• a number line

▶ **GOAL**

Calculate the difference between integers using a number line.

Learn about the Math

Miguel, Romona, and Bonnie found these data.

City	Lowest recorded temperature (°C)	Highest recorded temperature (°C)
Ottawa	−37	+39
Toronto	−32	+41
Montreal	−38	+38
Calgary	−36	+34
Vancouver	−17	+35
Winnipeg	−44	+42
Churchill	−49	+37

They wanted to determine which city had the greatest difference between lowest and highest recorded temperatures.

Bonnie said, "It's easy to find the difference between positive temperatures, such as +42 and +4. All you have to do is subtract: $(+42) - (+4) = (+38)$."

Miguel commented, "But all the lowest temperatures are negative. For Ottawa, we have to calculate $(+39) - (-37)$. We could use counters, but we need too many counters."

Bonnie replied, "We used number lines to add integers that were far from zero. Maybe we can use number lines to subtract, too."

? **How can you use a number line to calculate temperature differences?**

A. Mark the lowest and highest temperatures for Ottawa on a number line. Draw an arrow from the lowest temperature to the highest temperature. Write the subtraction question the arrow represents. How far away is the highest temperature from the lowest temperature?

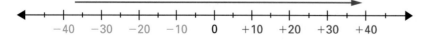

B. Why does the arrow in step A show $(+39) - (-37)$?

C. On the same number line, draw an arrow from the highest temperature to the lowest temperature. What does this arrow represent? Write the subtraction question this arrow represents.

D. Use the method you used in steps A to C to calculate the temperature range for the other cities in the table.

E. Use your calculations to identify the city with the greatest range of extreme temperatures.

F. Use the same method to calculate how much lower the $-49°C$ low for Churchill is than the $-17°C$ low for Vancouver.

Reflecting

1. When you want to calculate $(+39) - (-37)$, why do you start at (-37) on the number line?

2. a) How is the arrow going from the highest temperature to the lowest temperature like the arrow going from the lowest temperature to the highest temperature?

b) How is it different?

3. How does a number line model show whether the result of an integer subtraction will be positive or negative?

Work with the Math

Example 1: Modelling integer subtraction on a number line

Use a number line model to show that $(-15) - (-20) = (+5)$.

Romona's Solution

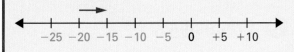

I marked -20 and -15 on the number line.

The order of the subtraction told me to start the arrow at (-20) and end it at (-15).
The arrow is 5 units long and goes right.

$(-15) - (-20) = (+5)$

The difference is $(+5)$.

Example 2: Modelling integer subtraction on a number line

Use a number line model to show that $(-20) - (+15) = (-35)$.

Miguel's Solution

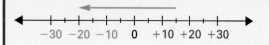

I marked −20 and +15 on the number line.

The order of the subtraction told me to start the arrow at (+15) and end it at (−20).
The arrow is 35 units long and goes left.

$(-20) - (+15) = (-35)$

The difference is (−35).

Example 3: Modelling integer subtraction on a number line

Use a number line model to calculate $(+15) - (-20)$.

Bonnie's Solution

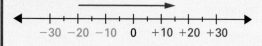

I marked −20 and +15 on the number line.

The order of the subtraction told me to start the arrow at (−20) and end it at (+15).
The arrow is 35 units long and goes right.

$(+15) - (-20) = (+35)$

The difference is (+35).

Ⓐ Checking

4. Write the subtraction question that each model represents.

a)
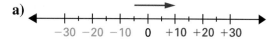

b)
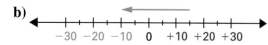

c)

d)
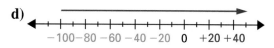

e)
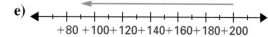

5. Calculate each difference in question 4.

6. An arrow is used on a number line model to represent $(-35) - (+40)$.

   a) Identify the starting point for the arrow.

   b) Identify the ending point for the arrow.

   c) Calculate the difference.

Ⓑ Practising

7. On a number line, how do the distance and direction from −2 to −4 compare with the distance and direction from −4 to −2?

8. The difference between two integers is −5. What does this tell you about their positions on a number line?

9. Use a number line to show that
$(-30) - (+40) = (-30) + (-40)$.
Explain why this is true.

10. Use a number line to explain why
$(+36) - (-34)$ and $(-34) - (+36)$ have
different integer values.

11. Mark the starting and ending points for
each subtraction on a number line.
Calculate the difference.

a) $(-20) - (-40)$

b) $(+30) - (+70)$

c) $(-23) - (-21)$

d) $(+35) - (+32)$

e) $(+10) - (-10)$

f) $(-20) - (-20)$

g) $(-20) - (+20)$

h) $(+100) - (-100)$

12. Imagine that an arrow on a number line is
used to represent each difference in the
following table. Copy and complete the
table. Part (a) is done for you.

	Start of arrow	End of arrow	Subtraction statement
a)	−5	−1	$(-1) - (-5)$
b)	+15	+10	
c)	−10	−16	
d)	0	+8	
e)	−8	0	
f)			$(-80) - (+20)$
g)			$(+15) - (-15)$
h)			$(-85) - (-15)$

13. Use a number line to calculate each
difference in question 12.

14. The subtraction $6 - 4$ represents the
difference between 6 and 4. Use a number
line to explain why $(+6) - (-4)$ and
$(-4) - (+6)$ do not represent the same
difference.

15. Ravi tracked the performance of several
investments. Copy and complete Ravi's
chart. Use a number line to model each
difference.

	Last year's gain/loss ($)	Current gain/loss ($)	Change in value this year ($)
a)	−300	−350	$(-350) - (-300) = (-50)$
b)	+200	−150	
c)	+150	+20	
d)	−595	+105	
e)	−1005	−950	
f)	+537	−111	
g)	−97	−121	
h)	−32	+128	

C Extending

16. Copy and complete the table.

	a	b	a − b
a)	−2150	+205	
b)	−1510		+103
c)		+237	−150

17. Use a number line to calculate each
answer.

a) $(+40) + (+20) - (+30)$

b) $(-45) + (-35) - (-20)$

c) $(+37) - (-85) + (-10)$

d) $(+120) - (-90) + (-10)$

e) $(-100) - (-510) + (-520)$

f) $(+301) - (-205) + (-153)$

18. a) Use a number line to show why
$(+15) - (-9) = (+15) + (+9)$.

b) Use a number line to show why
$(+15) - (-20) = (+15) + (+20)$.

c) Predict the difference $(+15) - (-35)$.
Explain your prediction.

Addition and Subtraction of Integers **215**

6.8 Solve Problems by Working Backwards

▶ **GOAL**

Use the strategy of working backwards to solve problems.

Learn about the Math

Meagan played a number trick with Yoshi.
She told him to follow these steps:

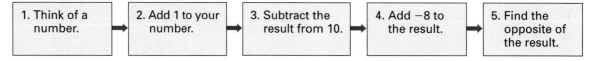

1. Think of a number.	2. Add 1 to your number.	3. Subtract the result from 10.	4. Add −8 to the result.	5. Find the opposite of the result.

Yoshi said that his result was −7.
Meagan said, "I think your number was −6."

? **How did Meagan know Yoshi's number?**

1 **Understand the Problem**

Yoshi wants to know how Meagan determined his number
from his result.

2 **Make a Plan**

Yoshi realizes that he needs to start with the result and
work backwards through the steps to find the original number.

3 **Carry Out the Plan**

Yoshi goes through the steps in two ways:
• He completes the original steps in order.
• He works backwards from the result.

Original Steps

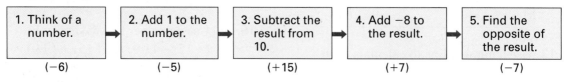

1. Think of a number.	2. Add 1 to the number.	3. Subtract the result from 10.	4. Add −8 to the result.	5. Find the opposite of the result.
(−6)	(−5)	(+15)	(+7)	(−7)

Working Backwards

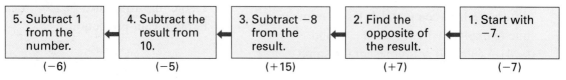

5. Subtract 1 from the number.	4. Subtract the result from 10.	3. Subtract −8 from the result.	2. Find the opposite of the result.	1. Start with −7.
(−6)	(−5)	(+15)	(+7)	(−7)

4 Look Back

Yoshi checks to see if this method works with −2 as the number.

Original Steps

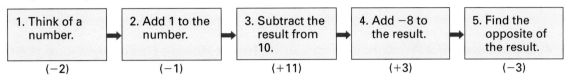

1. Think of a number.	2. Add 1 to the number.	3. Subtract the result from 10.	4. Add −8 to the result.	5. Find the opposite of the result.
(−2)	(−1)	(+11)	(+3)	(−3)

Working Backwards

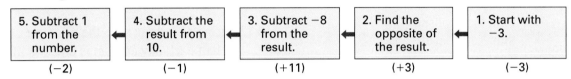

5. Subtract 1 from the number.	4. Subtract the result from 10.	3. Subtract −8 from the result.	2. Find the opposite of the result.	1. Start with −3.
(−2)	(−1)	(+11)	(+3)	(−3)

Reflecting

1. How does working backwards help to solve Meagan's number trick?

2. Is working backwards the only way to solve this problem? Explain.

Work with the Math

Example: Working backwards

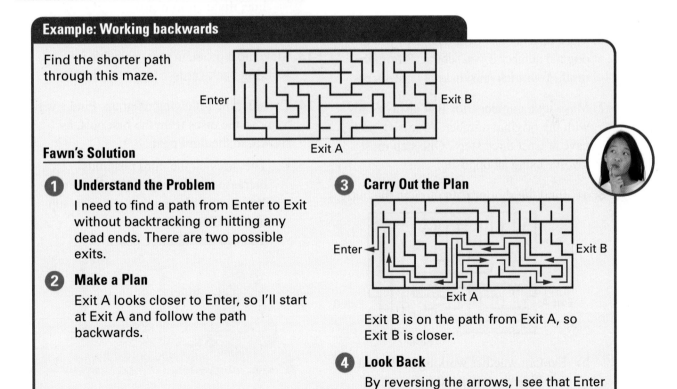

Find the shorter path through this maze.

Enter   Exit B

Exit A

Fawn's Solution

1 Understand the Problem

I need to find a path from Enter to Exit without backtracking or hitting any dead ends. There are two possible exits.

2 Make a Plan

Exit A looks closer to Enter, so I'll start at Exit A and follow the path backwards.

3 Carry Out the Plan

Enter   Exit B

Exit A

Exit B is on the path from Exit A, so Exit B is closer.

4 Look Back

By reversing the arrows, I see that Enter to Exit B is the shorter path.

Ⓐ Checking

3. Try Meagan's trick using other numbers. Is there a quick way to find the original number? Explain how it works.

4. Use Meagan's trick. Write the steps, in order, to find the original number.
- Subtract 7.
- Add −9.
- Find the opposite.
- Subtract from 12.
- The answer is +10.

Ⓑ Practising

5. Find the original number. State the steps, in order, that you will use.
- Add −31.
- Subtract −9.
- Add 18.
- Subtract from 12.
- The answer is −12.

6. Make up a number trick that gives you the original number if you subtract 3 from the result. Your trick must have at least 4 steps.

7. Make up a number trick that always ends with the original number. Your trick must have at least three steps. One step must involve using an opposite.

8. a) Find the shorter path through the maze.

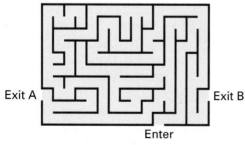

Exit A   Exit B

Enter

b) Explain whether working backwards is an efficient strategy to use.

9. Lloyd is lifting weights over a nine week training period. Every week, he lifts 2 kg more than he lifted the previous week. During the ninth week, he lifts 80 kg. How much was he lifting during his first week?

10. On Monday, Heidi bought some grapes. Each day, she ate half of them. On Friday, only eight grapes were left. How many grapes did Heidi buy?

11. During a clothing sale, the price goes down by half each day an item is not sold. If an item costs $2.50 after 8 days, what was the original price?

12. Romona takes a shape and cuts away half of it five times. The following triangle is what remains.

a) Draw the original shape.

b) Draw a polygon that is symmetric. With a single line, divide it into two identical pieces.

13. Make up a problem you can solve by working backwards. Show how to solve it.

14. Consider the following diagram. Find a way to move the discs from the first post, as shown, to the third post.
- The discs on the third post must increase in size from top to bottom.
- You cannot place a larger disc on top of a smaller disc.
- You can use all three posts.

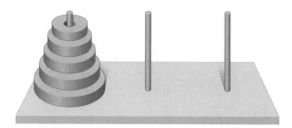

Chapter Self-Test

1. Draw a number line with integers from −5 to +5.

   a) Mark the position of a number that is 6 more than −2.

   b) Mark the position of a number that is 2 less than 1.

2. Use < or > to make each statement true.

   a) (+3) ▦ (−2)

   b) (−4) ▦ (−7)

   c) (−4) ▦ 0

3. Write the addition question that each model represents. Calculate each sum.

   a)

   b)

   c)

   d)

   e)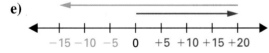

4. Use counters to model and add (−2) + (+3). Explain what you did.

5. Use a number line to model and add (+3) + (−2). Explain what you did.

6. Use counters to model and subtract (−3) − (+4). Explain what you did.

7. Write the subtraction question that each model represents.

   a)

   b)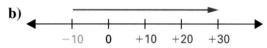

   c) (●●● ●●●●) − (●●)

   d) (●) + (●●) − (●●)

   e) (●●●●) − (●●)

8. What keys would you press to calculate the answer to each question using a calculator?

   a) (+125) − (−23)

   b) (−117) − (+89)

9. Calculate. Do not use a calculator.

   a) (+5) + (−8)

   b) (−10) + (−3)

   c) (−7) − (+2)

   d) (−8) − (−4)

   e) (−4) + (−5) + (+6)

   f) (+10) + (−15) + (−5)

10. Sam checks his investments every day. One Friday, he missed the news and did not know if there was a gain or a loss. His mother told him that the overall value had dropped $210 during the week. What was his gain or loss on the Friday?

Day	Mon.	Tues.	Wed.	Thu.	Fri.
Gain/loss ($)	+300	−193	−25	+51	

Mystery Integer

Select three integers. Make up a set of eight clues that will allow others to figure out the integers you chose. All eight clues must be necessary.

The clues must
- include both addition and subtraction of integers
- compare integers

For example, suppose your integers are -8, $+17$, and -33. Here are three possible clues:
- If you add all three integers, the result is less than -20.
- The difference between the least two integers is 25.
- The sum of the least and greatest integers is -16.

? **What eight clues can you write to describe your three integers?**

A. The three clues above do not give enough information for someone to figure out the integers. What five additional clues would give enough information?

B. Select three integers of your own, and make up eight clues. Remember that all the clues must be necessary. It should not be possible to figure out all the integers with only some of the eight clues.

Task Checklist

☑ Do your clues involve both addition and subtraction of integers?

☑ Does at least one of your clues compare integers?

☑ Did you include any information you did not need?

☑ Did you check to make sure that your clues give the correct integers?

Cross-Strand Multiple Choice

(4.2) **1.** Which sequence follows the rule "Each number is the previous number, multiplied by 2, and then reduced by 1"?

A. 1, 2, 3, 5, 9   **C.** 3, 5, 7, 9, 11

B. 0, 2, 3, 5, 9   **D.** 2, 3, 5, 9, 17

(4.3) **2.** What is the term value for the 8th term number?

Term number	Term value
1	3
2	7
3	11
4	15
5	19

A. 31   **B.** 32   **C.** 39   **D.** 23

(4.4) **3.** The table shows the gas usage for a motorcycle.

Time (h)	Gas (L)
1	0.5
2	1.0
3	1.5
4	2.0

What is the gas usage for 9 h?

A. 4 L   **B.** 5 L   **C.** 5.5 L   **D.** 4.5 L

(5.1) **4.** What is the area of this parallelogram?

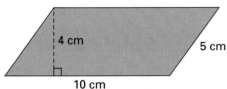

A. 50 cm^2 **C.** 40 cm^2

B. 50 cm   **D.** 20 cm^2

5. What is the area of this trapezoid?   (5.4)

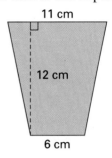

A. 204 cm^2 **C.** 102 cm^2

B. 41 cm^2 **D.** 30 cm^2

6. What is the area of one side of a picket in this fence?   (5.6)

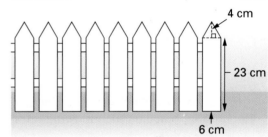

A. 150 cm   **C.** 62 cm

B. 162 cm^2 **D.** 150 cm^2

7. Which expression is *not* correct?   (6.1)

A. $(-3) > (-10)$ **C.** $(-4) > (-1)$

B. $(-2) < (+2)$ **D.** $(-5) < (-2)$

8. What is the sum of $(+4) + (-7)$?   (6.3)

A. -3 **B.** $+3$ **C.** $+11$ **D.** -11

9. What is the difference of $(+8) - (-5)$? (6.7)

A. $+3$ **B.** $+13$ **C.** -13 **D.** -3

Cross-Strand Investigation

You have been hired by a band called The Geometrics to help with marketing and sales.

10. a) Your first job is to design a cover for the band's new CD. The band has given you some general requirements for the cover:

- The CD cover is a 12 cm by 12 cm square.
- Because of the band's name, previous CD covers have had drawings of polygons. The band has decided that the following polygons must be included on this CD cover: one trapezoid, one triangle, and one parallelogram. These polygons are the only drawings that should appear on the cover.
- Each polygon must be a single colour and outlined in black.
- The polygons must take up 70% of the area of the cover.
- The band's name must be in the space not taken up by the polygons.
- The cover should be colourful and interesting, so it will attract attention and encourage sales.

Create a CD cover, based on these general requirements.

b) Calculate the cost for the printing company to print your CD cover, based on the following information. Remember to show your work.

- Each colour costs $0.02 per square centimetre.
- The black outline costs $0.03 per centimetre.
- The band's name costs $0.02 per letter.

c) The band created three different demos for radio stations. Each demo cost $5 to produce and was sold to a radio station for $2. Write a number sentence to express how much money the band spent on the demos.

d) To help sell the new CD, you have scheduled an outdoor concert for the band. The band is only able to perform in temperatures greater than $-3°C$, because the instruments do not work properly at lower temperatures. If the scheduled day starts out at $-10°C$, warms up by $8°C$ at lunch, cools off by $4°C$, and then finally warms up again by $6°C$, will the band be able to perform? Explain your answer, using a number sentence to illustrate your thinking.

CHAPTER 7

2-D Geometry

▶ GOALS

You will be able to
- perform rotations, reflections, and translations
- describe and use congruence and similarity
- work with tilings, and create tessellations
- use mathematical language to describe and identify 2-D shapes

Pentomino Tiles

A pentomino is a 2-D shape that is made up of five **congruent** squares.
Each square shares a complete edge with at least one other square.

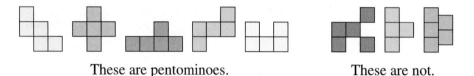

These are pentominoes. These are not.

If your family gets a new kitchen floor, you do not expect the installers to
leave any gaps between the tiles. Mathematicians are also interested in
tilings that have patterns of one repeated shape, with no gaps. These
tilings are called **tessellations**.

? **How can you tile an area using pentominoes?**

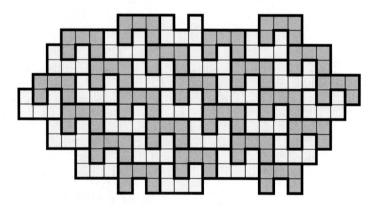

This drawing shows how one type of pentomino can be used to tile a
plane. The pink pentominoes are congruent to the yellow pentominoes,
but they have been turned.

A. Choose three different pentominoes. Which of your pentominoes
could you use to tile a plane?

B. Choose one pentomino, and sketch the tessellation on grid paper.
Colour the shapes so that you can see the tiling pattern.

C. If you were to use the yellow pentomino shown above to tile a
rectangular room, you would need to cut the tiles at the edges of the
room. Draw a large rectangle on grid paper. Could you tile the
rectangle with congruent pentominoes without cutting any tiles?

Do You Remember?

1. Show each integer on a number line.

  a) −2   **b)** +1   **c)** +4   **d)** −6

2. Arrange the following integers from least to greatest:

$$-4, -7, +5, -2, +3$$

3. Match the name of the shape to the correct picture.

  a) isosceles triangle   **d)** trapezoid

  b) equilateral triangle   **e)** hexagon

  c) parallelogram

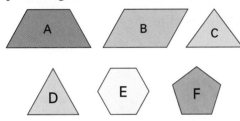

4. Match the name of the triangle to the correct picture.

  a) acute scalene triangle

  b) right isosceles triangle

  c) acute isosceles triangle

  d) obtuse scalene triangle

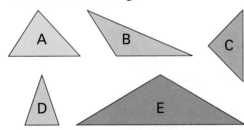

5. Draw an *x*-axis and a *y*-axis on grid paper. Label a scale from 0 to 10 on each axis, using the same spacing for both scales. Plot the following points. Join the points in order, and connect *D* to *A*. What shape have you drawn?

  $A(1, 1), B(2, 4), C(6, 4), D(7, 1)$

6. For each colour of figures, what **transformation** would make the figure in column A match the figure in column B?

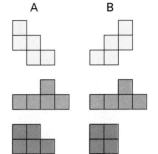

  • a **translation**

  • a **reflection**

  • a **rotation**

7. This snowflake has **rotational symmetry**. Trace the snowflake. Rotate the tracing clockwise about the centre. How many turns produce an exact copy of the original snowflake before you get back to where you started?

8. State whether or not each figure has rotational symmetry.

  a) scalene triangle   **d)** square

  b) equilateral triangle   **e)** regular hexagon

  c) rectangle

9. State whether or not each pentomino has symmetry. For each pentomino with symmetry, identify the type of symmetry as rotational symmetry or **reflection symmetry**.

  a)   **c)**   **e)**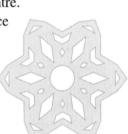

  b)   **d)**   **f)**

Comparing Positions on a Grid

You will need
• grid paper

▶ **GOAL**

Locate positions on a Cartesian grid with integer coordinates.

Learn about the Math

Katya and Susan use a map of their neighbourhood to plan a bicycle route. The girls want their route to start at Katya's house and end at Escarpment Park. They decide to mark Katya's house with a red star.

Katya draws a grid of squares on her copy of the map to help her find different locations. Susan uses a **Cartesian coordinate system**. She draws two axes that cross at Katya's house.

Cartesian coordinate system

a method (named after mathematician René Descartes) for describing a location by identifying the distance from a horizontal number line (the *x*-axis) and a vertical number line (the *y*-axis); the axes intersect at 0; the location is represented by an ordered pair of coordinates (*x*, *y*)

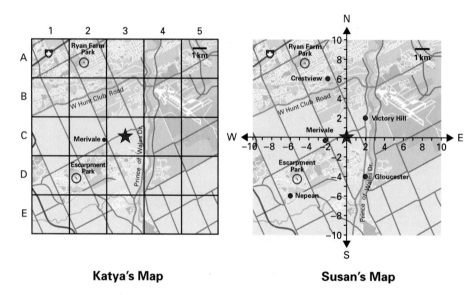

Katya's Map   **Susan's Map**

? **Whose method is easier for identifying locations on the map, Katya's or Susan's?**

A. Using Susan's map, identify the location that is represented by each set of coordinates below.

  a) (+2, +2)   **b)** (−6, −6)   **c)** (−2, +6)   **d)** (+2, −4)

B. Name the coordinates of each location using the grid system on Katya's map and Susan's map.

  a) Katya's house   **b)** Escarpment Park   **c)** Ryan Farm Park

C. Susan says that CD Express has the coordinates $(+3, -4)$ on her map and Pizza Shack has the coordinates $(-4, -2)$.
   a) Which location is farther west? How do you know?
   b) Which location is farther south? How do you know?

D. Susan rides her bicycle from her school, at $(-5, +4)$, to her grandmother's house, at $(-3, +6)$.
   a) Does she ride east or west?   **b)** Does she ride north or south?

Reflecting

1. You are given the coordinates of two points on a Cartesian coordinate system. How can you tell which point is farther north? Explain.

2. You are given the coordinates of two different points. How can you tell which point is farther west? Explain.

3. Which coordinate grid is easier for identifying locations, Katya's or Susan's? Explain your thinking.

Work with the Math

Example: Describing locations of points

Point A has coordinates $(4, -2)$ on a Cartesian coordinate system. Where is $B(-3, 3)$ in relation to A?

James's Solution

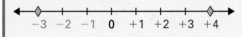

I compared the first coordinates of the two ordered pairs, 4 and –3, on a horizontal number line. Since –3 < 4, B is to the left of A.

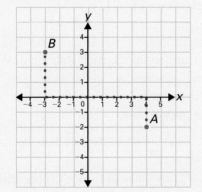

I compared the second coordinates of the ordered pairs, −2 and 3, on a vertical number line.

Since 3 > −2, B is above A.

I drew the horizontal number line to intersect the vertical number line, so they become the axes of a Cartesian coordinate system. When I plotted the points, $B(-3, 3)$ is to the left and above $A(4, -2)$.

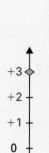

4. Name the coordinates for each point.

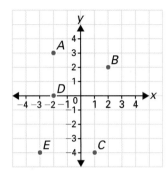

5. Plot the following points on a Cartesian coordinate system:

(2, 3), (−1, 4), (−3, −5), (4, −6), (0, 0)

6. Plot each set of points on a separate Cartesian coordinate system. Connect the points in order, and connect the last point to the first point. Write the name of the polygon you formed. Explain how you know the name.

a) A(0, 5), B(4, 5), C(4, 0)

b) D(−3, 1), E(−3, −3), F(−1, −4), G(−1, 3)

7. Use each inequality to complete the ordered pairs and make the statement true.

a) Since −5 < 6, (▢, 4) is to the left of (▢, 4).

b) Since −2 > −8, (10, ▢) is above (10, ▢).

8. Copy and complete each statement by filling in the boxes. (Replace ? with a number from the first ordered pair.)

a) (5, −3) is to the right of (▢, 2) because ? > ▢.

b) (2, −6) is below (2, ▢) because ? < ▢.

9. Name the coordinates for each point.

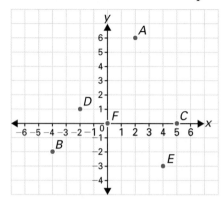

10. Plot each set of points on a separate Cartesian coordinate system. Connect the points in order, and connect the last point to the first. Write the name of the polygon you formed. Explain how you know the name.

a) R(−5, 2), S(0, 0), T(0, −4)

b) K(0, 3), L(−5, 0), M(0, −3), N(5, 0)

c) U(−2, 3), V(4, 3), W(3, −1), X(−3, −1)

d) A(−3, 4), B(4, 5), C(1, −2), D(−2, −5), E(−6, 0)

11. Name three points for each description.

a) to the left of (−17, −12)

b) to the right of (−17, −12)

c) above (−17, −12)

d) below (−17, −12)

12. Copy and complete each statement. Write "above" or "below" in the green box. Write the correct numbers from the vertical axis in the grey boxes.

a) (4, −5) is ▢ (−4, −3) because ▢ < ▢.

b) (−10, −1) is ▢ (−4, −3) because ▢ > ▢.

13. Copy and complete each statement. Write "left" or "right" in the green box. Write the correct numbers from the horizontal axis in the grey boxes.

   a) $(4, -3)$ is to the ▢ of $(-5, -2)$ because ▢ $>$ ▢.

   b) $(-10, -1)$ is to the ▢ of $(-5, -2)$ because ▢ $<$ ▢.

14. Copy and complete each statement. Write $>$ or $<$ in the grey boxes. Write "above" or "below" in the green box. Write "right" or "left" in the blue box.

   a) Since -15 ▢ -42 and 31 ▢ -16, $(-15, 31)$ is ▢ and to the ▢ of $(-42, -16)$.

   b) Since 8 ▢ -5, -10 ▢ -18, $(8, -10)$ is ▢ and to the ▢ of $(-5, -18)$.

15. Use the points $(-21, 38)$, $(-38, 25)$, $(-25, -38)$, and $(-38, -25)$.

   a) Which point is farthest left?

   b) Which point is farthest right?

   c) Which point is highest up?

   d) Which point is lowest down?

16. a) List five points where the y-coordinate is the opposite integer of the x-coordinate.

   b) Plot your points on a Cartesian coordinate system.

   c) What pattern do these points form?

   d) As the first coordinate increases, does the point move up, down, to the right, or to the left?

17. (a, b) is below and to the left of (c, d).

   a) Is a greater or less than c?

   b) Is b greater or less than d?

18. Each unit on this map represents 5 km.

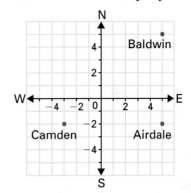

   a) Find the distance between Baldwin and Airdale.

   b) Find the distance between Camden and Airdale.

   c) Can you accurately find the distance between Baldwin and Camden? Explain.

19. A right triangle has these two vertices: $A(-2, 2)$ and $Z(-8, -3)$. What are the coordinates of the other vertex? Draw two possible answers.

20. Draw a parallelogram that has $(-5, -3)$ as its top right corner. State the possible coordinates of the other corners. Explain how you determined these coordinates.

ⓒ Extending

21. Suppose that $c < d$ and $f > g$.

   a) Describe the positions of (c, f) and (d, g) in relation to each other.

   b) Describe the positions of (c, d) and (f, g) in relation to each other.

22. The diagonals of a rectangle intersect at $(0, 0)$. The rectangle is 6 units long and 4 units wide. Find the coordinates of the four corners of the rectangle.

23. On a Cartesian coordinate system, draw two different triangles that each have an area of 30 square units.

Translations

You will need
• centimetre grid paper
• a ruler

▶ **GOAL**

Recognize the image of a 2-D shape after a translation.

Learn about the Math

Jody is tiling a floor using square black and white tiles. She wants the floor to have a chessboard pattern. All the tiles are in a pile at $(-2, 0)$. Jody takes one tile at a time from the pile and slides it into position.

? **How can Jody slide the tiles into position?**

Communication Tip

The new shape that is created when a shape is transformed is called the *image*. The original shape is called the *pre-image*. The vertices of the image are often labelled using the same letters as the pre-image, but with primes. (For M', say "M prime.") This shows which vertices of the image match which vertices of the pre-image. When M' is transformed, its image is M''.

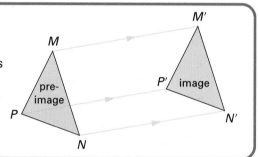

Example 1: Describing translation images using a coordinate grid

Describe how Jody slides the tiles into position.

Jody's Solution

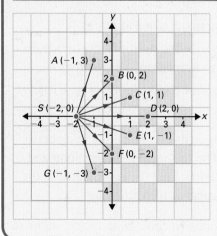

I began by moving some of the black tiles into position from their starting point at $S(-2, 0)$. I moved them along the arrows shown in the diagram. Here's how I moved each tile and the tile's final position:

A: 1 unit to the right and 3 units up to $A(-1, 3)$
B: 2 units to the right and 2 units up to $B(0, 2)$
C: 3 units to the right and 1 unit up to $C(1, 1)$
D: 4 units to the right to $D(2, 0)$
E: 3 units to the right and 1 unit down to $E(1, -1)$
F: 2 units to the right and 2 units down to $F(0, -2)$
G: 1 unit to the right and 3 units down to $G(-1, -3)$

Then I moved the rest of the tiles into position until the floor was complete.

Example 2: Determining translations using the coordinates of the starting point and endpoint

Determine how Jody slid each tile into position, using just the coordinates of the starting point and endpoint of the slide.

Indira's Solution

starting point: $S(-2, 0)$
endpoint: $A(-1, 3)$

$(-1) - (-2) = 1$ or $(-1) - (-2) = +1$

$3 - 0 = 3$ or $(+3) - 0 = +3$

The endpoint, A, has coordinates $(-1, 3)$.
The starting point, S, has coordinates $(-2, 0)$.

The difference between the x-coordinates is $+1$ or 1.

The difference between the y-coordinates is $+3$ or 3.

So, Jody slid the tile 1 unit to the right and 3 units up.

Reflecting

1. For the translation of a tile, draw an arrow from each vertex of the pre-image to the matching vertex of the image. What do the arrows have in common?

2. What properties of a pre-image and its image are the same after a translation? What properties are different? Include **orientation** in your answer.

orientation

the direction that a shape or an object is facing; for example, $\triangle ABC$ and $\triangle A'B'C'$ have the same orientation

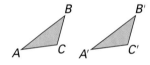

Work with the Math

Example 3: Describing translation images using a coordinate grid

Describe the effects of the translation shown below.

Kwami's Solution

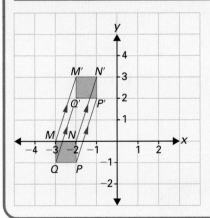

From the diagram, the coordinates of M are $(-3, 0)$ and the coordinates of M' are $(-2, 3)$.

The difference between the x-coordinates is $(-2) - (-3) = +1$.

The difference between the y-coordinates is $(+3) - 0 = +3$.

So, M moved 1 unit to the right and 3 units up to the image point M'.

Also, the other three vertices moved the same way, 1 unit to the right and 3 units up.

A Checking

3. Point *A* has coordinates (2, 4) on centimetre grid paper. It is translated 12 cm to the right and 3 cm up. What are the coordinates of the new location?

4. Describe the transformation that moved quadrilateral *DEFG* to quadrilateral *D′E′F′G′*.

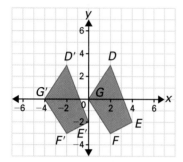

5. Which of figures B, C, and D is not a translation of figure A? Explain.

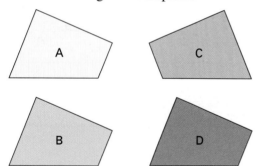

B Practising

Use centimetre grid paper for the following questions.

6. The vertices of △*ABC* have coordinates *A*(1, 2), *B*(3, 5), and *C*(3, −1). △*ABC* is translated 2 units to the right and 1 unit down. Determine the coordinates of the image triangle.

7. a) Describe the transformation that moved △*RST* to △*R′S′T′*.

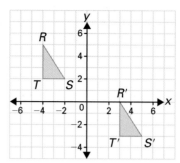

b) Describe the transformation that moved rectangle *JKLM* to rectangle *J′K′L′M′*.

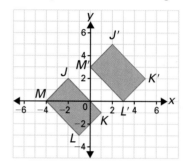

8. Copy each shape and translate it to determine the image coordinates.

a) Translate parallelogram *ABCD* 3 cm down.

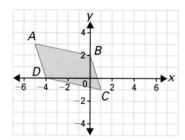

b) Translate △*EFG* 2 cm to the left and 1 cm up.

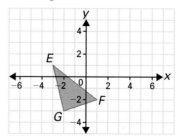

9. The vertices of square $ABCD$ are $A(-1, -1)$, $B(1, -1)$, $C(1, 1)$, and $D(-1, 1)$. Square $ABCD$ is translated 3 cm to the left and 2 cm up. Determine the coordinates of all the vertices of the image square.

10. a) Draw any triangle, $\triangle DEF$, on a coordinate grid. Record the coordinates of each vertex.

b) Transform $\triangle DEF$ 5 cm to the left and 4 cm down. Label the image $\triangle D'E'F'$.

c) Determine the coordinates of D', E', and F'.

11. $\triangle DEF$ has a base, DE, that is 7 cm long. Describe how you would translate $\triangle DEF$ so that D' is the same point as E. Draw a diagram to illustrate your description.

12. Square $WXYZ$ has sides that are 2 cm long. Squares A and B are translation images of square $WXYZ$. Describe how square $WXYZ$ was translated to create squares A and B.

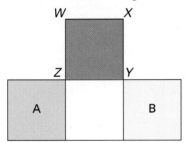

13. The vertices of $\triangle ABC$ have coordinates $A(3, -2)$, $B(0, 0)$, and $C(2, 2)$.

a) $\triangle ABC$ is translated 3 cm to the right and 2 cm down. Determine the coordinates of the image triangle. Label the image triangle $\triangle A'B'C'$.

b) $\triangle A'B'C'$ is translated 1 cm to the right and 3 cm up. Determine the coordinates of the image triangle. Label the image triangle $\triangle A''B''C''$.

c) Describe a single translation that moves $\triangle ABC$ directly to $\triangle A''B''C''$.

14. Nathan has a paper route in a part of town where all the streets run either north-south or east-west. From his home, he travels 4 blocks north, 3 blocks east, 7 blocks south, 2 blocks east, 5 blocks north, 3 blocks west, and 2 blocks south. At the end of his route, how many blocks is he from his home, and in what direction?

15. The vertices of $\triangle ABC$ have coordinates $A(2, 0)$, $B(3, 0)$, and $C(2, 2)$.

a) Translate $\triangle ABC$ 2 cm to the right and 1 cm down. Then translate the resulting image 1 cm to the right and 2 cm up.

b) Start again with $\triangle ABC$. Translate $\triangle ABC$ 1 cm to the right and 2 cm up. Then translate the resulting image 2 cm to the right and 1 cm down.

c) Compare your results in parts (a) and (b). If you apply two translations, one after the other, does the order in which you apply them matter? Write a hypothesis, and explore it using several examples.

Ⓒ Extending

16. $\triangle ABC$ is translated 3 units to the left and 2 units down. The vertices of the image triangle, $\triangle A'B'C'$, have coordinates $A'(1, -1)$, $B'(3, -2)$, and $C'(2, 1)$. Determine the coordinates of the vertices of $\triangle ABC$.

17. $\triangle ABC$ is translated 2 units to the right and 1 unit up to produce $\triangle A'B'C'$. Then $\triangle A'B'C'$ is translated 3 units to the right and 2 units down to produce $\triangle A''B''C''$. The coordinates of three of the vertices are $A(0, 0)$, $B'(6, 0)$, and $C''(6, 0)$. Determine the coordinates of all the other vertices of the three triangles.

7.3 Reflections

▶ **GOAL**

Explore the properties of reflections of 2-D shapes.

You will need
- centimetre grid paper
- a ruler
- a transparent mirror
- a protractor

Learn about the Math

Andrea is decorating the door of her room.

? **How can you draw the reflection of your name without a mirror?**

A. On a piece of paper, print your name in capital letters on a slant.

B. Draw a reflection line that is not horizontal or vertical.

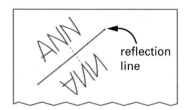

reflection line

C. Use a transparent mirror to trace the image of your name.

D. Connect three different points on your name with their image points.

E. Use a ruler and a protractor to draw a more accurate reflection image of your name.

Reflecting

1. How does the size and shape of a reflected image compare with the size and shape of the pre-image?

2. What do you notice about the distances on each side of the reflection line?

3. What do you notice about the angle between the reflection line and the line segments you drew?

4. Why does using a ruler and protractor allow you to draw a more accurate reflection image?

Example 1: Constructing a reflected image

Reflect quadrilateral *ABCD* in reflection line *IJ*.

James's Solution

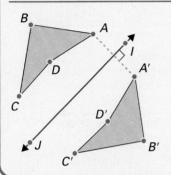

I drew a line segment from point *A*, perpendicular to *IJ*, and continued my line segment beyond *IJ*. I measured the distance from *A* to *IJ* along my line segment. I measured the same distance on the other side of *IJ*, and labelled the point *A'*.

I used this method to find the other three image vertices.

I joined *A'*, *B'*, *C'*, and *D'* to form the image. I noticed that the orientation of the image is opposite to the orientation of the pre-image.

Example 2: Drawing a reflected image using coordinates

Reflect △*ABC* in the *y*-axis.

Simon's Solution

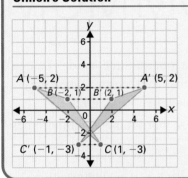

A is 5 units to the left of the *y*-axis, so the image point *A'* is 5 units to the right of the *y*-axis at (5, 2).

B is 2 units to the left of the *y*-axis, so the image point *B'* is 2 units to the right of the *y*-axis at (2, 1).

C is 1 unit to the right of the *y*-axis, so the image point *C'* is 1 unit to the left of the *y*-axis at (−1, −3).

I joined the image points to form the image △*A'B'C'*.

Ⓐ Checking

5. Which of figures A, B, and C is not a reflection of figure G? Explain.

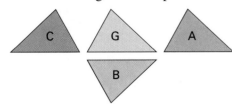

6. a) Reflect quadrilateral *ABCD* in the *y*-axis. Determine the coordinates of the image quadrilateral.

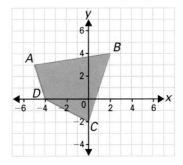

6. b) Reflect △*RST* in the *x*-axis. Determine the coordinates of the image triangle.

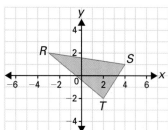

B Practising

7. Which of figures R, S, T, and U are not reflections of figure M? Explain.

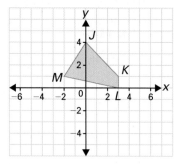

8. a) Reflect quadrilateral *JKLM* in the *x*-axis. Determine the coordinates of the image.

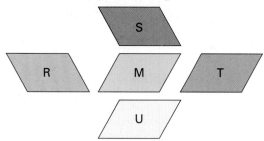

b) Reflect pentagon *DEFGH* in the *y*-axis. Determine the coordinates of the image.

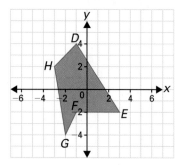

9. Triangles M and N are images of △*ABC*. Describe the transformation(s) that created each image.

a) image M   **b)** image N

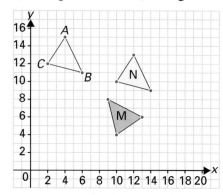

10. The word MOM is a **palindrome**. This means that it is the same word when it is read backward. As well, MOM is the same word when it is reflected in a vertical line.

a) Write the longest three words you can think of that are the same when they are reflected in a vertical line.

b) Must these words also be palindromes? Explain.

11. The word BOB is also a palindrome. As well, it is the same word when it is reflected in a horizontal line.

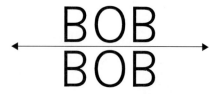

a) Write the longest three words you can think of that are the same when they are reflected in a horizontal line.

b) Must these words also be palindromes? Explain.

12. The vertices of △*ABC* have coordinates *A*(−3, 0), *B*(1, 3), and *C*(2, −1). Determine the coordinates of the image of △*ABC*.

   a) after a reflection in the *y*-axis

   b) after a reflection in the *x*-axis

13. Figure X is the image of figure W.

   a) Can you tell whether figure W was translated or reflected? Explain.

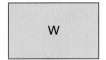 

   b) Sketch the shapes in part (a). Label the vertices so that figure X is a reflection of figure W.

14. a) Use centimetre grid paper to draw a trapezoid. Label the trapezoid *WXYZ*.

   b) Draw a reflection line so that *W′* (the image point of vertex *W*) is 6 cm from *W*. Draw the reflected image of *WXYZ*, and label it *W′X′Y′Z′*.

   c) What is the distance from *W* to the reflection line?

15. The reflected image of △*CDE* is △*C′D′E′*. Is line segment *WY* the reflection line? Explain, giving at least two reasons for your answer.

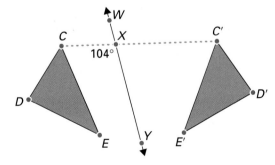

16. Determine the greatest four-digit number that is a lesser number when you reflect it in a vertical line.

ⓒ Extending

17. a) Reflect △*ABC* in the *y*-axis to produce the image △*A′B′C′*.

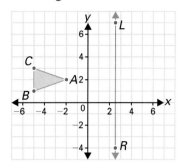

   b) Reflect △*A′B′C′* in the line *LR* to produce the image △*A″B″C″*.

   c) Determine the coordinates of △*A″B″C″*.

   d) Is there a single transformation that moves △*ABC* to the image △*A″B″C″*? If so, describe the transformation. If not, explain why not.

   e) Will the result in part (d) be true for reflections in every pair of parallel lines? Write a hypothesis, and explore it using a number of examples.

18. The vertices of △*ABC* have the coordinates *A*(−2, 0), *B*(0, 0), and *C*(0, 3). Draw a line through the points *S* (−3, 3) and *T* (3, 3).

   a) Reflect △*ABC* in the *y*-axis. Then reflect the resulting image in the line *ST*. Determine the coordinates of the final image triangle.

   b) Reflect △*ABC* in the line *ST*. Then reflect the resulting image in the *y*-axis. Determine the coordinates of the final image triangle.

   c) Compare your results in parts (a) and (b). If you apply two reflections, one after the other, does the order in which you apply them matter? Write a hypothesis, and explore it using several examples.

Rotations

You will need
- centimetre grid paper
- a compass
- a protractor
- a ruler

▶ **GOAL**

Identify the properties of a 2-D shape that stay the same after a rotation.

Learn about the Math

Kwami sees this starburst quilt in an art gallery and notices the rotated pattern.

? **How can Kwami create his own starburst pattern?**

Example 1: Constructing rotation images

Use geometry tools to construct rotations.

Kwami's Solution

I saw that the quilt's starburst is made of 8 rotated copies of a square design. I made 8 copies of it on separate pieces of grid paper.

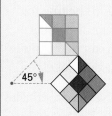

A circle has 360°, so I figured the 8 copies had to be 45° apart. On a large piece of paper, I used a protractor to draw a 45° angle.

I lined up the diagonal of a square along one of the angle arms. Then I moved the squares back and forth along the lines until they met at a corner. I glued the squares in place.

I set my compass to the distance that the corner of the square had to be from the turn centre.

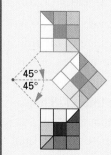

I marked another 45° angle at the centre from the previous angle and used my compass to mark where the corner of the next square had to go. Then I glued the square in place.

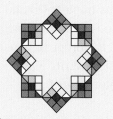

I kept drawing new 45° angles, marking the corners, and gluing the squares until I finished the starburst.

Reflecting

1. Why was Kwami correct when he said that the copies of the square design had to be 45° apart? Where is the **centre of rotation**?

2. What properties of the squares changed after each rotation? What properties did not change?

3. Suppose that Kwami wants to make a quilt in which the square pattern has 10 rotated copies. What should he do differently?

centre of rotation

a fixed point around which other points in a shape rotate in a clockwise (cw) or counterclockwise (ccw) direction; the centre of rotation may be inside or outside the shape

Work with the Math

Example 2: Recognizing rotation images

Which figures could be rotated images of figure 1?

Jody's Solution

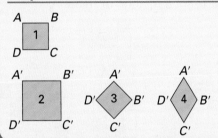

Rotations do not change the size of a figure, so figure 2 cannot be a rotated image of figure 1.

Figure 3 could be a rotated image of figure 1, since it is congruent to figure 1.

Rotations do not change the shape of a figure, so figure 4 cannot be a rotated image of figure 1.

Example 3: Rotating a shape on a coordinate grid

Plot points $A(5, 6)$, $B(8, 6)$, $C(10, 4)$, and $D(3, 4)$ on a Cartesian grid. Join the points to form a quadrilateral. Rotate the quadrilateral 90° cw about the origin, O.

Indira's Solution

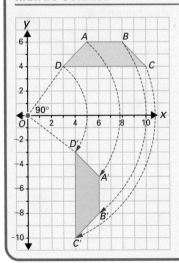

I plotted the points and joined them to form a trapezoid. I drew a line from O to D. Using a protractor, I measured a 90° angle from line segment OD and drew another line segment, perpendicular to OD. I placed the point of a compass on O and the pencil on D. I drew an arc from D to meet the perpendicular line. I marked this point D' (the image of D after rotation).

I used the same method to find the other image points.

I joined the image points to form trapezoid $A'B'C'D'$.

The coordinates of the image are $A'(6, -5)$, $B'(6, -8)$, $C'(4, -10)$, and $D'(4, -3)$.

Ⓐ Checking

4. a) Which figures below could be rotated images of figure 1? Explain.

   b) Look at the figures that cannot be rotated images of figure 1. Could they be images of figure 1 after a different transformation? Explain.

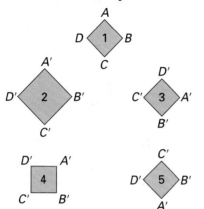

5. Your teacher tells you that the hour hand on an analog clock will rotate 45° while you write a math quiz. If the quiz starts at 9:00, what time does it end? What is the angle between the starting and ending positions of the minute hand?

6. △A′B′C′ is the image of △ABC after a rotation about the centre of rotation D. Suppose that you use a compass to draw a circle with centre D, and B is on the circle. Which other point must be on the circle? Explain.

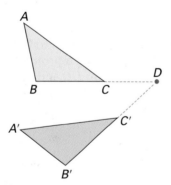

7. The vertices of △ABC have coordinates A(2, 3), B(1, 5), and C(−1, 1). Determine the coordinates of the image of △ABC after a 90° ccw rotation about the origin.

Ⓑ Practising

8. Rotate rectangle QRST 90° cw about the origin. Label the coordinates of the vertices of the image.

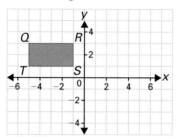

9. a) Which figure below is not the result of a rotation of figure A? Explain.

   b) What transformation created the image that is not a rotation of figure A?

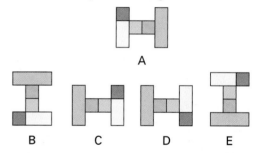

10. One of the rotated images in question 9 can also be formed by reflecting figure A. How can you form the image by reflection?

11. △D′E′F′ is the image of △DEF after a rotation about centre K. Suppose that you use a compass to draw a circle with centre K. E is on the circle. Which other point must be on the circle? Explain.

Use centimetre grid paper for questions 12 to 15.

12. a) Rotate $\triangle RST$ 270° cw about the origin. Label the coordinates of the vertices of the image.

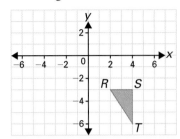

b) Predict a different rotation that would move $\triangle RST$ to the same image as in part (a). Justify your prediction.

13. a) Rotate quadrilateral $ABCD$ 180° ccw about vertex D. Label the coordinates of the vertices of the image.

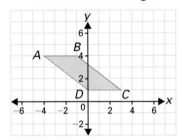

b) Predict a different rotation that would move quadrilateral $ABCD$ to the same image as in part (a). Justify your prediction.

14. a) Draw any triangle, and label its vertices X, Y, and Z.

b) Rotate $\triangle XYZ$ 360° cw about point X. Label the points of the image X', Y', and Z'.

c) What do you notice? Explain why this happens.

15. $\triangle E'F'G'$ with $E'(1, -1)$, $F'(1, 3)$, and $G'(4, 0)$ is the rotation image of $\triangle EFG$ after a 90° cw rotation about F. Determine the vertices of $\triangle EFG$.

16. a) Draw rhombus $PQRS$ with a 30° angle at vertex R and sides 2 cm long. How many degrees must you rotate $PQRS$ about vertex R so that the image touches the pre-image along an edge?

b) Rotate the rhombus about vertex R so that all the images touch along an edge, without overlapping. How many figures are in the design? What are the angles of rotation? What pattern do you see?

c) If the rhombus were drawn with an angle of 60° at vertex R, how many figures would be in the starburst? What would the angles of rotation be? Show how you know.

17. a) Plot $A(4, 5)$, $B(7, 5)$, $C(10, 2)$, $D(2, 2)$, and $R(5, 4)$. Join points A, B, C, and D to form a quadrilateral.

b) Rotate quadrilateral $ABCD$ 180° about point R. What are the coordinates of the vertices of the rotated image, $A'B'C'D'$?

C Extending

18. Use $A(0, 0)$, $B(0, 3)$, and $C(1, 3)$.

a) Draw and rotate $\triangle ABC$ 90° cw about point A to produce $\triangle A'B'C'$.

b) Rotate $\triangle A'B'C'$ 90° cw about point A to produce $\triangle A''B''C''$.

c) Is there a single rotation that will move $\triangle ABC$ directly to $\triangle A''B''C''$? If so, what is the angle of rotation? What is the centre of rotation?

19. Use $A(-5, 2)$, $B(-2, 3)$, and $C(-2, 1)$.

a) Reflect $\triangle ABC$ in the y-axis to produce $\triangle A'B'C'$. Then reflect $\triangle A'B'C'$ in the x-axis to produce $\triangle A''B''C''$.

b) Determine the coordinates of $\triangle A''B''C''$.

c) Is there a single transformation that will move $\triangle ABC$ to $\triangle A''B''C''$? Explain.

Congruence and Similarity

You will need
• a ruler
• a protractor

▶ **GOAL**

Investigate the conditions that make two shapes congruent or similar.

Explore the Math

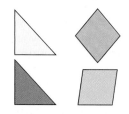

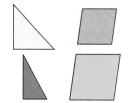

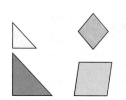

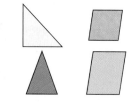

These pairs of shapes are congruent.   These are not.   These pairs of shapes are similar.   These are not.

? **What information do you need to know if you want to construct congruent and similar shapes?**

A. Look at the diagrams of congruent shapes. Explain what you think "congruent" means.

B. Look at the diagrams of similar shapes. Explain what you think "similar" means.

C. For each group of triangles described in the table, construct as many triangles as you can that fit the description.

Group of triangles	How are the triangles the same?	How are the triangles different?	Are the triangles congruent?	Are the triangles similar?
equilateral triangles with sides that are 4 cm				
triangles with sides that are 3 cm, 6 cm, and 4 cm				
isosceles triangles with an angle of 40° between two equal sides of 2 cm each				
isosceles triangles with equal sides that are 4 cm long				
right triangles with one side that is 4 cm long				
triangles with angles of 30°, 80°, and 70°				
triangles with a 5 cm side between angles of 60° and 70°				
triangles with a base of 3 cm and a height of 4 cm				

D. Draw an irregular quadrilateral (with unequal sides). Describe the quadrilateral in terms of its sides and angles. Test your description by having someone follow it to draw the same shape.

Reflecting

1. Look at your table for step C.

   a) Choose a group with triangles that are congruent. Why do you think they are congruent?

   b) Choose a group with triangles that are not congruent. What other information would you need to give so that the triangles for this group would be congruent?

2. Suppose that you have two hexagonal paving tiles that you think might be congruent, but they are too heavy for you to move. What tools would you need to check whether the tiles are congruent? How would you use these tools?

3. a) How can you tell when two shapes are congruent?

   b) How can you tell when two shapes are similar?

   c) Is it easier to tell whether two shapes are congruent or whether they are similar? Explain.

4. Read each statement about the relationship between congruence and similarity. Decide whether the statement is true or false, and use examples to explain why.

   a) All similar geometric figures are also congruent.

   b) Congruent geometric figures are not always similar.

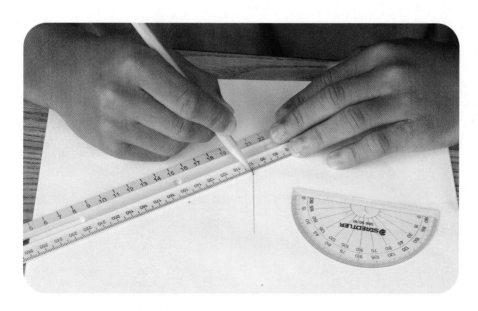

Frequently Asked Questions

Q: How can you tell whether a shape is the image of another shape after each transformation?

- a translation
- a reflection
- a rotation

A: Each of these transformations produces an image that is congruent to the pre-image.

Transformation	Relationship between vertices of pre-image and vertices of image	Example
translation	Each pre-image vertex is moved the same direction and the same distance. The image is oriented in the same way as the pre-image.	
reflection	The line segments that join each pre-image vertex to its image vertex are • perpendicular to the line of reflection • cut in half by the line of reflection The orientation of the image is opposite to the orientation of the pre-image.	
rotation	There is a common centre of rotation, O, from which a circle can be drawn to pass through each pre-image vertex and its image. Angles AOA', BOB', COC', and so on, are all the same. The orientation of the image depends on the angle of rotation.	

Q: How can you tell whether figures are congruent?

A: Congruent figures have exactly the same size and shape. This means that the measures of matching sides and angles are equal.

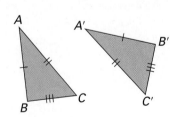

Q: How can you tell whether figures are similar?

A: Similar figures have the same shape, but not necessarily the same size. This means that matching angles are equal.

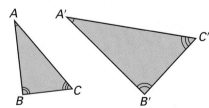

Practice Questions

(7.1) **1.** Complete each statement by filling in the boxes. (Replace ? with a number from the first ordered pair.)

  a) $(-1, -2)$ is to the right of ($\blacksquare$, 0) because ? > $\blacksquare$.

  b) $(0, 0)$ is below $(-4, \blacksquare)$ because ? < $\blacksquare$.

(7.2) **2. a)** Draw square *CDEF* with vertices $C(1, 1)$, $D(3, 2)$, $E(2, 4)$, and $F(0, 3)$.

  b) Translate square *CDEF* 5 units to the right. Determine the coordinates of the image vertices.

(7.2) **3. a)** Draw $\triangle ABC$ with vertices $A(0, 0)$, $B(4, 0)$, and $C(0, 3)$.

  b) Translate $\triangle ABC$ 6 units to the left and 2 units up. Determine the coordinates of the image vertices.

(7.2) **4. a)** Draw trapezoid *CDEF* with vertices $C(-1, 3)$, $D(4, 3)$, $E(6, -1)$, and $F(-3, -1)$.

  b) Translate *CDEF* so that C' has coordinates $(-10, 7)$. What are the coordinates of D', E', and F'?

(7.2) **5.** Rhombus *RSTU* is congruent to rhombus *WXYZ*. Could rhombus *RSTU* be a translation image of rhombus *WXYZ*? Why or why not? Can you be sure that rhombus *RSTU* is a translation image of rhombus *WXYZ*? Why or why not?

(7.3) **6. a)** Draw $\triangle ABC$ with vertices $A(-4, 8)$, $B(-8, 3)$, and $C(-3, 3)$.

  b) Reflect $\triangle ABC$ in the *y*-axis. What are the coordinates of the reflected image?

7. Quadrilateral *WXYZ* is reflected in line *AB*. Determine the distances between vertices *W*, *X*, *Y*, and *Z* and their images. (7.3)

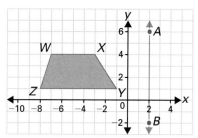

8. a) Draw $\triangle LMN$ with vertices $L(-5, 2)$, $M(-2, 4)$, and $N(0, 1)$. (7.4)

  b) Rotate $\triangle LMN$ 180° cw about point *L*. Label the coordinates of the image.

9. a) Draw $\triangle ABC$ with vertices $A(4, 2)$, $B(3, -3)$, and $C(1, -2)$. (7.4)

  b) Rotate $\triangle ABC$ 90° ccw about the origin. What are the coordinates of the vertices of the image?

10. Describe a single transformation that could move parallelogram *HIJK* to each image. If possible, give more than one answer. (7.4)

  a) image A **b)** image B

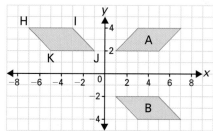

11. Draw a triangle with two angles of 70° and one angle of 40°. Draw another triangle with the same angle measurements. Must the second triangle be congruent to the first triangle? Must the second triangle be similar to the first triangle? Use diagrams to illustrate your answers. (7.5)

7.6 Tessellations

▶ **GOAL**

Create and analyze designs that tessellate a plane.

You will need
- models of 12 pentominoes
- coloured pencils
- grid paper

Explore the Math

The following diagram shows 12 different pentominoes.
These pentominoes can be identified by the letters they resemble.

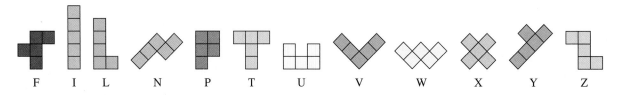

F   I   L   N   P   T   U   V   W   X   Y   Z

Amelia says that all pentominoes can be used to make a tessellation.
Is she correct?

❓ How can you use each pentomino to make a tessellation?

A. This diagram shows a tessellation that has the I pentomino in one orientation. Draw three other tessellations that have pentominoes in one orientation.

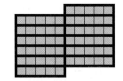

B. This diagram shows a tessellation that has the T pentomino in two orientations. What transformation is needed to move from a green T to a yellow T?

C. The diagram below shows a tessellation that has the W pentomino transformed into different orientations by rotations. How many orientations of the W pentomino can you see in the tessellation? Sketch the W pentomino in each of these orientations, and colour each pentomino to match the orientation in the tessellation. Complete the table to describe how you would rotate the yellow pentomino (the pre-image) into the other pentominoes.

Image tile	Rotation
blue	

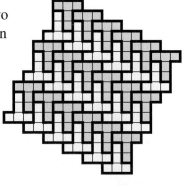

D. This diagram shows a tessellation with the N pentomino. How many different orientations can you see in the tessellation? Complete the table to describe how you would transform the light green pentomino (the pre-image) into the other pentominoes. The first one is done for you.

Image tile	Rotation	Reflection(s)
red	90° cw	in a vertical line
blue		

E. Draw a tessellation with a pentomino you did not use in steps A to D. Record the number of orientations in the tessellation. Repeat this for three other pentominoes.

F. Use the table to summarize what you have discovered about how each pentomino can make a tessellation.

Pentomino shape	Was this transformation needed to make a tessellation with the shape?			Properties of the shape
	Translation	Rotation	Reflection	

Reflecting

1. Which pentominoes required the fewest transformations to tessellate? Explain.

2. Which pentominoes required the most transformations to tessellate? Explain.

3. Was it possible to tessellate every pentomino? Explain.

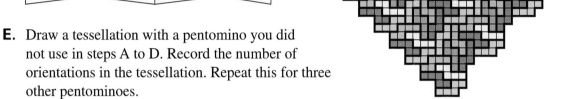

You will need
• coloured pencils
• centimetre grid paper

7.7 Communicating about Geometric Patterns

▶ **GOAL**

Describe designs in terms of congruent, similar, and transformed images.

Communicate about the Math

Mohammed used computer software to draw this design, which he plans to stencil on homemade ceramic tiles. This is Mohammed's description of his design:

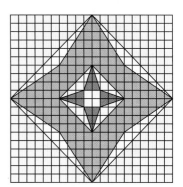

It has a white square in the middle, inside a red star. The red star is in a white square, which is in a green square. The green square is in a blue star. Then there is a white square around the blue star and, finally, a yellow square around everything. The whole design is 20 cm high and 20 cm wide.

This is what Samantha drew, based on Mohammed's description.

❓ How can Mohammed's description be improved?

Samantha looked at Mohammed's design and the Communication Checklist, and asked him these questions:

• Did you completely describe both the shape and the orientation of each part of your design?
• Did you describe all the transformations you used?
• Did you describe any equal sides and use appropriate measurements?

A. Identify the parts of Mohammed's description that were accurate but not clear enough to allow Samantha to draw the design correctly.

B. What did Samantha think was missing from Mohammed's description?

C. What other questions would you have asked Mohammed to help him clarify his design for Samantha?

> **Communication Checklist**
>
> ☑ Was your description clear?
>
> ☑ Was your description complete and thorough?
>
> ☑ Did you use necessary and appropriate math language?

Reflecting

1. How could Mohammed have improved his description so that Samantha's design would have been closer to his design?

2. Which questions in the Communication Checklist were covered in Samantha's questions?

Work with the Math

Example: Describing a design

Improve Zach's description of this diagram.

Zach's description:

There are three similar right triangles.

The bases of the triangles are 1 cm apart, on the same line.

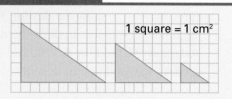

1 square = 1 cm²

The triangle on the left has a 90° angle between a side of 6 cm and a base of 9 cm.

The triangle in the middle has a 90° angle between a side of 4 cm and a base of 6 cm.

The triangle on the right has a 90° angle between a side of 2 cm and a base of 3 cm.

Kwami's Solution

1. Draw a right triangle with a 9 cm base and a 6 cm height. The right angle should be at the bottom left vertex. The triangle should point up.

2. On the same base, draw a second triangle that is similar to the first, but has a 6 cm base and a 4 cm height. Make the second triangle oriented the same way as the first triangle.

3. Line up the base of the second triangle with the base of the first triangle.

4. Translate this triangle to the right until its right angle is 1 cm to the right of the bottom right vertex of the first triangle.

5. Draw a third similar triangle with a 3 cm base and a 2 cm height, oriented the same way as the other two triangles.

6. Translate the third triangle so that its base lines up with the other bases and is 1 cm to the right of the base of the second triangle.

I wrote step 1 because I couldn't tell where Zach wanted the right angle to go. From his description, it could have been on the left or the right. I also couldn't tell whether the triangle should point up or down.

I wrote the other steps to describe how the triangles line up. When Zach said "in the middle," I knew there would be a third triangle. He didn't say whether the smaller triangle should go to the right or the left of the original. Also Zach didn't say whether the triangles should point in the same direction, or how far apart to space them.

Ⓐ Checking

3. Consider the following design. Assume that each square in the grid is 1 cm by 1 cm.

a) Describe the design, keeping the Communication Checklist in mind.

b) Test your description by asking someone else to draw the design. Was the other person able to draw the design accurately? Could the description be used to draw a different design?

4. This design is made up of four triangles. Assume that each square in the grid is 1 cm by 1 cm.

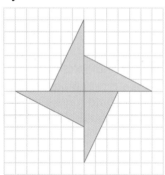

a) Describe the design, keeping the Communication Checklist in mind.

b) Test your description by asking someone else to draw the design. Was the other person able to draw the design accurately? Could the description be used to draw a different design?

Ⓑ Practising

5. The symbol for radioactive materials includes some simple shapes and some complex shapes. The following is Heidi's description of the symbol:

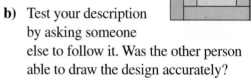

"There is a black triangle, and the inside of it is all yellow. In the middle of the triangle is a black dot, surrounded by three equally spaced shapes."

a) Carefully explain and correct any errors in Heidi's description.

b) Use the Communication Checklist to help you assess Heidi's description.

c) Use your answers to parts (a) and (b) to help you write an improved description.

d) Test your description by asking someone else to follow it. Was the other person able to draw the symbol accurately?

6. The log cabin square is a very popular quilt design.

a) Carefully describe the design.

b) Test your description by asking someone else to follow it. Was the other person able to draw the design accurately?

7. A design called "steps to the courthouse square" is often used with the log cabin square.

a) Carefully describe the design.

b) Test your description by asking someone else to follow it. Was the other person able to draw the design accurately?

8. Design your own quilt square using geometric shapes. Write a description of your design, keeping the Communication Checklist in mind. Can someone else recreate your design from your description? Review and revise your description until it is clear.

1. A 5-by-5 square can be tiled with one, two, three, four, or five different types of pentominoes. Here is one possibility for each tiling. Can you find others? Use grid paper to show your work.

One type of pentomino:

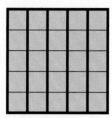

Four types of pentominoes:

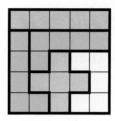

Two types of pentominoes:

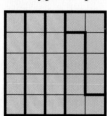

Five types of pentominoes:

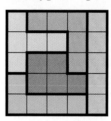

Three types of pentominoes:

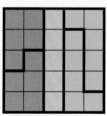

2. A rectangle can be covered using each of the 12 pentominoes once, if the rectangle has dimensions 4 by 15, 5 by 12, or 6 by 10. Show how to create one of these designs. (A tiling of a 4-by-15 rectangle is started below.)

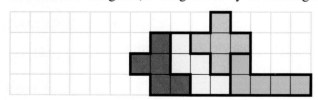

3. Create your own design with at least four different pentominoes.

Investigating Pattern Blocks

▶ **GOAL**

Use transformations and properties of congruent shapes to solve problems.

Explore the Math

The shapes of the six different pattern blocks are shown below.

? **What are the angle measurements and the area relationships for each pattern block?**

regular polygon
a polygon with all sides equal and all angles equal

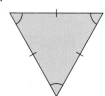

A. Which blocks are **regular polygons** ? Explain how you can find out, using only the blocks and measuring tools. Explain how you can find out by folding a paper copy of each block.

B. The diagram at the right shows a tessellation of three yellow hexagons around a single point. How does this diagram help you calculate each vertex angle?

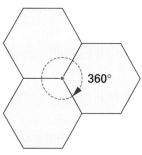

360°

C. Trace each block to show tessellations around a single point for all the pattern blocks. Use these tessellations to determine each block's vertex angles.

D. Show how to tile the hexagon with the green triangle. What fraction of the area of the hexagon is the area of the triangle?

E. Repeat step D with the red trapezoid, and then with the blue rhombus.

F. Try step E with the orange square and the beige rhombus. What do you notice?

G. Summarize your findings in steps A to F by completing the following table. If the area of a block cannot be compared with the area of the hexagon block by using a fraction with the numerator 1, write "larger than the hexagon" or "smaller than the hexagon" in the third column.

Pattern block	Regular or irregular?	Relationship of area to area of hexagon	Number of tessellated blocks that meet at a vertex	Angle measures
yellow hexagon			3	
red trapezoid		$\frac{1}{2}$ hexagon		
blue rhombus				
orange square				
green triangle				
beige rhombus				

H. Use a protractor to check the accuracy of the angle measures you calculated.

Reflecting

1. a) How did you use properties of congruent shapes to find the measures of the vertex angles in the pattern blocks?

  b) For which blocks did you use a transformation to help you find the angle measures?

2. What is the relationship between the vertex angles in a pattern block and the ability of the pattern block to tessellate?

3. a) A regular octagon has eight sides, and each vertex angle is 135°. Use the relationship you described in question 2 to explain why a regular octagon cannot be used to create a tessellation.

  b) What figure can be used to fill the gaps in a tiling created with regular octagons?

4. a) Identify several other polygons (regular and irregular) that can be used to tessellate a plane.

  b) For each polygon you identified in part (a), verify that the relationship you described in question 2 is satisfied.

DIVIDING SHAPES INTO CONGRUENT PARTS

1. Copy the parallelogram onto square dot paper. Divide the parallelogram into these shapes.

a) two congruent obtuse triangles   **c)** four congruent parallelograms

b) four congruent obtuse triangles   **d)** two congruent trapezoids

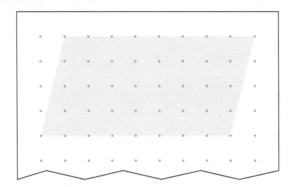

2. Copy the rectangle onto square dot paper. Divide the rectangle into these shapes.

a) two congruent right triangles   **d)** four congruent rectangles

b) eight congruent right triangles   **e)** two congruent squares

c) sixteen congruent right triangles   **f)** eight congruent squares

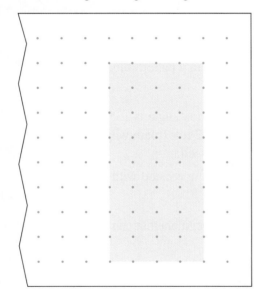

Tessellating Designs

▶ **GOAL**

Create irregular tiles, and determine whether irregular tiles can be used to tessellate a plane.

Learn about the Math

Kwami sees a copy of this M.C. Escher drawing on a poster. He wonders how the artist managed to get all the birds to fit together so well.

Escher chose the square as the basic shape for his artwork.

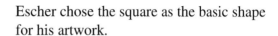

The black section from the top of the square can be cut out and translated to the bottom of the square. The black section from the right of the square can also be cut out and translated to the left.

? How can you make an irregular tessellating tile?

Example 1: Creating an irregular tessellating tile from a hexagon

Create a tessellating lizard from a hexagon.

Simon's Solution

A: I started with a hexagon because it tessellates.

B: I changed half of one side.

C: I rotated the changed part 180° to the other half of the same side. (I used the midpoint of the side as the centre of rotation.)

D: Using rotations and translations, I transferred the changed side to all the other sides.

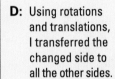

E: I added some details to finish the lizard.

F: I cut out the lizard.

G: Then I used the tile to create a tessellation.

Example 2: Creating an irregular tessellating tile from a square

Create an irregular tessellating tile that looks like a horse and rider.

Jody's Solution

A: I began with a square.

B: I changed the left side with a curve.

C: I translated the curve to the right side.

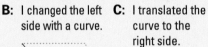

D: I changed the bottom with a curve, and then I translated the curve to the top.

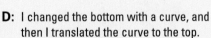

E: I cut out the tile.

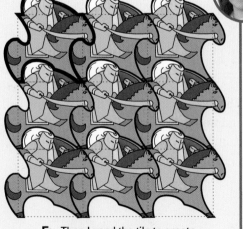

F: Then I used the tile to create a tessellation.

Reflecting

1. Look at the bird and lizard tessellations. Explain which transformations were used in each case to create the tessellations.

2. Each tessellating design started with a polygon that can tessellate. What needs to be true about the vertex angles in a polygon to ensure that it can tessellate?

3. The next step in creating a tessellating design is to change the sides of the polygons that make up the tessellation. What kinds of changes are allowed?

Work with the Math

Example 3: Creating a tessellating tile by modifying half of each side

Create a tessellating tile by modifying half of each side of a square.

James's Solution

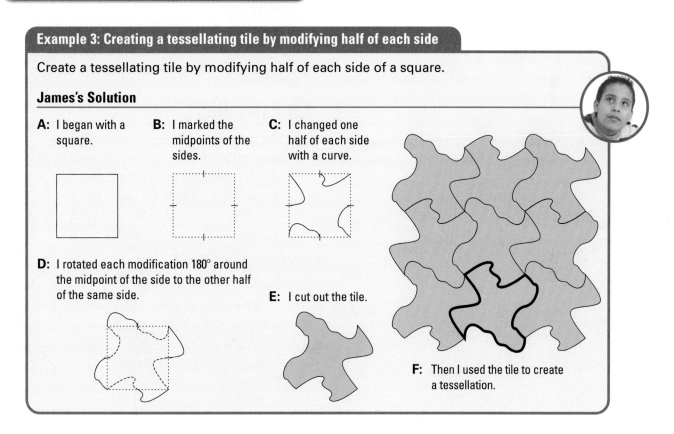

A: I began with a square.

B: I marked the midpoints of the sides.

C: I changed one half of each side with a curve.

D: I rotated each modification 180° around the midpoint of the side to the other half of the same side.

E: I cut out the tile.

F: Then I used the tile to create a tessellation.

Ⓐ Checking

4. This diagram shows a tessellation of four squares. What property of the angles at the common vertex, *A*, allows the squares to tessellate?

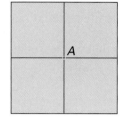

5. This diagram shows the beginning of a change to a triangle.

a) Would the triangle have tessellated without being changed? Explain.

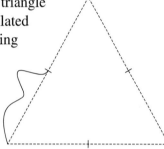

b) Show how you could finish changing the triangle so that it will tessellate as a modified shape.

Ⓑ Practising

6. a) Create a tile by following this description:

Start with a square.

Change the left side, and translate the change to the right.

Use a different change for the top. Translate this change to the bottom.

b) Cut out the tile, and use it to create a design.

c) Will the tile you created tessellate a plane?

7. Draw a square. Change each side by drawing a curve from each vertex to the adjacent vertex. Is it likely that this shape will tessellate? Explain.

8. a) Create a tessellating tile by following this description:

Start with a triangle.

Change one half of one side. Rotate this change 180° about the midpoint of the side to complete the side.

Change the other sides the same way.

b) Cut out the tile, and use it to create a design.

c) What kind of triangle did you need to start with?

d) Will the tile you created tessellate a plane?

e) Can you rotate a tile in the tessellation so that it has the same orientation as another tile in the tessellation? Explain.

Ⓒ Extending

9. The following drawing is a complex tessellation by M.C. Escher. It is made up of two fish that always appear together. One fish has a single fin on its tail, and the other fish has a double fin on its tail.

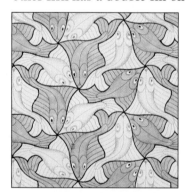

a) What is the basic shape that Escher began with?

b) What transformations can you see in Escher's drawing? Explain.

c) Look at the side of the basic shape that has the fish with the double fin tail. How did Escher change this side of the basic shape?

TRANSFORMATIONAL GOLF

Your task is to design a hole for a golf course. Here are the requirements:

- The design must fit on the piece of paper provided by your teacher.
- The hole must have a tee and a green.
- The hole must have hazards, such as water, sand, and trees.
- The hole must have a "par" value of 3, 4, or 5. The par value is the number of transformations needed to get the ball from the tee to the green, while following the rules of the game.

Rules

1. The object of the game is to get the ball from the tee to the green.

2. The ball is moved using translations, reflections, and/or rotations.

3. Rotations and translations travel along the ground. Reflections travel through the air.

4. There are four restrictions for hazards:
- You may not rotate through a hazard.
- The centre of rotation cannot be in a hazard.
- You may not translate through a hazard.
- You may reflect over a hazard, but the line of reflection cannot pass through a hazard.

5. The path of the ball, the centres of rotation, and the lines of reflection must remain on the paper.

A. The following hole was completed in two moves: a reflection and a translation. Is it a par 2 hole or a par 3 hole?

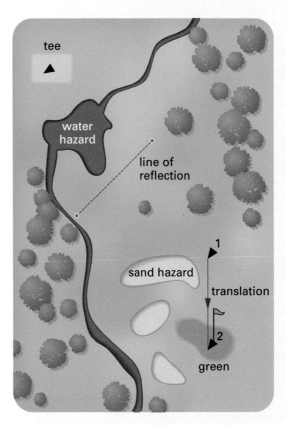

B. Once you have designed your hole, exchange holes with a partner.
- Can you complete your partner's hole at par?
- Can you go under par?

C. Put together several holes to make a course. Can you complete the course under par?

1. △*DEF* has a base, *DE*, that measures 9 cm between angles of 60° and 70°. Describe how you would translate △*DEF* so that *D'* is the same point as *E*. Draw a diagram to illustrate your description.

2. a) On centimetre grid paper, draw a triangle and label it △*XYZ*. Translate △*XYZ* 3 cm to the right and 2 cm down to produce the image, △*X'Y'Z'*.

   b) Extend line segment *XY* to form line *LM*.

   c) Reflect △*X'Y'Z'* in line *LM*.

3. a) Draw a polygon with vertices *D*(2, 2), *E*(−1, 4), *F*(−2, −1), and *G*(3, −2).

   b) Rotate the polygon 90° cw about the origin.

4. A trapezoid has coordinates *J*(1, 1), *K*(6, 1), *L*(5, −2), and *M*(2, −2). The image of the trapezoid has coordinates *J'*(1, −1), *K'*(1, −6), *L'*(−2, −5), and *M'*(−2, −2).

   a) Draw the figure and the image.

   b) Describe two different transformations that move trapezoid *JKLM* to trapezoid *J'K'L'M'*.

5. For each of the following, state whether the two shapes appear congruent, similar, or neither. Explain.

   a)

   b)

   c)

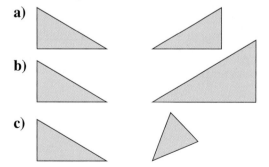

6. Determine whether each tile can tessellate. Explain your answers.

   a)

   c)

   b)

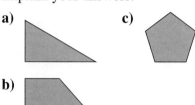

7. The figure in the diagram is formed from congruent shapes. Describe it so that someone else could draw it.

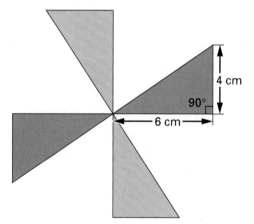

4 cm
90°
6 cm

8. Anthony has five pattern blocks. Two of the blocks are squares and three are triangles.

   a) Can Anthony place the five blocks on a table such that one vertex from each block meets at a single point? Draw a sketch to illustrate your answer.

   b) Is it possible to predict the answer to part (a) by using the measures of the angles in the blocks? If so, explain how.

9. Use a square to create an irregular tessellating tile. Explain how you created your tile. Does your tile have rotational or reflection symmetry?

Chapter Review

Frequently Asked Questions

Q: **What is a tessellation?**

A: A tessellation is a geometric pattern made up of one repeated shape that covers a flat area (plane) without overlapping or gaps.

Q: **Why does the orientation of shapes have to change in some situations, but not others, to create a tessellation?**

A: If a shape is simple, like a rectangle, you can use translations to complete a tessellation. If a shape is complex, you may need to turn it or reflect it to fill gaps.

Q: **Why must the angles at any common vertex in a tessellation add up to 360°?**

A: If the angles at a vertex add up to less than 360°, then there will be a gap. If the angles add up to more than 360°, then there will be overlapping. (There are 360° in a circle.)

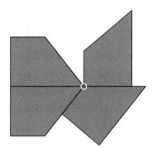

The angles at the common vertex add up to less than 360°.

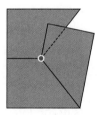

The angles at the common vertex add up to more than 360°.

Q: **How can you change a tile so that it will tessellate a plane?**

A: You must start with a shape that will tessellate. You must change more than one side the same way, using appropriate transformations.

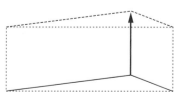

Practice Questions

(7.1) **1.** Use the points $(-14, 0)$, $(0, 0)$, $(14, 0)$, $(0, 14)$, and $(0, -14)$. State the point that matches each description. Explain how you know.

    a) farthest right   **c)** lowest down

    b) farthest left   **d)** highest up

(7.2) **2. a)** Draw $\triangle PQR$ with vertices $P(-1, 3)$, $Q(0, 1)$, and $R(2, 4)$.

    b) Transform $\triangle PQR$ by translating it 6 units to the left and 4 units up.

    c) Determine the coordinates of the image vertices.

(7.3) **3.** Which of these images is not a reflection of the red triangle? Explain.

(7.3) **4.** The coordinates of quadrilateral $ABCD$ are $A(0, 2)$, $B(3, 1)$, $C(3, -2)$, and $D(-1, -1)$.

    a) Draw quadrilateral $ABCD$.

    b) Reflect $ABCD$ in the y-axis to produce the image $A'B'C'D'$.

    c) Determine the coordinates of the vertices of $A'B'C'D'$.

    d) Translate $A'B'C'D'$ 2 units to the left and 1 unit up to produce $A''B''C''D''$.

    e) Determine the coordinates of $A''B''C''D''$.

(7.4) **5.** How would you transform triangle 1 to match triangle 2? Describe as many ways as you can.

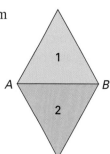

6. Determine whether or not the shape on the right is the image of the shape on the left after a transformation. If it is, describe the transformation.   (7.4)

7. How would you transform shape A to match shape B?   (7.4)

8. Shape C was transformed twice to form image shape D. What transformations were used? Describe the transformations.   (7.4)

9. $\triangle ABC$ has coordinates $A(-1, 3)$, $B(-2, 0)$, and $C(1, -1)$.   (7.4)

    a) Draw $\triangle ABC$.

    b) Rotate $\triangle ABC$ $90°$ cw about B.

10. A figure has coordinates $A(2, 0)$, $B(6, 0)$, $C(7, 3)$, and $D(3, 3)$. The image of the figure has coordinates $A'(2, -4)$, $B'(6, -4)$, $C'(7, -7)$, and $D'(3, -7)$.   (7.4)

    a) Draw the figure and the image on a coordinate grid.

    b) How was figure $ABCD$ transformed to form figure $A'B'C'D'$?

(7.4) **11.** The transformation of $\triangle PQR$ is a 360° ccw rotation. Explain what you know about the coordinates of $\triangle PQR$ and $\triangle P'Q'R'$. Describe a different rotation that would also result in image $\triangle P'Q'R'$.

(7.4) **12.** Matthew says that figure $J'K'L'M'$ is the image of figure $JKLM$ rotated 180° about the origin, O. Malik says that $J'K'L'M'$ is the translated image of $JKLM$. Are they both right? Is there another possible transformation? Explain.

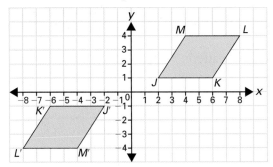

(7.4) **13.** Use this quilt design. Describe examples of the transformations.

  a) translations

  b) reflections

  c) rotations

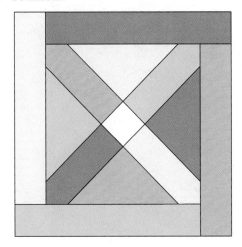

14. Tien drew two triangles. Both triangles had a side of 5 cm between angles of 60° and 70°. Are the two triangles similar, congruent, or neither? Explain with a diagram. (7.5)

15. $\triangle XYZ$ has two 45° angles. What type of triangle is it? What other piece of information do you need to construct a triangle that is congruent to it? (7.5)

16. Create a tessellation that has only two orientations of the T pentomino. How are the two orientations related? (7.6)

17. The two green triangles in the following diagram are congruent, and the hexagon is regular. (7.8)

  a) How many green triangles would you need to cover the hexagon?

  b) What is the sum of the angles at the centre of the hexagon?

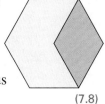

18. a) What fraction of the regular hexagon is the blue rhombus?

  b) How many rhombuses would you need to cover the hexagon?

  c) What transformation will move one rhombus to another rhombus? (7.8)

19. Start with a square. Change the sides to make a tessellating tile. Create a tessellation. (7.9)

Floor Designs

Your family has asked you to design a new kitchen floor. You want to make it interesting, so you have decided to use at least two different tiles. As well, you want to include several transformations in your pattern.

? **How can you create an interesting pattern to tile your kitchen floor?**

A. Use your creativity to make the pattern. There are only four restrictions:
- You must include some type of symmetry in your pattern.
- You must include at least two shapes and their transformed images.
- You must use translations, rotations, and reflections.
- You must use congruent and similar shapes.

B. Write a paragraph that clearly describes your pattern.

Task Checklist

☑ Did you use more than one shape?

☑ Did you use both congruent and similar shapes?

☑ Did you use translations, rotations, and reflections?

☑ Did you use symmetry?

☑ Did you clearly describe your pattern?

Math in Action

Computer Animator

A Canadian company called Side Effects is leading the way in three-dimensional computer animation. Side Effects has developed PRISMS™ software, which is used by clients to create special effects in movies, such as *Apollo 13*, and in rock videos, video games, and commercials.

Mark Mayerson, a computer animator at Side Effects, says, "Everything I do as an animator involves moving objects around in space. When working on my computer, I might have to move something 5 positions over on an *x*-axis and 10 positions up on a *y*-axis. I'm always thinking in terms of coordinates. I might copy, flip, slide, and turn figures to create exciting visual effects."

Problems, Applications, and Decision Making

Mark created a snowflake figure for a snow scene in one computer animation sequence. "I made a figure, and then I flipped it and started rotating it. Then I began copying the figure I created."

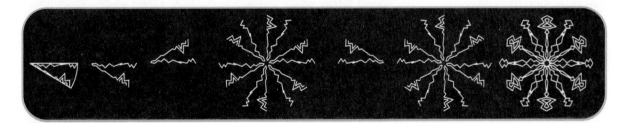

1. Mark described the first two transformations he performed. Describe the rest of the transformations he performed.

2. On a sheet of graph paper, follow Mark's steps, performing each transformation to create a snowflake.

3. Work with a partner. Describe how you performed each transformation to create your snowflake.

4. How might a computer make the transformations easier?

Mark explains how he uses a computer: "I draw and set up the first and last positions in the animation sequence. Say I want the robot's knee to turn 80°. I set up the position I want at the start of the turn and the position I want at the end of the turn. The computer calculates and draws all the positions in between."

5. Describe the transformations that occur from frame to frame.

Mark explains how the computer does the transformations: "Suppose there are 10 positions of a ball shown on the screen. The computer must perform two calculations per position: one for how far the ball moves on the *x*-axis and one for how far the ball moves on the *y*-axis."

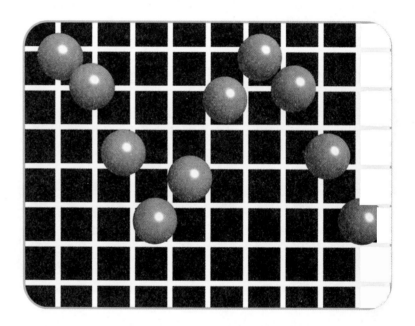

6. What transformation(s) can you identify?

7. What other transformation(s) might happen as the ball moves?

Advanced Applications

8. Create your own animation figure. Plan the moves it will make. Will the moves be reflections, translations, or rotations? Develop an animation sequence on a piece of graph paper.

9. Describe, to a classmate, the transformations you performed to create your animation sequence.

CHAPTER 8

Variables, Expressions, and Equations

▶ GOALS

You will be able to

- use variables to represent and describe mathematical relationships
- evaluate expressions
- solve equations and verify the solutions
- solve problems that lead to equations

Getting Started

You will need
- grid paper
- a calculator
- coloured pencils

Finding Checkerboard Patterns

Omar and Kaitlyn are doing
a project about patterns in
board games. They want to
find a way to figure out the
number of squares on an
8-by-8 checkerboard.

? **How many squares
are on an 8-by-8
checkerboard?**

A. Copy the table.
Complete the row
for 1-by-1 squares.

Dimensions of each square	Number of squares in a row	Number of rows of squares	Number of squares
1-by-1			
2-by-2			
3-by-3			

B. Omar started this pattern for the 2-by-2 squares in one row. Use colour to copy and continue Omar's pattern on an 8-by-8 grid. Count and record the number of 2-by-2 squares in a row.

C. Kaitlyn started this pattern for the 2-by-2 squares in one column. Use colour to copy and continue Kaitlyn's pattern on an 8-by-8 grid. Count and record the number of 2-by-2 squares in a column.

D. How many 2-by-2 squares are on the checkerboard? Explain how you know. Complete the table for 2-by-2 squares.

E. Repeat steps B to D for 3-by-3 squares, and then for 4-by-4 squares.

F. Write pattern rules to describe the patterns in your table.

G. Use your pattern rules. Calculate the total number of squares on the checkerboard.

H. Predict the total number of squares on each size of grid. Show your work.

    a) 10-by-10 grid   **b)** 15-by-15 grid

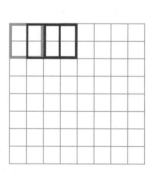

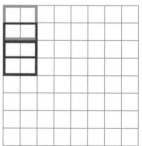

Do You Remember?

1. What is the 20th term in each sequence?

   a) 4, 9, 14, 19, …

   b) 5, 8, 6, 9, 7, 10, 8, …

2. Copy and complete each table of values. Describe a pattern rule that relates the term values to the term numbers.

a)

Term number	Term value
1	4
2	8
3	12
4	16
5	
6	
7	

b)

Term number	Term value
1	5
2	8
3	11
4	14
5	
6	
7	

3. a) In a table of values, record the number of pattern blocks you would need to build each of the first five figures.

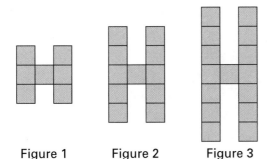

   Figure 1   Figure 2   Figure 3

   b) Describe the pattern rule in words.

   c) Predict the number of pattern blocks you would need to build the 20th figure in the pattern.

4. Use = or ≠ to make each statement true.

   a) 12×3 3×12

   b) $9 \times 7 + 3$ $7 \times 9 + 3$

   c) $19 - 6 \times 3$ $(19 - 6) \times 3$

   d) $23 + 8 \times 2$ $8 + 23 \times 2$

   e) $2 \times (5 + 6)$ $2 \times 5 + 2 \times 6$

5. Determine each missing number.

   a) $4 + \blacksquare = 11$ **d)** $45 \div \blacksquare = 5$

   b) $8 = 15 - \blacksquare$ **e)** $9 = \blacksquare \div 8$

   c) $7 \times \blacksquare = 42$ **f)** $\blacksquare - 6 = 9$

6. The following table of values shows Amanda's time in the 100 m race at a track-and-field meet.

Time (s)	Distance (m)
1	5
2	10
3	15
4	20

   a) Describe the pattern rule that shows the relationship between time and distance.

   b) Make a scatter plot using the ordered pairs in the table of values. Use the horizontal axis to show time and the vertical axis to show distance.

   c) How far will Amanda run in 10 s if her pace continues? Explain your thinking.

Exploring Pattern Representations

▶ **GOAL**

Explore different ways of describing a pattern.

You will need
- grid paper
- a ruler
- a calculator
- triangle dot paper
- coloured pencils
- square dot paper

Explore the Math

Both Colin and Kaitlyn made the same growing pattern of triangles. Then they used tables, rules, and graphs to represent the pattern. Both students were correct, but they were surprised to find out that their representations were different. They decided to find as many ways as they could to represent the pattern.

? **How can you represent this triangle pattern in different ways?**

  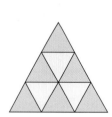 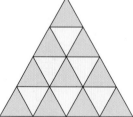

A. Look for different patterns. For example, find changes for these:
- number of rows
- number of small triangles
- number of yellow triangles
- number of green triangles
- lengths of sides

B. Choose one of the patterns. Make a table of values for the pattern.

C. Describe the pattern rule using words and numbers.

D. Make a scatter plot for the pattern. Use your graph to name the 10th term in the pattern.

E. Repeat steps B, C, and D for at least three other patterns.

F. Create your own pattern with squares, and represent it in different ways.

Reflecting

1. Which representation (the table of values, the pattern rule, or the scatter plot) would you use to determine the 57th term of the triangle pattern? Explain your reasoning.

2. Which representation for your square pattern would you find most helpful to show how the term values grow? Explain.

3. What are some advantages of each representation for describing a pattern?

4. What are some disadvantages of each representation for describing a pattern?

Mental Math

SUBTRACTING DECIMALS IN PARTS

You can mentally subtract hundredths from a whole number by subtracting in parts.

Example 1:

$6.00 - 0.65$
$= 6.00 - 0.60 - 0.05$
$\quad\quad 5.40$
$\quad\quad\quad\quad 5.35$

Example 2:

$6.00 - 4.65$
$= 6.00 - 4.00 - 0.60 - 0.05$
$\quad\quad 2.00$
$\quad\quad\quad\quad 1.40$
$\quad\quad\quad\quad\quad\quad 1.35$

1. Show how to add in parts to check the answer.

2. What number adds to 0.65 to make 1? What number adds to 1 to make 6?

3. Subtract in parts.

   a) $7.00 - 0.65$ **c)** $3.00 - 0.45$ **e)** $9.00 - 1.35$ **g)** $50.00 - 2.99$

   b) $6.00 - 0.35$ **d)** $5.00 - 0.85$ **f)** $10.00 - 5.25$ **h)** $100.00 - 3.95$

Using Variables to Write Pattern Rules

▶ **GOAL**

Use numbers and variables to represent mathematical relationships.

Learn about the Math

Rana described the pattern rule for her pattern as "Start with 2 blue rhombuses and place 1 on top. Then add 1 more block on top each time."

? **How can you use symbols to predict the number of pattern blocks needed to build any figure in this pattern?**

A. Copy and complete this table.

Figure number	Number of pattern blocks
1	3
2	4
3	

B. Write a word phrase to describe the relationship between the figure number and the number of blocks you would need to build the figure.

C. Use your word phrase to predict the number of blocks needed to build figure 5. Build the figure to check.

D. Use the **variable** n to represent the figure number. Write your word phrase from step B as an **algebraic expression** with the variable n. Explain how you know that your algebraic expression is correct.

variable

a letter or symbol, such as a, b, or x, that represents a number

Reflecting

1. a) Why is the variable n a good choice to represent the figure number in the expression you wrote in step D?

  b) What other choices could you have used for the variable in the expression? Give reasons for your suggestions.

2. How is writing the pattern rule in words like writing it as an algebraic expression? How are these two representations different?

3. What are the advantages of using an algebraic expression to represent a pattern rule, compared with describing the pattern in words?

algebraic expression

a combination of one or more variables; it may include numbers and operation signs; for example, $2 \times d + 5$ is an algebraic expression that could represent two times the figure number plus five

---

Communication Tip

When variables represent quantities that are multiplied together, they are often written without a multiplication sign between them; for example, the expression $2 \times a$ can be written as $2a$.

Work with the Math

Example 1: Using an algebraic expression to describe a pattern

Write an algebraic expression for the number of blocks you would need to build the figures in this pattern.

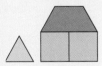
Figure 1

Figure 2

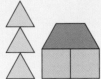
Figure 3

Omar's Solution

My algebraic expression is $t + 3$.

The barn shape does not change, so you always need three pieces to make it.

The number of triangles needed to build each tree is the same as the figure number in the pattern.

I used the variable t to represent the figure number.

Example 2: Using different expressions to describe the same pattern

Write an algebraic expression for the number of squares you would need to build the figures in this pattern.

Tynessa's Solution

My algebraic expression is $n + 2$.

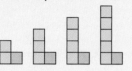

I coloured the two green squares on the bottom blue.

I used the variable n to represent the figure number.

The number of green squares is the same as the figure number.

Colin's Solution

My algebraic expression is $(n + 1) + 1$.

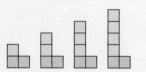

I coloured the vertical squares green, and the one remaining square blue.

I noticed that the number of green squares is always one more than the figure number.

Ⓐ Checking

4. Use words to describe the pattern rule for the total number of blocks in each figure below. Then use an algebraic expression to describe the pattern rule.

Figure 1   Figure 2   Figure 3

5. a) What stays the same and what changes in this pattern?

b) Describe the pattern rule in words.

c) Write an algebraic expression that describes the pattern.

Ⓑ Practising

6. a) Copy and complete the table of values for this triangle pattern.

Figure number	Number of triangles
1	4
2	5
3	
4	
5	

b) Describe how the number of triangles is related to the figure number.

c) Write an algebraic expression for the number of triangles. Use a variable to represent the figure number.

7. Anne, Sanjay, and Robert each wrote an algebraic expression for the pattern of squares. Explain each student's reasoning.

Anne: "My pattern rule is $n + 1 + n$."

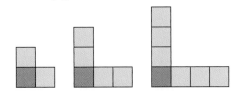

Sanjay: "My pattern rule is $n + (n + 1)$."

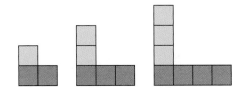

Robert: "My pattern rule is $2n + 1$."

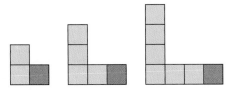

8. The two sets of diagrams below represent the same pattern rule in two different ways.

Kyle's pattern

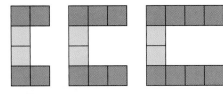

Meagan's pattern

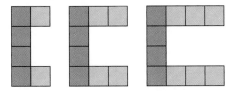

a) What stays the same and what changes in each representation of the rule?

b) Write the algebraic expression that describes Kyle's pattern.

c) Write the algebraic expression that describes Meagan's pattern.

9. Use each description to create an algebraic expression for the pattern rule.

a) The number of blocks increases by six each time.

b) The number of blocks doubles each time.

c) The number of blocks is the sum of the figure number and 10.

10. Each algebraic expression represents a pattern rule. Draw examples that show possible figures for the first three terms of each pattern.

a) $n + 2$ **c)** $3x + 1$

b) $4s$ **d)** $2b + 1$

11. Two people are looking at the same pattern of blue squares. They write different algebraic expressions to describe the pattern rule.

a) Use diagrams and words to show how this is possible.

b) Write the algebraic expression for each diagram.

12. Why is it helpful to use a variable, rather than just numbers, to describe a growing pattern?

Ⓒ Extending

13. Create a pattern for which each algebraic expression might be a pattern rule.

a) n^2 **b)** $2(n + 1)$

14. Explain whether you could build a figure with 257 squares using each of the patterns.

a) the pattern in question 7

b) the pattern in question 8

8.3 Creating and Evaluating Expressions

▶ **GOAL**

Translate statements into algebraic expressions, and evaluate the expressions.

Learn about the Math

Colin is helping to organize a school trip to Ottawa. He needs to figure out the cost of transportation and choose the least expensive bus company. A total of 260 students are going on the trip. He has prices from three bus companies for the cost of seven buses:

- CarryAll Bus Company charges $350 per bus.
- School Bus Transport charges $7 per student.
- Zim Transport charges $500 to cover the cost for the drivers plus $6 per student.

? **Which bus company should Colin choose?**

First Colin calculated the cost of using the CarryAll Bus Company.

$350 \times 7 = \$2450$

Then he calculated the cost of using School Bus Transport and Zim Transport.

Communication Tip

- Use brackets to show when you have substituted a number for a variable. This will help prevent errors caused by accidentally running numbers together. For example, to evaluate the expression $2a$, when $a = 10$, write $2(10)$.
- Write each step in a calculation directly under the previous step, and line up the equal signs one under the other. This makes the calculation easier to read and check later. For example,

$$2a + 5$$
$$= 2(10) + 5$$
$$= 20 + 5$$
$$= 25$$

Example 1: Evaluating an algebraic expression with one step

Create an algebraic expression to represent School Bus Transport's cost for any number of students. Evaluate the expression for 260 students.

Colin's Solution

My expression for School Bus Transport's cost:

$$7s$$

$$= 7(260)$$
$$= 1820$$

School Bus Transport would charge $1820 for the buses.

I used the variable s to represent the number of students going on the trip.

I wrote an expression to represent multiplying s by $7 (the charge per student).

I substituted 260 for s because 260 is the number of students going on the trip. Then I multiplied 7 by 260.

Example 2: Evaluating an algebraic expression with several steps

Create an algebraic expression to represent Zim Transport's cost for any number of students. Evaluate the expression for 260 students.

Colin's Solution

My expression for Zim Transport's cost:

$$500 + 6s$$
$$= 500 + 6(260)$$
$$= 500 + 1560$$
$$= 2060$$

Zim Transport would charge $2060 for the buses.

I used the variable s to represent the number of students going on the trip.

I wrote an expression to represent multiplying s by $6 (the charge per student) and then adding $500 (the cost of the drivers).

I substituted 260 for s because 260 is the number of students going on the trip.

I used the order of operations to evaluate the expression.

Colin chose School Bus Transport because it is the least expensive.

Reflecting

1. Why do you think Colin did not use a table of values to find the cost for School Bus Transport or Zim Transport?

2. Why do you think Colin did not use a scatter plot to find the cost for School Bus Transport or Zim Transport?

3. Explain how to create and evaluate an algebraic expression.

Variables, Expressions, and Equations **279**

Work with the Math

Example 3: Creating and evaluating an algebraic expression

A cafeteria charges $250 plus $5 per meal after the first 140 students.
Create an algebraic expression to calculate the cafeteria bill for 260 students.
Evaluate the expression.

Rana's Solution

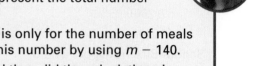

$250 + 5(m - 140)$

$= 250 + 5((260) - 140)$

$= 250 + 5(120)$

$= 250 + 600$

$= 850$

The cafeteria bill is $850.

I used the variable m to represent the total number of meals served.

The charge of $5 per meal is only for the number of meals greater than 140. I found this number by using $m - 140$.

I substituted 260 for m, and then did the calculations in the correct order.

A Checking

4. A bowl of chili costs $4. Which expression represents the cost of buying chili for b people?

   a) $7b - 4$ **b)** $b + 4$ **c)** $4b$

5. Evaluate each expression when $d = 5$.

   a) $6d$ **c)** $d + 1$

   b) $5d - 1$ **d)** $3(d + 2)$

6. Write an algebraic expression to represent the cost of renting a sleigh for $12 per hour plus $35.

B Practising

7. Asha skated the length of a frozen canal t times, except the last time she stopped 2 km from the end. The canal is 7 km long.

   a) Write an algebraic expression that represents how far Asha skated in kilometres.

   b) Suppose that $t = 4$. Use your expression to calculate the distance Asha skated.

8. Evaluate each algebraic expression when $a = 3$ and $b = 5$.

   a) $3a$ **e)** $3a + 2$

   b) $8b$ **f)** $4b - 6$

   c) $9a$ **g)** $5b + 7$

   d) $2(b - 1)$ **h)** $5(a + 5)$

9. Evaluate $6(b - 1) + 3$ when $b = 4$. Show and explain all the steps.

10. Write an algebraic expression for each cost.

    a) $4 a pair for skate sharpening

    b) hamburgers at $3 per person

    c) $2 per hour plus $5 for renting skates

    d) hats on sale for $10 each

    e) cost of a pizza shared equally by four students

11. Samantha works in the snack bar at a community centre. She earns $8 an hour. On her last day, she is paid a bonus of $50.

    a) Choose a variable to represent the number of hours Samantha works.

    b) Write an algebraic expression that describes Samantha's earnings.

    c) Use your expression to calculate how much Samantha would earn if she works 15 h and receives the bonus. Show your work.

12. Jerry sells toques at a kiosk. He is paid $25 a day plus $2 for each toque he sells.

    a) What part of his salary never changes, no matter how many toques he sells?

    b) Write an algebraic expression that describes Jerry's daily salary.

    c) Use your algebraic expression to calculate how much Jerry will earn in one day if he sells 17 toques.

13. Winnie can use up to 10 coupons when buying a box of DVDs. Each coupon is worth $3. She bought a box of DVDs for $56 less the value of her coupons.

    a) Choose a variable to represent the number of coupons.

    b) Write an algebraic expression that describes the amount Winnie paid for the box of DVDs.

    c) Use your expression to calculate how much Winnie would pay if she had 5 coupons.

14. In some parts of question 5, you needed to use the order of operations. Show how you could have made a mistake if you had not used the correct order of operations.

15. Create a problem, based on a real-life situation, that can be represented using the algebraic expression $2x + 6$. Solve your problem.

C Extending

16. Write an algebraic expression for each description.

    a) the cost of pizza at $4 per student added to the cost of drinks at $2 per student plus $200 for the bus

    b) four times the combined number of apples and pears

    c) the cost of hot chocolate for six people plus the cost of muffins for eight people

17. Popcorn and drinks are sold at the school's movie night. Popcorn is $0.75 per bag, and drinks are $1.25 each.

    a) Write an algebraic expression that represents the total amount of money received by selling popcorn and drinks.

    b) Predict which of the three movie nights received the most money for selling popcorn and drinks.

Movie night	Number of bags of popcorn sold	Number of drinks sold
in November	103	76
in February	70	85
in April	68	75

    c) Calculate the amounts received to check your prediction.

Mid-Chapter Review

Frequently Asked Questions

Q: What is a variable?

A: A variable is a quantity that varies, or changes. In the algebraic expression 8*h*, the symbol *h* is a variable. In the example 8*h*, the variable might be number of hours worked, and the expression might represent the number of hours worked at $8 per hour.

Q: How do you choose a symbol for a variable?

A: You can choose any letter as a symbol for a variable. The variable is often the first letter of the word the variable represents. This makes the meaning of the symbol easy to remember. In the expression 2(*l* + *w*), for example, *l* stands for length and *w* stands for width.

To avoid confusion with *x*, which is a symbol commonly used in algebra, the multiplication sign is sometimes not written in algebraic expressions describing a situation. So instead of writing $3 \times n$, we write 3*n*.

Q: How do you evaluate an algebraic expression?

A: First you substitute a number for the variable. Then you use the order of operations to calculate the answer. For example, to evaluate the expression 2*n* + 1 when *n* = 4:

$$2n + 1$$
$$= 2(4) + 1$$
$$= 8 + 1$$
$$= 9$$

To prevent copying and arithmetic errors, use brackets to show the substitution, and write each step below the step before it.

Q: How can you represent a pattern rule given in words as an algebraic expression?

A: You use one or more variables, and possibly numbers and operations signs. For example, the rule for this pattern can be described as "Start with one red square and one column of three green squares, and add another column of green squares each time." An algebraic expression for this is 3*n* + 1, where *n* represents the figure number.

You can check by substituting the figure number into the expression and calculating the total number of squares.

Practice Questions

(8.2) **1. a)** Copy and complete the table of values for this pattern.

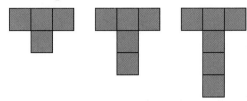

Figure number	Number of squares
1	4
2	5
3	
4	
5	

b) Use words to describe the pattern rule for the relationship between the figure number and the number of squares.

c) Write your pattern rule as an algebraic expression.

d) What did you use as the variable in your algebraic expression?

e) What does the variable represent?

f) Which value in your algebraic expression does not change?

(8.2) **2.** The two sets of diagrams show two different ways of representing the same pattern rule. Write an algebraic expression that represents each set of diagrams.

a)

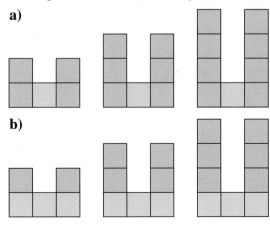

b)

3. How many toothpicks would you need to make the 10th figure in the following pattern? Show three different ways of solving this problem. (8.2)

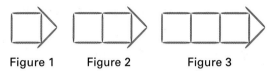

Figure 1   Figure 2   Figure 3

4. Match each phrase with the correct algebraic expression. (8.3)

a) a number increased by seven   **A.** $7n$

b) a number decreased by seven   **B.** $n + 5$

c) a number multiplied by seven   **C.** $n - 7$

d) five subtracted from a number   **D.** $n + 7$

e) the sum of five and a number   **E.** $n - 5$

5. Use the algebraic expressions $5n$ and $n + 5$. (8.3)

a) Explain why these expressions represent different calculations.

b) Suppose that you select a whole number and use it to evaluate each expression. Which expression is more likely to have a greater result? Why?

6. Write an algebraic expression for each. (8.3)

a) Angela's age five years from now

b) Raji's height increased by eleven centimetres

c) four times the figure number

d) the number of books decreased by nine

7. Explain how to evaluate the expression $4b + 3$ when $b = 8$. Show your work. (8.3)

8.4 Solving Equations by Inspection

▶ **GOAL**
Write equations and solve them by inspection.

Learn about the Math

John and Heidi are building this toothpick pattern.
They have 28 toothpicks.

Figure 1   Figure 2   Figure 3

? **How many squares can they build for this pattern with 28 toothpicks?**

A. Copy and complete the first four rows of the table of values for the toothpick pattern. Write the pattern rule for the relationship between the number of squares and the number of toothpicks.

Number of squares	Number of toothpicks
1	4
2	
3	
4	

B. Use the variable s to represent the number of squares. Write an algebraic expression that describes the relationship in step A.

C. Complete the **equation** $t = $ ▓, where t represents the number of toothpicks you need, and ▓ is your algebraic expression from step B.

D. When the number of toothpicks is 28, the value of t in your equation is 28. Rewrite your equation with 28 substituted for t.

E. Part of your equation should be ▓s. Explain how you can figure out the value of ▓s in your equation. Then explain how you can figure out the value of s.

F. Check your **solution to the equation** by building the squares.

Reflecting

1. Why would continuing the table of values not be an efficient strategy for solving this problem?

equation

a mathematical statement in which the value on the left side of the equal sign is the same as the value on the right side of the equal sign; for example, $6d + 1 = 13$ is an equation for 6 times d plus 1 equals 13

solution to an equation

the value of a variable in an equation that makes the equation true

2. The method you used to solve the equation in step E is called "solving by inspection." The word "inspect" means "examine carefully." Explain how your solution strategy used a careful examination of the equation.

Work with the Math

Example 1: Solving an equation by inspection

Use the method of inspection to solve the equation $2a - 1 = 13$.

Kaitlyn's Solution

$$2a - 1 = 13$$
$$2a = 14$$
$$a = 7$$

Check.

Left side	Right side
$2a - 1$	13
$= 2(7) - 1$	
$= 13 \checkmark$	

I looked at the equation.
Since you subtract 1 from $2a$ to get 13, $2a$ must be 14.
Since 2 times a number is 14, the number must be 7.

I checked my solution by substituting it into the equation to see if it works.
My solution $a = 7$ works, so it is correct.

Example 2: Solving a problem by solving an equation

The cost to rent skates is $3 plus $2 per hour. Omar has $21. What is the greatest number of hours he can rent the skates?

Omar's Solution

My equation for the rental cost is

$$c = 3 + 2h$$
$$21 = 3 + 2h$$

$$18 = 2h$$
$$9 = h$$

Check.

Left side	Right side
21	$3 + 2h$
	$= 3 + 2(9)$
	$= 3 + 18$
	$= 21 \checkmark$

I used h to represent the number of hours I'll rent the skates and c to represent the cost.

I want to find out how many hours I can rent the skates with $21, so I substituted $c = 21$ into my equation.

What number plus 3 gives 21? $2h$ must be 18.

What number multiplied by 2 gives 18? h must be 9.

I checked to make sure that this solution works by substituting it into the equation.

My solution $h = 9$ is correct.

I can rent the skates for 9 h with $21.

Variables, Expressions, and Equations **285**

A Checking

3. Use inspection to solve the equations. Justify your reasoning, and check your solutions.

   a) $n + 6 = 13$

   b) $w - 11 = 22$

   c) $9p = 63$

   d) $2n + 3 = 15$

   e) $7r - 1 = 20$

4. Rana wrote the expression $4n + 3$ to describe the pattern below, where n represents the figure number.

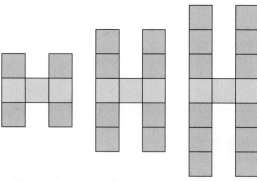

   Figure 1   Figure 2   Figure 3

   a) Write an equation that describes the number of squares (s) needed to build each figure in this pattern.

   b) Rana built the figure in this pattern that has 23 squares. Write the equation that you must solve to determine the figure number.

   c) Solve your equation.

   d) Check your solution by using the equation.

   e) Check by drawing the figure and counting the squares.

B Practising

5. Solve each equation by inspection.

   a) $7b = 84$ **d)** $11 = q - 4$

   b) $8 + z = 30$ **e)** $2w + 1 = 17$

   c) $22 = m + 2$ **f)** $9n - 4 = 32$

6. a) Copy the following solution. Explain each step.

$$6 + 5m = 16$$
$$5m = 10$$
$$m = 2$$

   b) Explain how you can check to make sure the solution $m = 2$ is correct.

7. Can you build a figure that belongs in Rana's pattern in question 4 using 39 squares? Explain.

8. a) Write an equation that describes the number of squares needed to build each figure in this pattern.

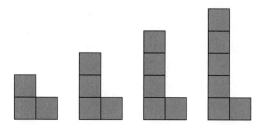

   b) Suppose that you want to build a figure in this pattern using 24 squares. Write the equation you must solve to determine the figure number.

   c) Solve your equation.

   d) Check your solution by using the equation, and by drawing the figure and counting the squares.

   e) Use your equation to determine which figure number you could build using 21 squares. Describe this figure.

9. a) Suppose that you want to build the figure in this pattern that uses 28 pattern blocks. Write an equation you can solve to determine the figure number.

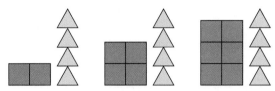

b) Solve your equation, and check your solution.

10. a) Suppose that you want to build the figure in this pattern that uses 30 squares. Write an equation you can solve to determine the figure number.

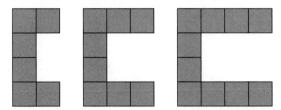

b) Solve your equation, and check your solution.

11. To rent a movie, a company charges $2 plus $1 per day.

a) Write an equation that describes the cost (c) of renting a movie in terms of the number of days (d).

b) Write an equation that represents the following question: "How many days can you rent a movie for $9?"

c) Solve your equation.

d) Check that your solution solves the problem.

12. Susan has $25. She is going to spend $4 on a book, then $3 per day on lunch.

a) Write an equation that represents the following question: "How many days can Susan buy lunch with this plan?"

b) Solve your equation, and check your solution.

13. Kevin and Zach are playing a number guessing game. Kevin says, "I am thinking of a number. If you double it and then subtract 1, the result is 7."

a) Write an equation that Zach could solve to find Kevin's number.

b) Explain the steps Zach could use to solve the equation.

C Extending

14. The sum of three consecutive whole numbers is 33. Create an algebraic equation, and solve to determine the numbers.

15. What is the greatest number of squares you can build with 100 toothpicks using this pattern?

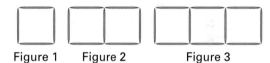

Figure 1   Figure 2   Figure 3

16. What is the greatest number of triangles you can build with 100 toothpicks using this pattern?

Figure 1   Figure 2   Figure 3

Variables, Expressions, and Equations **287**

Solving Equations by Systematic Trial

▶ **GOAL**

Write equations and solve them using systematic guessing and testing.

Learn about the Math

Tonya and Becky work at a provincial park. Tonya's job is to plant trees around groups of picnic tables to provide a windbreak. She uses a pattern like the one shown. Becky assembles picnic tables from kits and puts them in position. Tonya has 30 trees to plant. Becky wonders how many picnic-table kits she should get from storage.

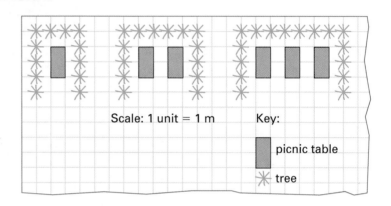

Scale: 1 unit = 1 m   Key:

picnic table

✳ tree

? **How many picnic tables can be sheltered by 30 trees?**

Example 1: Using a table of values to solve a problem

Extend the pattern of trees around picnic tables using a table of values.

Rana's Solution

Number of tables	Number of trees
1	12
2	14
3	16
4	18
5	20
6	22
7	24
8	26
9	28
10	30

I set up a table of values to record the information in the drawing of the trees around the picnic tables.

From my table of values, I noticed that the number of trees goes up by 2 from one row to the next.

I used this pattern to fill in the rest of my table.

Ten picnic tables can be sheltered by 30 trees.

Example 2: Using systematic trial to create and solve an equation

Write an equation to figure out how many picnic tables can be sheltered by 30 trees. Solve the equation.

Tynessa's Solution

Predict p.	Evaluate $2p + 10$.	Is it equal to 30?
7	$2(7) + 10$ $= 14 + 10$ $= 24$	No, it's too low.
12	$2(12) + 10$ $= 24 + 10$ $= 34$	Now it's too high.
10	$2(10) + 10$ $= 20 + 10$ $= 30$	Yes!

I used p for picnic tables. Then I created the expression $2p + 10$ to represent the number of trees in terms of the number of picnic tables. There are 30 trees, so I substituted 30 for t in my equation $2p + 10 = t$ and wrote $2p + 10 = 30$.

I started with a guess of 7 and changed my prediction if the result was too high or too low.

The solution shows that you can shelter 10 picnic tables using 30 trees.

Reflecting

1. Who do you think used the more efficient strategy? Explain your answer.

2. After you solve an equation using inspection, you should check your answer. Why is this step not necessary when using systematic trial?

3. Describe how systematic trial requires both an organized approach to the problem and estimation skills.

Work with the Math

Example 3: Using systematic trial to solve an equation

Use systematic trial to solve $6c - 4 = 92$.

Kaitlyn's Solution

Predict c.	Evaluate $6c - 4$.	Is it equal to 92?
20	$6(20) - 4$ $= 120 - 4$ $= 116$	No, it's too high.
15	$6(15) - 4$ $= 90 - 4$ $= 86$	Now it's too low, but I'm getting closer.
16	$6(16) - 4$ $= 96 - 4$ $= 92$	I've solved it!

I guessed possible values for the variable and evaluated the expression. I kept track of my guessing and testing until I solved the equation.

A Checking

4. Which solution is correct? Explain your thinking.

  a) $4r = 16$ $r = 4$ or $r = 64$

  b) $p - 9 = 15$ $p = 6$ or $p = 24$

  c) $n + 12 = 20$ $n = 8$ or $n = 32$

  d) $5w - 2 = 73$ $w = 15$ or $w = 75$

5. Copy each table, and use it to solve the equation. In the second column, show your work. In the third column, write "too high," "too low," or "correct."

  a) $n - 12 = 122$

Predict n.	Evaluate $n - 12$.	Is this the correct solution?
144		
124		
134		

  b) $15b = 315$

Predict b.	Evaluate $15b$.	Is this the correct solution?
15		
21		
25		

  c) $4x + 9 = 73$

Predict x.	Evaluate $4x + 9$.	Is this the correct solution?
10		
15		
16		

6. Use systematic trial to solve the equation $3p + 7 = 82$.

  a) What value would you try first? Explain your choice.

  b) Solve the equation.

B Practising

7. a) Write an equation that relates the number of black counters to the total number of counters used to make a figure in this pattern.

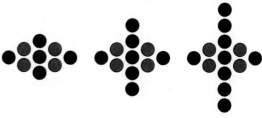

Figure 1   Figure 2   Figure 3

  b) Write an equation that represents the following question: "How many black counters would you need to make a figure with 73 counters?"

  c) Use systematic trial to solve your equation.

  d) Would you prefer to use systematic trial or inspection to solve your equation? Justify your choice.

8. Use systematic trial to find the value of each variable.

  a) $p - 8 = 33$

  b) $97 + h = 201$

  c) $9r = 792$

  d) $5 + 4c = 105$

  e) $78 = 3y + 3$

9. Write each sentence as an equation. Solve each equation.

  a) The sum of a number and 19 is 55.

  b) Eight times a number is 192.

  c) Nine times a number subtract 26 is 136.

10. The sum of two consecutive whole numbers, n and $n + 1$, is given by the equation $n + (n + 1) = 171$. What are the two numbers?

11. a) Build this pattern with coloured counters.

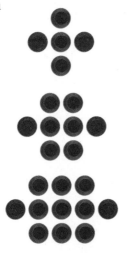

b) Write and solve an equation to answer the following question: "If you use a total of 32 counters, how many of them will be blue?"

c) Is systematic trial an efficient method for solving your equation? Explain your thinking.

12. How do you decide when systematic trial will be more efficient than inspection for solving an equation?

13. The formula for the area of a parallelogram is $A = bh$.

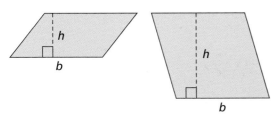

a) Determine the value of b, in centimetres, if A is 52 cm² and h is 4 cm. (*Hint*: Solve $52 = 4b$.)

b) Determine the value of h, in centimetres, if A is 171 cm² and b is 9 cm.

c) Determine the value of h, in metres, if b is 12 m and A is 156 m².

d) Determine the value of b, in metres, if h is 11 m and A is 121 m².

14. A scientist measured the temperature of a hot liquid using a thermometer that was marked in degrees Kelvin. The temperature was 403 K. Change the temperature to degrees Celsius by solving the equation $K = C + 273$. (*Hint*: Solve the equation $403 = C + 273$.)

❸ Extending

15. Rosa and her brother worked last summer weeding their neighbour's garden. Rosa earned $8 more than twice the amount earned by her brother. If their earnings totalled $71, how much did Rosa earn? Use an algebraic equation and systematic trial to solve this problem.

16. Which of the following relationships does the equation $2(10 + w) = 28$ describe? Explain your choice.

A. The area of a rectangle is 28 square units.

B. The length of a rectangle is 12 times its width.

C. Twice the length of a rectangle is 28 units.

D. The perimeter of a rectangle is 28 units.

17. Solve the equation in question 16 to find the width of the rectangle. Show your thinking.

18. Create a real-life problem based on the equation $28n + 75 = 215$. (*Hint:* You might consider 28 people in a group, or 28 days in February, or $28 as the cost of something.) Show how you would use systematic trial to solve the equation.

19. Look back at question 10. Is it possible for two consecutive whole numbers to have each sum below? Use equations to explain your answer.

a) an odd number   **b)** an even number

Variables, Expressions, and Equations **291**

Communicating the Solution for an Equation

You will need
• red and blue counters

▶ **GOAL**

Communicate about solving equations using correct mathematical language.

Communicate about the Math

Omar and Tynessa are trying to figure out the number of counters in the container on this pan balance.

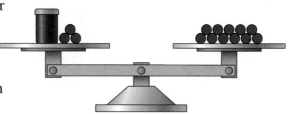

Omar says that he knows how many counters are in the container. He tries to show his thinking using algebra. Tynessa asks him to explain what he means.

? **How can Omar improve his explanation?**

Omar's Explanation	Tynessa's Questions
$c + 3 - 3 = 11 - 3 = 8$ ◀——	I see where $c + 3$ comes from, but how do you get 8?
$c = 8$	
I subtracted 3 from 11. ◀——	Why did you subtract?
There are 8 red counters in the container. ◀——	How do you know your solution is correct?

A. Would you have asked any of the same questions? Explain why or why not.

B. Which parts of Omar's solution and explanation were not clear?

C. What other questions would you ask about Omar's explanation?

D. Write your explanation. (You can use any method that makes sense to you.) Make sure that Tynessa's questions and your questions from step C are answered.

Communication Checklist

- ☑ Did you show each step of your thinking?
- ☑ Did you express yourself clearly?
- ☑ Were you convincing?

Reflecting

1. Which parts of the Communication Checklist did Omar cover well? Explain.

2. How could Omar have made his solution easier to understand?

3. Why is it important to show your steps in a logical order when solving an algebraic equation?

Work with the Math

Example: Explaining the solution for an equation

Omar wants to help Mei make her algebraic solutions easier to understand. He shows his solution to this balance problem.

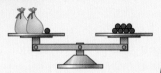

Omar's Solution

n represents the number of counters in each bag.

$$2n + 1 = 7$$
$$2n + 1 - 1 = 7 - 1$$
$$2n = 6$$
$$n = 3$$

Check.

Left side	Right side
$2n + 1$	7
$= 2(3) + 1$	
$= 6 + 1$	
$= 7 ✓$	

There are 3 counters in each bag.

I wrote the balance problem as an algebraic equation using the variable n.

When I take away 1 counter from each side, both sides still balance.

I solved for n.

I made sure that there was only one equal sign in each step.

I checked my solution by substituting the answer for the variable and checking that it made the equation true.

A Checking

4. Explain what is happening in each step of Tynessa's solution.

$$2c = 12$$
$$2c \div 2 = 12 \div 2$$
$$c = 6$$

Check.

Left side	Right side
$2c$	12
$= 2(6)$	
$= 12 ✓$	

Tynessa said, "When I remove one of the groups from each side, both sides still balance."

ⓑ Practising

Use the Communication Checklist to help you write clear solutions.

5. Explain what is happening in each step of the following solution.

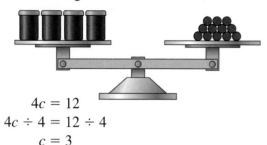

$$4c = 12$$
$$4c \div 4 = 12 \div 4$$
$$c = 3$$

Check.

Left side	Right side
$4c$	12
$= 4(3)$	
$= 12 \checkmark$	

6. Use pictures and words to explain each step in the solution to this balance problem.

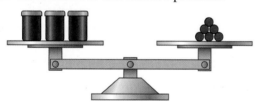

7. Explain why the steps in this solution for the equation $a + 4 = 13$ are confusing. Correct the steps.

$$a + 4 = 13 = 13 - 4 = 9$$

8. Improve Tynessa's solution and explanation.

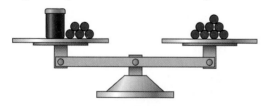

$$x + 5 = 8$$
$$x = 8 - 5$$
$$x = 3$$

"I subtracted 5 because of the zero principle, so $8 - 5$ equals 3."

9. Find and explain any errors in each solution to this balance problem. Rewrite each solution correctly, and check your work.

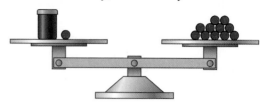

a) Mei's method using inspection:

What number added to 1 gives 11?

$$11 + 1 = 12$$

b) Tynessa's method using operations that balance:

$$n + 1 = 11$$
$$n + 1 - 1 = 11 - 1$$
$$n + 0 = 10$$

10. Use words, numbers, and diagrams to explain how you would solve this balance problem.

Step 1:

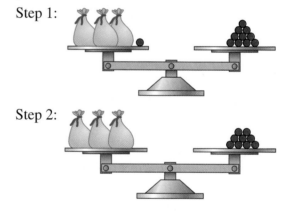

Step 2:

11. Write a clear algebraic solution for each balance problem. Check that your solution makes the equation true.

a) $4x = 8$

b) $x + 4 = 9$

c) $2x + 5 = 21$

d) $x - 2 = 9$

Math Game

This is a matching game. The goal is to find the pairs of matching Alge-cards before your partner does. Each time you make a match, you try to win the cards by correctly solving the equation on the cards.

You will need
• Alge-cards

Number of players: 2 or 3

Rules

1. Place your Alge-cards face down in rows on the table.

2. Take turns turning over a pair of cards.

3. If the code numbers do not match, place both cards face down in their original positions. If the code numbers do match, try to solve the equation shown on the cards.

4. If you solve the equation correctly, you take both cards. If you do not solve the equation correctly, your partner takes the pair of cards.

5. The game ends when all the cards are taken. The player with the most cards is the winner.

USING ALGEBRA TO SOLVE NUMBER TRICKS

Use blocks and counters to model algebraic expressions.

A. Try this trick.	**Get the picture.**	**Here's the algebra!**
Think of a one-digit number.		n
Multiply by 6.		$6n$
Add 10.		$6n + 10$
Divide by 2.		
		$3n + 5$
Subtract 5.		$3n$
Divide by 3.		n

What's the answer? Always your starting number!

B. Try this trick.	**Get the picture.**	**Here's the algebra!**
Think of a one-digit number.		a
Multiply by 6.		$6a$
Add 10.		$6a + 10$
Subtract the original number.		$5a + 10$
Divide by 5.		
		$a + 2$
Subtract the original number.		2

What's the answer? Always 2!

1. Finish the trick below so that the answer is always the original number.

Try this trick.	**Get the picture.**	**Here's the algebra!**
Think of a one-digit number.		y
Multiply by 10.		$10y$
Add 10.		$10y + 10$
Divide by .		
Subtract .		

What's the answer?

2. Make up a trick of your own. Use pictures and algebra to show how it works. Try it with a classmate.

1. Explain how these two algebraic expressions can represent the same pattern rule.

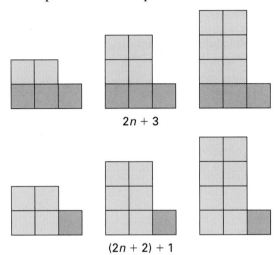

2n + 3

(2n + 2) + 1

2. Katya wants to build a figure in the following pattern. She has 61 toothpicks. What is the figure number of the figure that requires all her toothpicks? Use an algebraic expression to solve this problem.

3. Write an algebraic expression for each phrase.

a) a number doubled

b) six subtracted from five times a number

c) four times a number plus nine

4. At a bookstore, paperbacks sell for $8 each, including tax.

a) Write an algebraic expression for the cost of *n* paperbacks.

b) A membership in the store's Book Club costs $10. Members can buy paperbacks for only $6 each. Write an algebraic expression for the cost of paperbacks for a Book Club member, including the cost of a membership.

c) What is the difference between the cost of 12 books purchased with and without a membership? Show your work.

5. Substitute for the variable and evaluate the algebraic expression $9a + 7$ when $a = 5$. Show your work.

6. Solve the equation $2x - 1 = 9$ by inspection.

7. Solve the equation $29n + 43 = 391$ using systematic trial. Explain each step.

8. Determine the figure number that you could draw using 25 squares in this pattern. Show your work.

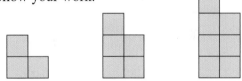

9. a) Write an equation for this balance problem.

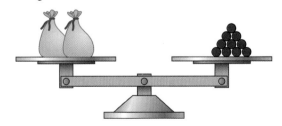

b) Solve your equation.

c) Explain your solution.

Chapter Review

Frequently Asked Questions

Q: **How are an algebraic expression and an algebraic equation different?**

A:

An algebraic expression... Examples: $3n$ $b + 4$ $2p - 7$	An algebraic equation... Examples: $3n = 6$ $b + 4 = 13$ $2p - 7 = 37$
... is like a word phrase	... is like a word sentence
... may contain one or more operation signs but does not have an equal sign	... may contain one or more operation signs and does have an equal sign
... can be evaluated by substituting a number for each variable, then calculating	... can be solved by determining the value of the variable that makes the equation true, or the left side equals the right side

Q: **How do you solve an equation by inspection?**

A: Solve for the variable by thinking about what steps are suggested by the equation. For example, solve $3b + 5 = 26$ by inspection.

- Since you add 5 to $3b$ to get 26, $3b$ must be 21.
- Use your mental math skills to solve for the variable. Since 3 times a number is 21, the number must be 7.
- Check your answer by substituting the value you got for the variable into the original equation. Both sides of the equation should have the same value. If not, try again.

$$3b + 5 = 26$$
$$3b = 21$$
$$b = 7$$

Check.

Left side   Right side
$3b + 5$ 26
$= 3(7) + 5$
$= 21 + 5$
$= 26$ ✓

Q: **How do you use systematic trial to solve an equation?**

A: Follow these steps to solve the equation $2p - 7 = 37$.

- Use your estimating skills to guess the value of the variable.
- Test your prediction by substituting for the variable.
- Calculate to see if your solution makes the equation true. If not, use the value of your last guess to help you make another guess.

Predict p.	Evaluate $2p - 7$.	Is this the correct solution?
30	$2(30) - 7$ $= 60 - 7$ $= 53$	No, it's way too high.
25	$2(25) - 7$ $= 50 - 7$ $= 43$	I'm getting closer.
22	$2(22) - 7$ $= 44 - 7$ $= 37$	Correct!

Practice Questions

(8.2) **1. a)** Make a table of values for this toothpick pattern.

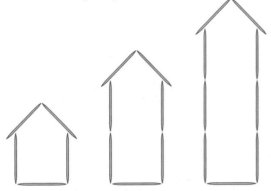

b) Explain the relationship between the figure number and the number of toothpicks in each figure.

c) Write an algebraic expression to represent the pattern rule.

(8.3) **2.** Write each phrase as an algebraic expression.

a) a number decreased by 38

b) the sum of 83 and a number

c) four times a number

d) 79 added to twice a number

(8.4) **3. a)** Use the equation $4 + n = c$ to determine the number of the figure you could build with 19 counters in this pattern.

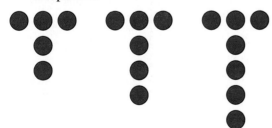

b) How may counters do you need to build the 21st figure in the pattern?

4. Solve each equation by inspection. Explain your thinking. (8.4)

a) $3n = 33$

b) $9 + w = 10$

c) $r - 15 = 30$

d) $2y + 3 = 11$

5. Solve each equation by systematic trial. (8.5)

a) $x - 68 = 192$

b) $g + 33 = 251$

c) $3z - 16 = 74$

d) $15h + 9 = 84$

6. Match each description with the correct equation. Then solve for the variable in each equation. (8.5)

a) When four is added to a number, the sum is twenty.

b) When a number is multiplied by four, the product is forty.

c) When four is subtracted from a number, the difference is forty.

A. $n - 4 = 40$

B. $n + 4 = 20$

C. $4n = 40$

7. Which solution is correct? Explain. (8.5)

a) $7n = 84$ $n = 12$ or $n = 77$

b) $k - 9 = 79$ $k = 70$ or $k = 88$

c) $p + 14 = 40$ $p = 26$ or $p = 54$

d) $4w + 4 = 52$ $w = 12$ or $w = 13$

8. Solve the equation $83 = 7m - 1$. (8.5)

9. Describe one advantage of each strategy for solving the equation $5y - 20 = 20$. (8.5)

a) inspection

b) systematic trial

Letters, Patterns, and Algebra

Kaitlyn and her classmates are creating patterns based on letters in their names. Their teacher has challenged them to "find the algebra" in their patterns.

? **What algebra can you find in a letter pattern?**

A. Use linking cubes or coloured square tiles to build one of the letters in your name. (This will be the figure in your pattern.)

B. Make up a pattern rule. Use your pattern rule to build the next two figures in a pattern that starts with the figure you built in step A.

C. Write a sentence that describes the relationship between the figure number and the number of cubes or squares in each figure.

D. Use diagrams with different colours to show which parts of your pattern change with every figure and which parts stay the same.

E. Create an algebraic expression that represents your pattern rule.

F. Use a different method to represent your pattern rule.

G. Write a question about your pattern rule that can be answered by solving an equation. Write the equation. Solve the equation to answer your question.

Task Checklist

☑ Did you use models?

☑ Did you include diagrams and a table of values?

☑ Did you explain your pattern using algebra and words?

☑ Did you write a clear algebraic solution for your equation?

Math in Action

Business

A business can make money by selling a product or providing a service. Here you will learn about businesses that sell products.

The selling price of a product is more than it costs the owner of the business. Generally, an item is priced using the following relationship:

Selling price = Cost + Markup

$$S = C + M$$

price paid by customer   price paid by seller   amount cost is increased

Problems, Applications, and Decision Making

The owner of Yeung's Supermarket buys grocery items for his store and then sells them to his customers.

1. The following table lists some of the items found in Yeung's Supermarket. Determine the missing amounts in the table.

Item	Selling price ($)	Cost ($)	Markup ($)
bag of cookies		1.98	0.98
box of crackers	2.58		1.09
butter	3.29	1.68	
loaf of bread	2.49		1.15
dozen eggs		1.05	0.73
flour	7.59	3.29	

2. Mr. Yeung buys a cash register from a local supplier for $739.55. If the markup on the cash register is $273.00, what was the cost of the cash register for the seller?

3. Mr. Yeung needs a new refrigeration unit. Determine the markup on the unit if the selling price is $1357.35 and the cost is $846.29.

4. Mr. Yeung buys tins of tuna in cases of 24 tins. He pays $20.16 for a case. If the markup on each tin of tuna is $0.12, how much do you pay for a tin of tuna?

Variables, Expressions, and Equations   **301**

In the following questions, you will look at other businesses that sell products.

5. A store buys shirts for $14.99 each. The selling price of the shirts is twice the cost for the store.

a) Write an equation for the selling price of a shirt.

b) What is the selling price?

c) What is the markup on each shirt?

6. A store buys jackets at a cost of $49.99. The markup on the jackets is $18.49. During a sale, the jackets are sold for $57.99. What is the markup on the jackets during the sale?

7. Your family wants to buy a used car. The asking price of the car is $4799. The dealer originally bought the car for $3800. For the dealer to make a profit, she must have a markup of at least $500 when the sale is complete.

a) If the dealer wants to make a profit, what is the lowest price your family can pay for the car?

b) If the dealer wants to make a profit, by how much can your family decrease the asking price?

Advanced Applications

The markup on products is supposed to cover all the expenses of a business and also earn the owner some money. Thus, markup can be thought of as expenses plus profit.

8. Rewrite the relationship for selling price using expenses and profit in place of markup.

9. A door is purchased by a lumber store for $184. The owner of the store calculates $75 in expenses for the door and wants to make a profit of $39 when the door is sold. Determine the selling price of the door.

10. A customer buys a bookshelf for $49.48. The bookshelf costs the store $21.98, and the estimated expenses are $7.00. What is the store's profit on the bookshelf?

CHAPTER 9

Fraction Operations

▶ GOALS

You will be able to

- add and subtract fractions using models, drawings, and symbols
- multiply a fraction by a whole number using models, drawings, and symbols
- estimate sums and differences of fractions

You will need
- pattern blocks
- triangle dot paper

Missing Measuring Cups

Susan is making her grandmother's brownie recipe. The recipe uses different units of measure: cups (c.) and teaspoons (tsp.).

Brownies

1/3 c. soft butter or margarine	1 1/3 c. flour
1 c. brown sugar	1/2 c. cocoa
2 eggs	3/4 tsp. baking powder
1/4 c. milk	1/2 tsp. salt
1/2 tsp. vanilla	2/3 c. chopped nuts, optional

Cream butter and sugar.
Add the eggs, milk, and vanilla.
Mix together the dry ingredients, and
add to the creamed ingredients. Mix well.
Pour into a greased square pan.
Bake at 350°F for 25 to 30 minutes.

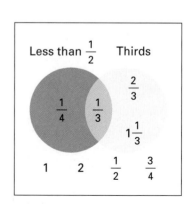

? **How do you measure ingredients if some of your measuring cups are missing?**

A. Susan can only find three measuring cups: $\frac{1}{4}$ c., $\frac{1}{3}$ c., and $\frac{2}{3}$ c. How can she measure $\frac{1}{2}$ c. of cocoa?

B. Which measuring cups can Susan use to measure $1\frac{1}{3}$ c. of flour? Explain.

C. How can knowing that $1\frac{1}{3} = \frac{4}{3}$ help Susan measure the flour?

D. Which ingredients can Susan measure using the $\frac{1}{4}$ cup? Explain.

E. The cocoa and vanilla measurements both start with the fraction $\frac{1}{2}$. Are the amounts the same? Explain your thinking.

F. One way to sort the fractions in the recipe is shown. Show two other ways to sort the same fractions.

G. How can sorting fractions be helpful when measuring cups are missing? Explain.

Less than $\frac{1}{2}$ Thirds

$\frac{2}{3}$

$\frac{1}{4}$ $\frac{1}{3}$

$1\frac{1}{3}$

1   2 $\frac{1}{2}$ $\frac{3}{4}$

Do You Remember?

1. Copy and complete the table of equivalent percents and fractions.

	Percent	Fraction
a)	10%	
b)		$\dfrac{1}{5}$
c)	25%	
d)		$\dfrac{2}{5}$
e)	50%	
f)		$\dfrac{3}{4}$

2. Write the letter that shows each fraction on the number line.

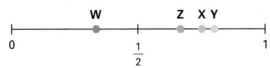

a) $\dfrac{2}{3}$ **b)** $\dfrac{3}{4}$ **c)** $\dfrac{1}{3}$ **d)** $\dfrac{4}{5}$

3. Draw a picture to show each fraction or **mixed number**.

a) $2\dfrac{1}{2}$ **b)** $1\dfrac{2}{3}$ **c)** $\dfrac{3}{5}$ **d)** $\dfrac{9}{4}$

4. Use the pictures. Replace each ▢ below with >, <, or =.

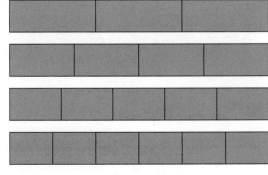

a) $\dfrac{1}{3}$ ▢ $\dfrac{2}{6}$ **c)** $\dfrac{4}{5}$ ▢ $\dfrac{2}{3}$

b) $\dfrac{3}{5}$ ▢ $\dfrac{3}{4}$ **d)** $\dfrac{2}{4}$ ▢ $\dfrac{4}{6}$

5. a) If a hexagon is a whole, what fraction does each pattern block represent?

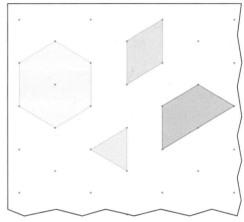

b) Which pattern block(s) would you use to model each fraction?

i) $\dfrac{5}{6}$ **ii)** $\dfrac{1}{3}$ **iii)** $1\dfrac{1}{2}$

c) Which fraction in part (b) is closest to 1?

6. a) Rewrite each **improper fraction** as a mixed number.

i) $\dfrac{3}{2}$ **ii)** $\dfrac{5}{3}$ **iii)** $\dfrac{8}{3}$ **iv)** $\dfrac{15}{12}$

b) Which fraction in part (a) is the greatest?

7. a) Rewrite each mixed number as an improper fraction.

i) $1\dfrac{3}{5}$ **ii)** $4\dfrac{2}{3}$ **iii)** $5\dfrac{3}{4}$ **iv)** $2\dfrac{5}{9}$

b) Which fraction in part (a) is the least?

8. a) Write each fraction in **lowest terms**, and draw it.

i) $\dfrac{20}{30}$ **ii)** $\dfrac{8}{10}$ **iii)** $\dfrac{15}{24}$ **iv)** $\dfrac{16}{12}$

b) Write two more equivalent fractions for each fraction in part (a).

9. List the first five **multiples** of each number.

a) 4 **b)** 5 **c)** 6 **d)** 8 **e)** 12

Adding Fractions with Pattern Blocks

You will need
- pattern blocks
- large triangle dot paper

▶ **GOAL**

Add fractions that are less than 1 using concrete materials.

Explore the Math

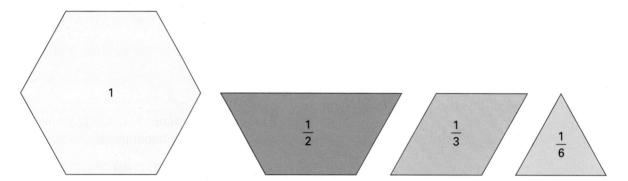

1   $\frac{1}{2}$ $\frac{1}{3}$ $\frac{1}{6}$

Yuki and Ryan are playing a fraction game. Working individually, they cover a hexagon using any of these pattern blocks. Then they write an equation to describe each combination. The player who writes more equations wins.

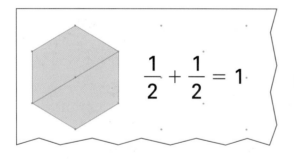

$$\frac{1}{2} + \frac{1}{2} = 1$$

? **How many different combinations of shapes can you use to cover a hexagon?**

A. Play Yuki and Ryan's game with a partner, using the same pattern blocks they used. Draw all your combinations on dot paper.

B. Write an equation to describe each combination.

C. What is the fewest number of blocks you can use to cover the hexagon? Explain how you know that this is the fewest number of blocks.

Communication Tip

You can remember the names for the parts of a fraction by thinking "the denominator is down."

D. What is the greatest number of blocks you can use? Explain how you know that this is the greatest number of blocks.

E. Can you use every number of blocks between the least number and greatest number to cover the hexagon? Explain.

F. Look at your combinations and your partner's combinations. How many different combinations do you have? Have you used all the possible combinations to cover the hexagon? Explain how you know.

Reflecting

1. Yuki put together two trapezoids and wrote $\frac{1}{2} + \frac{1}{2} = 1$. Explain why she could have written $\frac{1}{2} + \frac{1}{2} = \frac{2}{2}$.

2. Ryan wrote two different equations, like Yuki's, that involved adding fractions with the same denominator. What were these equations?

Mental Math

MULTIPLYING A DECIMAL CLOSE TO A WHOLE NUMBER

You can multiply decimals close to a whole number by multiplying and then subtracting.

To multiply 2×0.9, think of 2 groups of 1s instead of 2 groups of 9 tenths.

Calculate $2 \times 1 = 2$. Then subtract 2 tenths or 0.2.

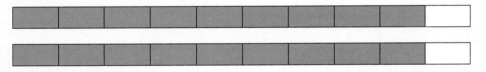

$$2 \times 0.9 = (2 \times 1) - (2 \times 0.1)$$
$$= 2 - 0.2$$
$$= 1.8$$

1. Why do you subtract 0.2?

2. Suppose that you used this strategy to multiply 4×2.98. What would you subtract after multiplying 4×3? Explain your thinking.

3. Multiply.

a) 4×0.9	**c)** 3×1.9	**e)** 8×2.9	**g)** 3×1.99
b) 7×0.8	**d)** 5×1.8	**f)** 6×3.9	**h)** 5×2.98

9.2 Adding Fractions with Models

You will need
- fraction strips
- number lines

▶ **GOAL**

Add fractions that are less than 1 using fraction strips and number lines.

Learn about the Math

Sandra is reading a mystery novel. Last weekend, she read $\frac{1}{3}$ of the book. Yesterday, she read $\frac{1}{4}$ more of the book.

? What fraction of the book has Sandra read?

You can use fraction strips or number lines to make models of fractions. A fraction strip shows rectangles that are the same size. The whole length of each fraction strip should be the same, no matter what fraction the strip represents.

The $\frac{1}{3}$ strip and the $\frac{1}{4}$ strip show different denominators.

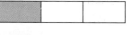

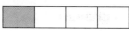

A number line is like a thin fraction strip.

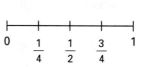

Example 1: Estimating using fraction strips

Use fraction strips to estimate $\frac{1}{3} + \frac{1}{4}$.

Sandra's Solution

I divided each fraction strip into the number of parts shown in the denominator. Then I coloured the number of parts shown in the numerator.

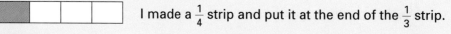

I made a $\frac{1}{3}$ strip.

I made a $\frac{1}{4}$ strip and put it at the end of the $\frac{1}{3}$ strip.

I made a $\frac{1}{2}$ strip to compare with $\frac{1}{4} + \frac{1}{3}$. It looks like the sum is a bit more than $\frac{1}{2}$.

Example 2: Adding using fraction strips

Use fraction strips to add $\frac{1}{3} + \frac{1}{4}$.

Ravi's Solution

If my fraction strips had the same number of parts, I could count the parts in the sum. Both the $\frac{1}{3}$ strip and the $\frac{1}{4}$ strip can be made into twelfths. 12 is a **common denominator** for $\frac{1}{3}$ and $\frac{1}{4}$ because 12 is a common multiple of 3 and 4.

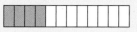 The $\frac{1}{3}$ strip becomes $\frac{4}{12}$ because $\frac{1}{3} = \frac{1 \times 4}{3 \times 4}$, which is $\frac{4}{12}$.

 The $\frac{1}{4}$ strip becomes $\frac{3}{12}$ because $\frac{1}{4} = \frac{1 \times 3}{4 \times 3}$, which is $\frac{3}{12}$.

I added $\frac{4}{12} + \frac{3}{12}$ to get $\frac{7}{12}$.

Example 3: Adding using a number line

Use a number line to add $\frac{1}{3} + \frac{1}{4}$.

Chang's Solution

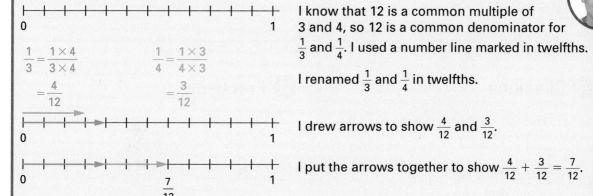

I know that 12 is a common multiple of 3 and 4, so 12 is a common denominator for $\frac{1}{3}$ and $\frac{1}{4}$. I used a number line marked in twelfths.

$\frac{1}{3} = \frac{1 \times 4}{3 \times 4}$ $\frac{1}{4} = \frac{1 \times 3}{4 \times 3}$

$= \frac{4}{12}$ $= \frac{3}{12}$

I renamed $\frac{1}{3}$ and $\frac{1}{4}$ in twelfths.

I drew arrows to show $\frac{4}{12}$ and $\frac{3}{12}$.

I put the arrows together to show $\frac{4}{12} + \frac{3}{12} = \frac{7}{12}$.

Reflecting

1. In Example 1, how did Sandra know that the answer was a bit more than $\frac{1}{2}$?

2. In Example 3, how could Chang use a number line to estimate that the answer must be more than $\frac{1}{4}$ but less than $\frac{2}{3}$?

3. Explain how using a common denominator helped Ravi and Chang add fractions using models.

Example 4: Estimating and adding using models

Estimate and then add $\frac{1}{3} + \frac{2}{5}$.

Solution A

Estimate. $\frac{1}{3} + \frac{2}{5}$ looks like about $\frac{3}{4}$.

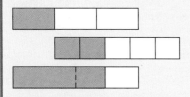

To add, first find the common denominator.

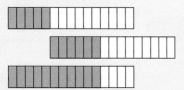

$$\frac{1}{3} = \frac{5}{15}$$

$$\frac{2}{5} = \frac{6}{15}$$

$$\frac{5}{15} + \frac{6}{15} = \frac{11}{15}$$

Solution B

Use a number line marked in fifteenths, since 15 is a common denominator for $\frac{1}{3}$ and $\frac{2}{5}$.

$$\frac{1}{3} = \frac{5}{15} \qquad \frac{2}{5} = \frac{6}{15}$$

0   1

Put the arrows together to model addition.

0                                    $\frac{11}{15}$ 1

$$\frac{1}{3} + \frac{2}{5} = \frac{5}{15} + \frac{6}{15}$$

$$= \frac{11}{15}$$

A Checking

4. a) How do you know that the sum of $\frac{3}{4}$ and $\frac{1}{6}$ is less than 1?

b) Use fraction strips to add $\frac{3}{4} + \frac{1}{6}$. Show your work.

5. Use a number line to add $\frac{2}{5} + \frac{7}{10}$. Show your work.

6. a) Describe how you would use fraction strips or a number line to estimate the sum for $\frac{1}{5} + \frac{1}{4}$.

b) Describe how you would use fraction strips or a number line to add $\frac{1}{5} + \frac{1}{4}$.

B Practising

7. Use fraction strips to estimate and then add. Show your work.

a) $\frac{2}{3} + \frac{1}{3}$ **d)** $\frac{2}{3} + \frac{1}{2}$

b) $\frac{1}{4} + \frac{1}{2}$ **e)** $\frac{2}{3} + \frac{3}{5}$

c) $\frac{1}{8} + \frac{1}{4}$ **f)** $\frac{5}{6} + \frac{3}{4}$

8. Use a number line to add. Show your work.

a) $\frac{3}{5} + \frac{1}{4}$ **d)** $\frac{1}{3} + \frac{4}{5}$

b) $\frac{2}{3} + \frac{1}{6}$ **e)** $\frac{5}{6} + \frac{1}{3}$

c) $\frac{1}{6} + \frac{1}{4}$ **f)** $\frac{5}{6} + \frac{1}{4}$

Use fraction strips or a number line to model each addition in questions 9 to 18.

9. Determine each sum.

a) $\dfrac{1}{8} + \dfrac{3}{4}$ b) $\dfrac{4}{5} + \dfrac{1}{2}$

10. The recipe for a cheese sauce requires $\dfrac{1}{3}$ c. of flour at the beginning and another $\dfrac{1}{8}$ c. of flour later. How much flour is required?

11. In a Grade 7 class, $\dfrac{1}{5}$ of the students have two pets and $\dfrac{1}{20}$ have three pets.

a) Estimate the fraction of the class that has either two or three pets.

b) Calculate the fraction of the class that has either two or three pets.

c) How many students do you think are in the class? Why?

12. a) Rewrite $\dfrac{1}{5}$ and $\dfrac{1}{20}$, from question 11, as percents. Add the percents.

b) How does your answer for part (a) relate to your answer for question 11, part (b)?

13. In the fall of 2003, the population of Ontario was about 39% of the population of Canada. The population of the western provinces was about $\dfrac{3}{10}$ of the population of Canada.

a) What fraction of Canadians live in Ontario and the western provinces?

b) What percent of Canadians live in Ontario and the western provinces?

14. Jane watched one television program for $\dfrac{1}{4}$ of an hour and then changed channels to watch another program for 20 min. Write an equation to describe the fraction of an hour that Jane watched television.

15. What denominators make this equation true? Write four possible answers.

$$\dfrac{1}{\blacksquare} + \dfrac{2}{\blacksquare} = \dfrac{3}{\blacksquare}$$

C Extending

16. Yan has three measuring cups filled with sugar.

a) Can Yan empty all three measuring cups into a 1 c. measuring cup? Explain.

b) How much sugar does he have in total?

17. a) Add each pair of fractions. Describe the model you used, and look for a pattern in the sums.

i) $\dfrac{1}{3} + \dfrac{1}{4}$ iii) $\dfrac{1}{5} + \dfrac{1}{6}$

ii) $\dfrac{1}{4} + \dfrac{1}{5}$ iv) $\dfrac{1}{6} + \dfrac{1}{7}$

b) Describe a rule for adding fractions in the form $\dfrac{1}{\blacksquare} + \dfrac{1}{\blacksquare}$. Justify your rule.

18. a) Copy and complete the table. Describe a pattern in the sums.

$\dfrac{1}{2} + \dfrac{1}{4} =$	$\dfrac{3}{4}$
$\dfrac{1}{3} + \dfrac{1}{6} =$	$\dfrac{3}{\blacksquare}$
$\dfrac{1}{4} + \dfrac{1}{8} =$	$\dfrac{3}{\blacksquare}$
$\blacksquare + \blacksquare =$	$\dfrac{3}{10}$

b) Use the pattern to predict the answer to $\dfrac{1}{20} + \dfrac{1}{40}$.

9.3 Multiplying a Whole Number by a Fraction

▶ **GOAL**

Use repeated addition to multiply fractions by whole numbers.

Learn about the Math

Leah is having a party. After a couple of hours, she notices that six lemonade pitchers are each only $\frac{3}{8}$ full. She decides to combine the leftovers to use fewer pitchers.

? **How many pitchers will be filled with the left-over lemonade?**

A. Sketch the pitchers. Estimate how many pitchers you think the lemonade will fill completely. Explain your thinking.

B. Use grid paper. Draw six 4-by-2 rectangles. Put counters on three of the eight squares in each rectangle to model the six $\frac{3}{8}$ full pitchers.

C. How many squares, in total, are covered by counters? If a full pitcher represents $\frac{8}{8}$, what improper fraction represents the total amount of lemonade?

D. Describe how you could move the counters to create as many full pitchers as possible. What part of a full pitcher would be left?

E. Use fraction strips or a number line to find the total amount of lemonade remaining.

Reflecting

1. How could you have predicted the numerator of your improper fraction in step C? Explain.

2. How could you have predicted that the amount left in the last pitcher would be a fraction with a denominator of 8?

3. Why could you write either $\frac{3}{8} + \frac{3}{8} + \frac{3}{8} + \frac{3}{8} + \frac{3}{8} + \frac{3}{8}$ or $6 \times \frac{3}{8}$ to describe the total amount of lemonade in the pitchers?

Work with the Math

Example 1: Multiplying a fraction by a whole number using grid paper

Multiply $4 \times \frac{5}{6}$ using grids and counters.

Chang's Solution

$4 \times \frac{5}{6}$ is 4 sets of $\frac{5}{6}$.

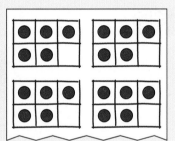

$$4 \times \frac{5}{6} = \frac{4 \times 5}{6}$$
$$= \frac{20}{6}$$

I used 3-by-2 rectangles, since I want to show sixths and $3 \times 2 = 6$. (I could have used 6-by-1 rectangles instead.)

I showed four sets of $\frac{5}{6}$ by putting counters on 5 out of 6 squares in each of the 4 rectangles.

$4 \times 5 = 20$ squares are covered.

Since each square represents $\frac{1}{6}$, the 20 covered squares represent $\frac{20}{6}$.

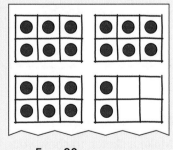

$$4 \times \frac{5}{6} = \frac{20}{6}$$
$$= 3\frac{2}{6} \text{ or } 3\frac{1}{3}$$

I'll write the improper fraction as a mixed number.

I moved 3 counters from the last rectangle to complete the other 3 rectangles. That means I filled 3 rectangles and there are 2 squares in another rectangle. So, $\frac{20}{6} = 3\frac{2}{6}$.

The fraction $\frac{2}{6}$ can be written in lowest terms as $\frac{1}{3}$.

Example 2: Multiplying a fraction by a whole number using fraction strips

Multiply $3 \times \frac{2}{3}$ using fraction strips.

Yuki's Solution

$3 \times \frac{2}{3}$ is 3 sets of $\frac{2}{3}$.

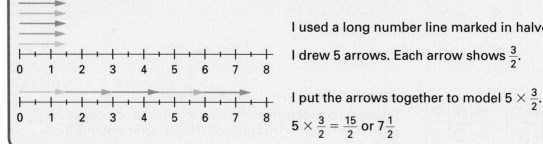

I made three $\frac{2}{3}$ strips.

I put the strips together.

$3 \times \frac{2}{3} = \frac{6}{3}$ or 2

Example 3: Multiplying a fraction by a whole number using a number line

Multiply $5 \times \frac{3}{2}$ using a number line.

Ryan's Solution

$5 \times \frac{3}{2}$ is 5 sets of $\frac{3}{2}$.

I used a long number line marked in halves.

I drew 5 arrows. Each arrow shows $\frac{3}{2}$.

I put the arrows together to model $5 \times \frac{3}{2}$.

$5 \times \frac{3}{2} = \frac{15}{2}$ or $7\frac{1}{2}$

Ⓐ Checking

4. Jennifer pours $\frac{2}{3}$ of a cup of water into a pot and repeats this seven times. How many full cups of water, in total, does she pour into the pot? Write your answer as a mixed number.

5. a) Multiply $4 \times \frac{5}{12}$ using a grid and counters. Make a sketch to show your work.

b) Write your answer as an improper fraction and as a mixed number.

6. a) Write $5 \times \frac{3}{4}$ as a repeated addition sentence.

b) Use fraction strips or a number line to calculate the answer.

c) Write your answer as an improper fraction and as a mixed number.

Ⓑ Practising

7. Multiply using grids and counters. Show your work.

a) $2 \times \frac{1}{3}$ **c)** $6 \times \frac{3}{8}$ **e)** $3 \times \frac{6}{7}$

b) $5 \times \frac{3}{5}$ **d)** $4 \times \frac{2}{5}$ **f)** $8 \times \frac{4}{2}$

8. Write as a repeated addition. Use fraction strips or a number line to calculate each answer.

a) $2 \times \frac{1}{3}$ **c)** $4 \times \frac{5}{2}$ **e)** $6 \times \frac{3}{5}$

b) $2 \times \frac{5}{4}$ **d)** $5 \times \frac{1}{6}$ **f)** $7 \times \frac{7}{6}$

9. Replace each missing value with a single-digit number to make the sentence true.

a) $4 \times \dfrac{1}{\blacksquare} = 2$

b) $5 \times \dfrac{\blacksquare}{9} = 4\dfrac{\blacksquare}{9}$

c) $\blacksquare \times \dfrac{4}{5} = 4\dfrac{4}{5}$

d) $\blacksquare \times \dfrac{\blacksquare}{8} = 1\dfrac{7}{8}$

10. Art class is $\frac{5}{6}$ of an hour each school day. How many hours of art does a student have in five days?

11. Kevin needs $\frac{2}{3}$ c. of sugar to make his favourite brownie recipe. How many cups of sugar does he need to make six batches of brownies for a bake sale?

12. Katya says that multiplying $17 \times \frac{1}{4}$ will tell her how many dollars that 17 quarters is worth. Do you agree? Explain.

13. a) Multiply $4 \times \frac{3}{5}$.

b) Rewrite $\frac{3}{5}$ as a percent, and multiply by 4.

c) Explain how you can use the calculation in part (b) to check your answer to part (a).

14. At a party, Raj notices that 15 pitchers of lemonade are filled to the same level, but not to the top. He combines all the lemonade to fill six whole pitchers. What fraction of each of the 15 pitchers was full?

15. A whole number multiplied by $\frac{3}{5}$ is 9. What is the number?

Ⓒ Extending

16. $\blacksquare \times \frac{5}{8}$ can be written as a whole number. What possible numbers can replace $\blacksquare$? Explain.

17. a) Use fraction strips divided into thirds to represent $6 \times \frac{2}{3}$.

b) Use fraction strips to represent $\frac{2}{3}$ of 6.

c) Explain why $\frac{2}{3}$ of 6 is the same as $6 \times \frac{2}{3}$.

18. a) $2 \times \frac{4}{5}$ means the same as doubling $\frac{4}{5}$. Explain why.

b) What would $\frac{1}{2} \times \frac{4}{5}$ mean?

c) How can you calculate the answer for $\frac{1}{2} \times \frac{4}{5}$?

d) How can you check whether your answer for $\frac{1}{2} \times \frac{4}{5}$ is correct?

9.4 Subtracting Fractions with Models

You will need
• fraction strips
• number lines

▶ **GOAL**

Subtract fractions less than 1 using fraction strips and number lines.

Learn about the Math

Yuki notices that $\frac{3}{4}$ of the houses on Fox Street have garages and $\frac{1}{6}$ have sheds.

? How many more of the houses have garages than sheds?

A. Model $\frac{3}{4}$ and $\frac{1}{6}$ using fraction strips.

B. Determine the common denominator of $\frac{3}{4}$ and $\frac{1}{6}$. Redraw your models with this denominator.

C. Look at the two new strips. How many parts of the new $\frac{3}{4}$ strip represent the difference between $\frac{3}{4}$ and $\frac{1}{6}$?

D. How many more houses have garages than sheds?

Reflecting

1. How could you have predicted that the answer in step D would be greater than $\frac{2}{4}$?

2. What equivalent fractions did you use for $\frac{3}{4}$ and $\frac{1}{6}$ in step B?

3. How did using a common denominator help you make the fraction strips in step B?

4. Explain how you can use fraction strips and a common denominator to subtract fractions.

Work with the Math

Example 1: Estimating and subtracting using fraction strips

Estimate and then subtract $\frac{2}{3} - \frac{1}{5}$ using fraction strips.

Ryan's Solution

I used fraction strips to show both fractions.

Then I figured out the difference between the longer coloured part and the shorter coloured part.

It looks like the difference is about $\frac{1}{2}$.

---

15 is a common denominator of $\frac{2}{3}$ and $\frac{1}{5}$ because 15 is a common multiple of 3 and 5. So, I divided the fraction strips into fifteenths.

The $\frac{2}{3}$ strip becomes $\frac{10}{15}$ because $\frac{2}{3} = \frac{2 \times 5}{3 \times 5}$, which is $\frac{10}{15}$.

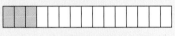

The $\frac{1}{5}$ strip becomes $\frac{3}{15}$ because $\frac{1}{5} = \frac{1 \times 3}{5 \times 3}$, which is $\frac{3}{15}$.

The difference is $\frac{10}{15} - \frac{3}{15} = \frac{7}{15}$.

Example 2: Subtracting using a number line

Subtract $\frac{5}{6} - \frac{1}{2}$ using a number line.

Yuki's Solution

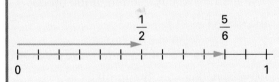

I used a number line showing twelfths, since 12 is a common denominator for $\frac{5}{6}$ and $\frac{1}{2}$.

$$\frac{5}{6} = \frac{5 \times 2}{6 \times 2} \qquad \frac{1}{2} = \frac{1 \times 6}{2 \times 6}$$

$$= \frac{10}{12} \qquad\qquad = \frac{6}{12}$$

I drew arrows to show $\frac{5}{6}$ and $\frac{1}{2}$.

I need to find the distance from $\frac{5}{6}$ to $\frac{1}{2}$.

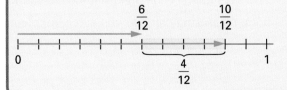

There are four spaces between $\frac{1}{2}$ and $\frac{5}{6}$.

Since each space is $\frac{1}{12}$, then $\frac{5}{6} - \frac{1}{2} = \frac{4}{12}$.

A Checking

5. a) How do you know that the difference for $\frac{4}{5} - \frac{1}{3}$ is about $\frac{1}{2}$?

b) Use fraction strips to subtract $\frac{4}{5} - \frac{1}{3}$.

6. Use a number line to model the difference for $\frac{4}{3} - \frac{1}{2}$. Show your work.

7. Suppose that $\frac{3}{5}$ of the students in your class have pets, and $\frac{1}{6}$ have more than one pet.

a) Use fraction strips or a number line to calculate the fraction of the students with only one pet.

b) How many students have no pets?

B Practising

8. Use fraction strips to estimate and then calculate each difference.

a) $\frac{5}{8} - \frac{1}{4}$ **d)** $\frac{11}{8} - \frac{3}{4}$

b) $\frac{7}{10} - \frac{1}{4}$ **e)** $\frac{3}{10} - \frac{1}{5}$

c) $\frac{7}{6} - \frac{2}{3}$ **f)** $\frac{5}{3} - \frac{5}{4}$

9. Use number lines to calculate each difference.

a) $\frac{3}{5} - \frac{1}{10}$ **d)** $\frac{7}{4} - \frac{2}{3}$

b) $\frac{5}{2} - \frac{3}{4}$ **e)** $\frac{3}{10} - \frac{1}{5}$

c) $\frac{8}{3} - \frac{7}{9}$ **f)** $\frac{9}{4} - \frac{7}{8}$

10. Draw and colour a shape so that $\frac{5}{8}$ is blue and $\frac{1}{3}$ is yellow.

a) What fraction describes how much more is blue than yellow?

b) What fraction describes the part of the shape that is not blue or yellow?

11. a) What fraction of the larger flag is red?

b) What fraction of the smaller flag is red?

c) What fraction more of the smaller flag is red than the larger flag?

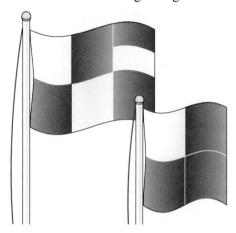

12. At Oakville School, $\frac{2}{10}$ of the students are in Grades 7 and 8, and $\frac{1}{4}$ of the students are in Grades 5 and 6.

a) Are there more Grade 5 and 6 students or Grade 7 and 8 students?

b) What is the difference between the sizes of the two groups of students? Give your answer as a fraction of the whole school.

13. Rosa does $\frac{1}{2}$ of her book report on Tuesday and another 20% of her book report on Wednesday. What fraction of the book report is left to do?

14. Anne needs $\frac{1}{6}$ c. of sugar to make a dessert. Ken says that she should fill a $\frac{1}{2}$ c. measuring cup with sugar and then pour out enough to fill a $\frac{1}{3}$ c. measuring cup. He says that $\frac{1}{6}$ c. of sugar will be left in the $\frac{1}{2}$ c. measuring cup. Do you agree? Explain.

15. Mohammed surveyed students in the school about their favourite activities. The results are shown in the following table. Use the fractions to answer the questions.

Activity	Fraction of students who prefer activity
swimming	$\frac{1}{4}$
tobogganing	$\frac{1}{6}$
skating	$\frac{1}{12}$
soccer	$\frac{1}{3}$

a) What fraction describes how many more students prefer playing soccer to swimming?

b) What fraction describes how many more students prefer swimming to skating?

c) What fraction describes how many more students prefer tobogganing to skating?

16. a) Subtract each pair of fractions. Look for a pattern in the differences.

i) $\frac{1}{3} - \frac{1}{4}$ iii) $\frac{1}{5} - \frac{1}{6}$

ii) $\frac{1}{4} - \frac{1}{5}$ iv) $\frac{1}{6} - \frac{1}{7}$

b) Use the pattern you saw in part (a) to help you calculate $\frac{1}{2} - \frac{1}{3}$.

17. Addition is not the same as subtraction. Yet, some of the steps for adding fractions are the same as the steps for subtracting fractions. Which steps are the same? Why are they the same?

ⓒ Extending

18. To subtract $\frac{4}{3} - \frac{3}{4}$, Ann decides to add $\frac{1}{4}$ to $\frac{1}{3}$.

a) Model $\frac{4}{3}$ and $\frac{3}{4}$ on a number line.

b) Explain Ann's method.

c) What is the difference for $\frac{4}{3} - \frac{3}{4}$?

19. A travel agency is finding volunteers for a travel group. Between $\frac{1}{4}$ and $\frac{1}{2}$ of the group must be between the ages of 15 and 21, and at least $\frac{1}{2}$ of the group must be over the age of 21.

a) What is the least fraction of the group that can be under 15? Show your work.

b) What is the greatest fraction of the group that can be under 15? Show your work.

20. Can the sum of two fractions equal the difference between the same two fractions? Explain.

21. What digit can you put in both boxes to make the following equation true?

$$\frac{\blacksquare}{7} - \frac{2}{\blacksquare} = \frac{11}{21}$$

Mid-Chapter Review

Frequently Asked Questions

Q: Why is adding fractions easy if the denominators are the same?

A: If the denominators are the same, all the pieces are the same size. You can add the numerators to count the pieces.

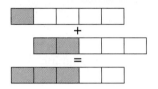

Q: What models are helpful to show adding and subtracting fractions?

A: You can use a hexagon pattern block as 1 and other pattern blocks to show sixths, thirds, and halves.

You can use a rectangle to show any fraction. Make the area of the rectangle the denominator of the fraction.

You can also use a number line or fraction strips to show any fraction. Use a ruler to divide a strip or number line into any number of equal pieces.

Q: Can you add or subtract two fractions if the denominators are different?

A: Expressing fractions as equivalent fractions makes adding or subtracting easier. After you have a common denominator, add or subtract the numerators to get the numerator for your answer. Use the common denominator as the denominator for your answer.

Adding:

$$\frac{3}{5} + \frac{1}{2} = \frac{6}{10} + \frac{5}{10}$$

$$= \frac{11}{10} \text{ or } 1\frac{1}{10}$$

Subtracting:

$$\frac{3}{5} - \frac{1}{2} = \frac{6}{10} - \frac{5}{10}$$

$$= \frac{1}{10}$$

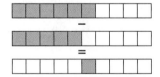

Q: What models can you use to multiply a whole number by a fraction that is less than 1?

A: You can use counters on a grid to model the fraction. Show the amount as many times as the whole number. Move counters around and count how many wholes and what fraction are covered in total. You can also use a number line or fraction strips to show the sets of fractions.

Practice Questions

(9.1) **1.** Use pattern blocks to show each fraction as a sum of other fractions in at least two different ways. Draw the blocks and record the addition sentences for each fraction.

a) $\dfrac{2}{3}$ b) $\dfrac{5}{6}$ c) 1

(9.2) **2.** Use fraction strips to estimate each sum and then to add. Show your work.

a) $\dfrac{5}{20} + \dfrac{1}{4}$ c) $\dfrac{7}{10} + \dfrac{2}{5}$

b) $\dfrac{7}{8} + \dfrac{1}{2}$ d) $\dfrac{2}{3} + \dfrac{1}{8}$

(9.2) **3.** Use a number line to add. Show your work.

a) $\dfrac{3}{4} + \dfrac{1}{5}$ c) $\dfrac{3}{8} + \dfrac{1}{2}$

b) $\dfrac{3}{4} + \dfrac{5}{6}$ d) $\dfrac{1}{4} + \dfrac{2}{3}$

(9.2) **4.** Write two fractions with different denominators that can be added to get the sum of $1\dfrac{1}{2}$. Explain your thinking.

(9.2) **5.** If the missing numbers are consecutive numbers, what are they?

$$\dfrac{A}{B} + \dfrac{C}{8} = \dfrac{7}{8}$$

(9.2) **6.** The area of Ontario is about $\dfrac{1}{7}$ the area of Canada. The area of Quebec is about $\dfrac{1}{5}$ the area of Canada. Approximately how much of the area of Canada is covered by the total area of these two provinces?

(9.3) **7.** Write as a repeated addition. Use fraction strips or a number line to add. Write each answer as an improper fraction and as a mixed number.

a) $6 \times \dfrac{1}{5}$ c) $8 \times \dfrac{3}{5}$

b) $4 \times \dfrac{5}{12}$ d) $5 \times \dfrac{4}{9}$

8. Use grid paper and counters to multiply. (9.3)

a) $3 \times \dfrac{3}{8}$ c) $5 \times \dfrac{5}{6}$

b) $2 \times \dfrac{5}{9}$ d) $4 \times \dfrac{2}{5}$

9. Give two possible values for each of the two missing numbers. Each missing number is different. (9.3)

$$8 \times \dfrac{\blacksquare}{9} = 6 \times \dfrac{\blacksquare}{9}$$

10. Use fraction strips or a number line to subtract. (9.4)

a) $\dfrac{4}{7} - \dfrac{1}{3}$ c) $\dfrac{5}{6} - \dfrac{2}{9}$

b) $\dfrac{6}{5} - \dfrac{3}{4}$ d) $\dfrac{7}{4} - \dfrac{3}{5}$

11. In the Yukon Territory, about $\dfrac{3}{4}$ of the population is between the ages of 15 and 65. About $\dfrac{1}{5}$ of the population is 14 or younger. What fraction describes how many more people are between 15 and 65 than are 14 or younger? (9.4)

12. Create a word problem for each number sentence. Then use a model to calculate the answer. Show your work. (9.4)

a) $\dfrac{1}{3} + \dfrac{1}{10}$ c) $\dfrac{4}{5} - \dfrac{2}{5}$

b) $\dfrac{1}{4} + \dfrac{2}{3}$ d) $7 \times \dfrac{1}{3}$

13. What fraction of April is taken up by non-holiday weekends? (9.4)

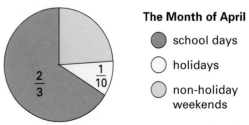

The Month of April

- ● school days
- ○ holidays
- ● non-holiday weekends

Subtracting Fractions with Grids

▶ **GOAL**

Subtract fractions using grids and counters.

You will need
• grid paper
• counters

Learn about the Math

Each school day, Mike is awake for $\frac{2}{3}$ of the time, and he spends $\frac{1}{4}$ of the time at school or on the school bus.

? **What fraction of the day does Mike have left for other activities?**

A. Outline a rectangle on grid paper with 3 rows of 4 to show the common denominator of $\frac{2}{3}$ and $\frac{1}{4}$.

B. How many rows of the rectangle show the fraction of the day Mike is awake? Use counters to cover the number of rows that model the fraction of the day Mike is awake.

C. Move some of the counters you placed in the rectangle so that your counters fill as many columns as possible.

D. How many columns of the rectangle show the fraction of the day Mike spends at school or on the school bus? Take away counters from the number of columns that model the fraction of the day Mike spends at school or on the school bus.

E. What fraction of the outlined rectangle has counters now?

F. Write a subtraction equation to describe your model.

G. What fraction of the day does Mike have left for other activities? Explain how you know.

Reflecting

1. Explain how the rectangle you outlined on grid paper in step A shows the common denominator of $\frac{2}{3}$ and $\frac{1}{4}$.

2. What fraction does each column of the rectangle you outlined represent? What fraction does each row represent?

3. Describe the rectangle you would outline on grid paper to subtract $\frac{3}{4} - \frac{1}{2}$. Explain how you would use counters to model the subtraction.

4. Explain how you decide how many rows and columns to outline on grid paper to show the common denominator for any two fractions.

Work with the Math

Example: Subtracting with grid paper

Subtract $\frac{2}{3} - \frac{2}{5}$ using a grid and counters.

Sandra's Solution

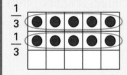

I used a 3-by-5 rectangle, since I want to show thirds and fifths. The rectangle has 15 squares since 15 is the common denominator of $\frac{2}{3}$ and $\frac{2}{5}$.

Each row shows $\frac{1}{3}$. To show $\frac{2}{3}$, I covered 2 rows with counters. That's 10 of the 15 squares, which represents $\frac{10}{15}$.

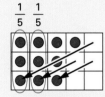

Next I want to subtract $\frac{2}{5}$. Each column shows $\frac{1}{5}$, so I moved the counters to fill as many columns as possible.

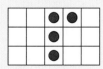

Then I removed 2 complete columns of counters to model subtracting $\frac{2}{5}$. That's 6 counters, which represents $\frac{6}{15}$.

4 of the 15 squares still have counters, so the difference is $\frac{4}{15}$.

$$\frac{2}{3} - \frac{2}{5} = \frac{10}{15} - \frac{6}{15}$$
$$= \frac{4}{15}$$

A Checking

5. Use a grid and counters to model and calculate each difference.

a) $\dfrac{2}{3} - \dfrac{1}{5}$ b) $\dfrac{5}{6} - \dfrac{1}{4}$

6. Susan has $\dfrac{7}{12}$ of a movie left to watch. She watched $\dfrac{1}{3}$ of the movie on Sunday. Use a grid and counters to show how much of the movie Susan still has to watch. Explain your thinking.

B Practising

7. Use a grid and counters to model and calculate each difference.

a) $\dfrac{4}{5} - \dfrac{2}{3}$ d) $\dfrac{7}{8} - \dfrac{2}{3}$

b) $\dfrac{1}{3} - \dfrac{1}{4}$ e) $\dfrac{3}{5} - \dfrac{1}{4}$

c) $\dfrac{1}{3} - \dfrac{2}{7}$ f) $\dfrac{3}{4} - \dfrac{2}{5}$

8. Subtract. Use a grid and counters to model.

a) $\dfrac{2}{3} - \dfrac{1}{2}$ c) $\dfrac{8}{9} - \dfrac{2}{3}$

b) $\dfrac{11}{12} - \dfrac{3}{4}$ d) $\dfrac{6}{7} - \dfrac{1}{3}$

9. Takumi is awake for $\dfrac{5}{8}$ of every Saturday. He spends $\dfrac{5}{24}$ of the day coaching gymnastics. What fraction of his Saturday does he have left for other activities?

10. Leanne puts some of her allowance in the bank to save for a bicycle, so she has $\dfrac{1}{2}$ of her allowance left. At the end of the week, she still has $\dfrac{1}{10}$ of her allowance left. What fraction of her allowance did she spend?

11. On any given day, all the cupcakes in a bakery have either blue, pink, or white icing. Copy and complete the table to determine the fraction with white icing.

	Blue icing	Pink icing	White icing
a)	$\dfrac{1}{4}$	$\dfrac{1}{3}$	
b)	$\dfrac{1}{6}$	$\dfrac{1}{4}$	
c)	$\dfrac{5}{12}$	$\dfrac{1}{6}$	

12. The gauge shows that the gas tank for a car is $\dfrac{3}{4}$ full. Lyle drives the car to the next town. When he looks at the gauge, it reads $\dfrac{1}{8}$ full. What fraction of the tank of gas did he use?

13. Make up your own problem that is similar to question 12. Then solve it.

14. Jeff bought a pie and cut it into equal-sized pieces. Jeff ate 2 pieces, so $\frac{3}{4}$ of the pie is left. What fraction of the pie is each piece?

15. An ice-cream parlour makes $\frac{1}{3}$ of its income from cones and 40% from sundaes. What fraction of its income comes from other items?

16. Explain why subtracting fractions with a common denominator $\left(\text{such as } \frac{3}{5} \text{ and } \frac{1}{5}\right)$ is usually easier than subtracting fractions without a common numerator $\left(\text{such as } \frac{3}{4} \text{ and } \frac{3}{5}\right)$.

17. When Marie arrived late for dinner, $\frac{1}{2}$ of the pan of lasagna had already been eaten. Marie ate $\frac{1}{10}$ of the pan of lasagna. How much lasagna was left in the pan?

18. Suppose that you subtract one fraction between $\frac{1}{2}$ and 1 from another fraction between $\frac{1}{2}$ and 1. Are the following statements true? Explain your thinking.

a) The difference can be less than $\frac{1}{2}$.

b) The difference can be greater than $\frac{1}{4}$.

❻ Extending

19. In the diagram below, the musical notes are described as fractions. The total of the fractions in each measure is 1. What notes can be added to the last measure to complete it?

20. Rachel is saving $100 to buy a CD player. She has saved half of the money.

a) If she earns another $\frac{1}{5}$ of the $100 by babysitting, what fraction does she have left to save?

b) How much money does she still need?

21. To calculate $\frac{7}{8} - \frac{2}{3}$, Anne calculated $1 - \frac{2}{3}$ and then subtracted $\frac{1}{8}$. Why is this correct?

22. Use a pattern and grouping to help you solve the following number sentence.
$$1 - \frac{1}{2} + \frac{3}{4} - \frac{1}{4} + \frac{5}{8} - \frac{1}{8} + \frac{9}{16} - \frac{1}{16} = \frac{\blacksquare}{\blacksquare}$$

23. In music, the time signature is written like a fraction. For example, $\frac{4}{4}$ time means there are 4 beats to each measure.

A quarter note gets one beat.

A half note gets two beats.

A whole note gets four beats.

a) How many beats does an eighth note get?

b) Create a diagram for a simple piece of music in $\frac{4}{4}$ time. For example,

"Are you sleep-ing?
Are you sleep-ing?
Broth-er John!
Broth-er John!
Time for break-fast.
Time for break-fast.
Please come on!
Please come on!"

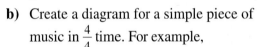

$$\frac{1}{4} + \frac{1}{8} + \frac{1}{8} + \frac{1}{8} + \frac{1}{8} + \frac{1}{16} + \frac{1}{16} + \frac{1}{16} + \frac{1}{16} \qquad \frac{1}{8} + \frac{1}{8} + \frac{1}{8} + \frac{1}{2} + \,?$$

Fraction Operations **325**

Adding and Subtracting Mixed Numbers

You will need
• grid paper
• counters
• fraction strips
• number lines

▶ **GOAL**

Add and subtract mixed numbers using different models.

Learn about the Math

Ashley just planted vegetables and a few flowers in her garden. She divided her garden into 11 rows. Since she loves tomatoes, she planted five rows of them.

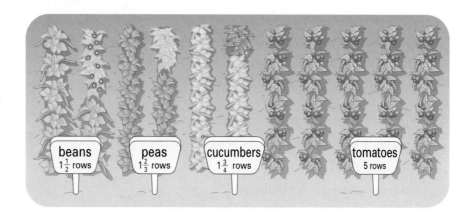

beans $1\frac{1}{2}$ rows   peas $1\frac{2}{3}$ rows   cucumbers $1\frac{3}{4}$ rows   tomatoes 5 rows

? **How many more rows of tomatoes did Ashley plant than rows of peas and cucumbers?**

A. Use fraction strips to model the rows of vegetables and flowers.

B. How many rows, in total, of peas and cucumbers did Ashley plant? Use counters to model the sum. Express your answer as a mixed number.

C. Write the subtraction equation that you need to solve to calculate how many more rows of tomatoes Ashley planted than rows of peas and cucumbers.

D. Use fraction strips to model the subtraction and calculate the difference. Express your answer as a mixed number.

Reflecting

1. In what step did you add $1\frac{2}{3}$ and $1\frac{3}{4}$? Explain why.

2. How is using fraction strips to model a difference of mixed numbers different from using them to model proper fractions? How is it the same?

3. Explain how you could estimate how many more rows of tomatoes Ashley planted than rows of peas and cucumbers.

Work with the Math

Add $3\frac{2}{3} + 1\frac{2}{5}$.

Chang's Solution

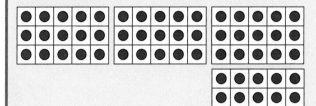

I used a grid and counters.
The common denominator for $\frac{2}{3}$ and $\frac{2}{5}$ is 15, so each rectangle has 15 squares.

First I added the whole numbers.

$3 + 1 = 4$

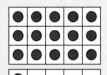

Then I added the fractions.

$\frac{2}{3} = \frac{10}{15}$

$\frac{2}{5} = \frac{6}{15}$

The sum of the fractions is $\frac{16}{15}$ or $1\frac{1}{15}$.

$4 + 1\frac{1}{15} = 5\frac{1}{15}$ To get the final answer, I added the whole number and fraction sums.

Yuki's Solution

First I used fraction strips to add the whole numbers.

$3 + 1 = 4$

Then I made a $\frac{2}{3}$ fraction strip and a $\frac{2}{5}$ fraction strip.

Since 15 is the common denominator for $\frac{2}{3}$ and $\frac{2}{5}$, I divided the fraction strips into fifteenths.

I put the fraction strips together to model adding.

I added $\frac{10}{15} + \frac{6}{15}$ to get $\frac{16}{15}$, which is $\frac{15}{15} + \frac{1}{15} = 1\frac{1}{15}$.

$4 + 1\frac{1}{5} = 5\frac{1}{15}$ I added the two sums to get the total.

Example 2: Subtracting mixed numbers using models

Subtract $7 - 5\frac{1}{3}$.

Ravi's Solution

$5\frac{1}{3} + \blacksquare = 7$

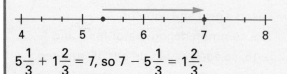

$5\frac{1}{3} + 1\frac{2}{3} = 7$, so $7 - 5\frac{1}{3} = 1\frac{2}{3}$.

I need to find out how much to add to $5\frac{1}{3}$ to get to 7. I used a number line divided into thirds to model the addition.

I added $\frac{2}{3}$ to get from $5\frac{1}{3}$ to the next whole number, 6, and then I added 1 more to get to 7. That's $\frac{2}{3}$ and 1.

Sandra's Solution

I used fraction strips to show both numbers, 7 and $5\frac{1}{3}$.

Then I figured out the difference between the numbers.

The difference is $1\frac{2}{3}$.

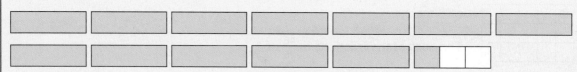

$7 - 5\frac{1}{3} = 1\frac{2}{3}$

A Checking

4. a) How many more rows of tomatoes did Ashley plant than rows of peas?

b) How many more rows of tomatoes did Ashley plant than rows of beans?

5. a) Use a model to add. Show your work.
$1\frac{1}{4} + 2\frac{7}{8}$

b) How can you add $1\frac{1}{4}$ and $2\frac{7}{8}$ by modelling only the fractions $\frac{1}{4}$ and $\frac{7}{8}$?

6. Use a model to subtract. Show your work.

a) $4 - 1\frac{4}{7}$ **b)** $3 - 2\frac{2}{3}$ **c)** $2 - \frac{1}{6}$

B Practising

7. Use grid paper and counters to add.

a) $1\frac{1}{2} + 1\frac{7}{8}$ **d)** $\frac{5}{6} + 1\frac{1}{3}$

b) $2\frac{2}{5} + 2\frac{2}{3}$ **e)** $1\frac{2}{3} + 5\frac{7}{9}$

c) $4\frac{2}{5} + \frac{3}{10}$ **f)** $4\frac{1}{12} + 3\frac{3}{4}$

8. Use a number line or fraction strips to add.

a) $2\frac{4}{5} + 3\frac{1}{10}$ **d)** $1\frac{5}{6} + 4\frac{1}{3}$

b) $3\frac{3}{4} + 1\frac{2}{3}$ **e)** $2\frac{3}{5} + 2\frac{1}{4}$

c) $2\frac{3}{4} + \frac{1}{2}$ **f)** $\frac{7}{9} + 1\frac{1}{3}$

9. Use a number line or fraction strips to subtract.

a) $8 - 1\frac{4}{5}$ **d)** $5 - \frac{7}{8}$

b) $7 - \frac{3}{7}$ **e)** $6 - 4\frac{2}{3}$

c) $3 - 1\frac{9}{10}$ **f)** $9 - 5\frac{5}{6}$

10. Dina is $11\frac{2}{3}$ years old. How old will she be after each number of years given below? Model the equation that answers each question. Express each answer as a whole number or a mixed number.

a) $3\frac{1}{3}$ years   **b)** $2\frac{1}{2}$ years

11. Express $\frac{8}{3}$ and $\frac{7}{5}$ as mixed numbers. Then add the mixed numbers. Show your work.

12. Malik and his friends ate $1\frac{1}{4}$ pepperoni pizzas, 2 cheese pizzas, and $\frac{2}{3}$ of a vegetarian pizza.

a) How many pizzas did they eat?

b) How many more cheese pizzas than vegetarian pizzas did they eat?

13. Gerald's garden has $1\frac{5}{8}$ rows of spinach, $2\frac{3}{4}$ rows of carrots, $\frac{9}{16}$ rows of corn, and 3 rows of turnips.

a) How many rows of spinach plus carrots are there?

b) How many more rows of turnips are there than carrots?

c) The garden has 10 rows altogether. Gerald says there are about 2 rows left for lettuce. Is he correct? Explain.

14. When can the sum of two mixed numbers be a whole number? Explain.

15. Yesterday, Jeff wallpapered $1\frac{2}{5}$ walls of a square room in his basement. How many more walls does he still have to wallpaper if a door takes up $\frac{1}{10}$ of one wall? (There are no windows in the room.) Model the equation that answers this question. Express your answer as a mixed number in lowest terms.

C Extending

16. It takes $9\frac{5}{6}$ h to show a movie five times. If $\frac{1}{6}$ h is needed to rewind the movie each time, how long is the movie? Write an equation or use a model to solve the problem.

17. There are five pies left at a bake sale. After Charlene and her friends buy pieces of pie, $3\frac{2}{3}$ pies are left.

a) How many pies did they buy, in total?

b) Each piece of pie is $\frac{1}{6}$ of a pie. How many pieces did they buy?

18. Use each digit from 1 to 5 once to make the following equation true.

$$\blacksquare - \blacksquare\frac{\blacksquare}{\blacksquare} = 3\frac{\blacksquare}{6}$$

19. The sum of three mixed numbers is $8\frac{2}{5}$. The difference between the first number and second number is 1. The difference between the first number and third number is 2. What is the least number?

20. Tori plays the tuba. She plays for $4\frac{1}{2}$ measures, rests for $8\frac{3}{8}$ measures, plays for another 16 measures, rests for $2\frac{1}{4}$ measures, and plays for the rest of a 36-measure song. How long was the last section that she played?

Communicating about Estimation Strategies

▶ **GOAL**

Explain how to estimate sums and differences of fractions and mixed numbers.

Communicate about the Math

Mr. Greene ordered seven pizzas for a math class party. The students ate all but $2\frac{1}{3}$ pizzas.

Sandra says, "Almost five pizzas have been eaten."

Ravi asks, "How do you know?"

Sandra explains, " When I estimate with fractions, I like to use whole numbers

that are easy to deal with. Here I estimated

with the closest whole numbers."

We started with 7 pizzas.
There are 2 left and
7 − 2 = 5.

? How can Sandra improve her explanation?

Sandra showed more detail in her explanation.

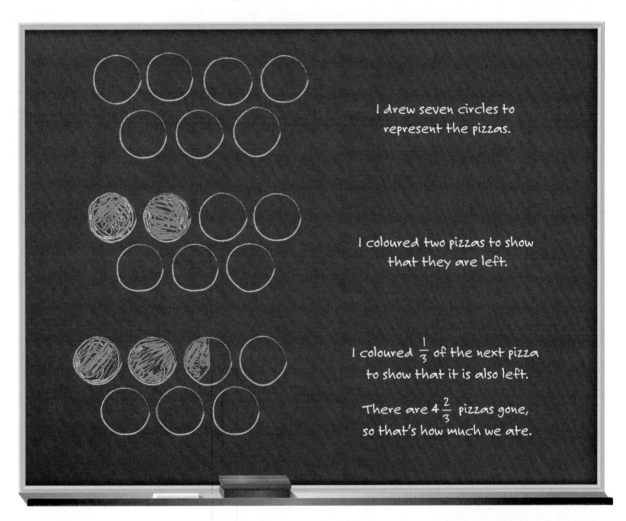

I drew seven circles to represent the pizzas.

I coloured two pizzas to show that they are left.

I coloured $\frac{1}{3}$ of the next pizza to show that it is also left.

There are $4\frac{2}{3}$ pizzas gone, so that's how much we ate.

A. Use the Communication Checklist to explain how Sandra improved her explanation.

B. Rewrite Sandra's explanation. Explain how your changes improve it.

Reflecting

1. Was Sandra's first estimate reasonable? Explain.

2. Why is a visual model helpful for explaining an estimation strategy to someone else?

3. Explain a different strategy to estimate the number of pizzas left.

> **Communication Checklist**
>
> ☑ Did you show all the necessary steps?
>
> ☑ Were your steps clear?
>
> ☑ Did you include words to describe your model, as well as pictures?
>
> ☑ Did your words support your use of the model?

Work with the Math

Example: Estimating a total

Ryan is building a dollhouse. He needs $2\frac{2}{3}$ boards for one part of the house and $3\frac{1}{2}$ boards for another part of the house. About how many boards does Ryan need, in total?

Ryan's Solution

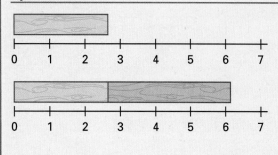

I estimated that $2\frac{2}{3}$ is a little more than $2\frac{1}{2}$.

If I add $2\frac{1}{2}$ and $3\frac{1}{2}$, I get 5 wholes and 2 halves. That's 6 whole boards.

I know the total is a little more than 6 boards, since $2\frac{2}{3}$ is a little more than $2\frac{1}{2}$.

A Checking

4. Mia has $4\frac{1}{4}$ packages of modelling clay. She wants to estimate how many packages of clay will be left if her brother uses $2\frac{1}{2}$ packages. The beginning of her explanation to her brother is given below. Complete her explanation. Use the Communication Checklist.

Mia explains, "$4\frac{1}{4}$ is a little more than 4. The distance from $2\frac{1}{2}$ to 3 is $\frac{1}{2}$."

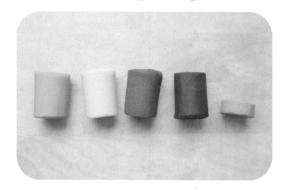

B Practising

Use models, words, and the Communication Checklist to explain how to estimate in the following questions.

5. George's family has $5\frac{1}{2}$ packages of crackers. On Sunday, their visiting cousins ate $1\frac{5}{6}$ packages of crackers. About how many packages are left?

6. Tony is painting his bedroom. He uses $\frac{3}{4}$ of a can of paint for the window frames and $\frac{1}{10}$ of a can for the baseboards. About how much more paint does he use for the window frames than for the baseboards?

7. Karen's father has 10 c. of flour. His brownie recipe requires $2\frac{1}{3}$ c. of flour. About how many batches of brownies can he bake with that much flour?

8. Braydon and Winnie are each sketching a bridge, to be constructed with straws for a science project. They have 9 bags of straws. Braydon thinks that he will use $3\frac{4}{5}$ bags of straws for his bridge. Winnie says that she will need $2\frac{3}{4}$ bags of straws for her bridge. About how many bags of straws will be left?

9. Suki, Lee, and Janice have the same number of pencils. When they put their pencils together, they have almost five full boxes of pencils. How many boxes of pencils does each person have?

Curious Math   EGYPTIAN FRACTIONS

The ancient Egyptians only used fractions with a numerator of 1 (called **unit fractions**). They used parts of the "eye of Horus" to represent these fractions, as shown below.

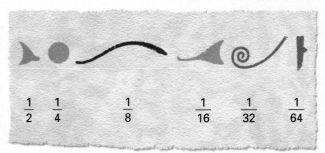

$$\frac{1}{2} \quad \frac{1}{4} \qquad \frac{1}{8} \qquad \frac{1}{16} \quad \frac{1}{32} \quad \frac{1}{64}$$

They wrote other fractions as sums of the unit fractions.

1. Show that $\frac{2}{3}$ equals $\frac{1}{2} + \frac{1}{6}$.

2. Write each fraction as a sum of unit fractions with different denominators.

a) $\dfrac{3}{4}$ **b)** $\dfrac{8}{15}$ **c)** $\dfrac{19}{24}$

3. Copy and complete the table to show that any unit fraction can be written as the difference of two other unit fractions.

4. Look for a pattern. Describe the pattern.

5. Use your pattern to write another fraction as the sum of unit fractions.

$\dfrac{1}{3}$	$=$	$\dfrac{1}{2} - \dfrac{1}{6}$
$\dfrac{1}{4}$	$=$	$\dfrac{1}{\blacksquare} - \dfrac{1}{\blacksquare}$
$\dfrac{1}{5}$	$=$	$\dfrac{1}{\blacksquare} - \dfrac{1}{\blacksquare}$
$\dfrac{1}{6}$	$=$	$\dfrac{1}{\blacksquare} - \dfrac{1}{\blacksquare}$
$\dfrac{1}{7}$	$=$	$\dfrac{1}{\blacksquare} - \dfrac{1}{\blacksquare}$
$\cdots$	$\cdots$	
$\dfrac{1}{50}$	$=$	$\dfrac{1}{49} - \dfrac{1}{2450}$
$\dfrac{1}{100}$	$=$	$\dfrac{1}{99} - \dfrac{1}{9900}$

9.8 Adding and Subtracting Using Equivalent Fractions

▶ **GOAL**

Develop a method for adding and subtracting fractions without using models.

Learn about the Math

Yuki wants to solve the following fraction puzzle. Each number in a purple box is the sum of the two numbers in the row or column in which it appears. The number in the blue box is the sum of the two numbers in the third column. It is also the sum of the two numbers in the third row.

Fraction Puzzle

$\frac{2}{5}$	A	$\frac{7}{10}$
B	$\frac{7}{20}$	$\frac{3}{5}$
C	$\frac{13}{20}$	$\frac{13}{10}$

? What are the missing fractions in the puzzle?

Example 1: Solving for the missing fractions

Calculate the missing fractions in the puzzle.

Yuki's Solution

$A = \dfrac{13}{20} - \dfrac{7}{20}$

$= \dfrac{6}{20}$

$\frac{2}{5}$	A	$\frac{7}{10}$
	$\frac{7}{20}$	$\frac{3}{5}$
	$\frac{13}{20}$	$\frac{13}{10}$

Since $\frac{7}{20} + A = \frac{13}{20}$, I subtracted $\frac{7}{20}$ from $\frac{13}{20}$ to calculate A.

$C = \dfrac{13}{10} - \dfrac{13}{20}$

$= \dfrac{13 \times 2}{10 \times 2} - \dfrac{13}{20}$

$= \dfrac{26}{20} - \dfrac{13}{20}$

$= \dfrac{13}{20}$

$\frac{2}{5}$	$\frac{6}{20}$	$\frac{7}{10}$
	$\frac{7}{20}$	$\frac{3}{5}$
C	$\frac{13}{20}$	$\frac{13}{10}$

Since $\frac{13}{20} + C = \frac{13}{10}$, I subtracted $\frac{13}{20}$ from $\frac{13}{10}$ to calculate the value of C.

I used an equivalent fraction with a denominator of 20 for $\frac{13}{10}$.

$B = \dfrac{13}{20} - \dfrac{2}{5}$

$= \dfrac{13}{20} - \dfrac{2 \times 4}{5 \times 4}$

$= \dfrac{13}{20} - \dfrac{8}{20}$

$= \dfrac{5}{20}$

$\frac{2}{5}$	$\frac{6}{20}$	$\frac{7}{10}$
B	$\frac{7}{20}$	$\frac{3}{5}$
$\frac{13}{20}$	$\frac{13}{20}$	$\frac{13}{10}$

To calculate the value of B, I subtracted $\frac{2}{5}$ from the value of C.

I used an equivalent fraction with a denominator of 20 for $\frac{2}{5}$.

Reflecting

1. Why was the value of A the easiest value to calculate?

2. Describe a different way that Yuki could have calculated the value of B. Which way do you prefer? Explain your reason.

3. Could Yuki have calculated the values of A, B, and C in a different order? Explain.

4. Why were equivalent fractions useful for determining the values of B and C?

Work with the Math

Example 2: Adding with equivalent fractions

Add $\frac{9}{10} + \frac{17}{20}$.

Solution

$$\frac{9}{10} + \frac{17}{20} = \frac{9 \times 2}{10 \times 2} + \frac{17}{20}$$
$$= \frac{18}{20} + \frac{17}{20}$$
$$= \frac{35}{20}$$
$$= 1\frac{15}{20}$$
$$= 1\frac{3}{4}$$

A common multiple of 10 and 20 is 20, so a common denominator for $\frac{9}{10}$ and $\frac{17}{20}$ is 20. The equivalent fraction for $\frac{9}{10}$ is $\frac{18}{20}$.

$\frac{15}{20}$, in lowest terms, is $\frac{15 \div 5}{20 \div 5} = \frac{3}{4}$.

Example 3: Subtracting with equivalent fractions

Subtract $\frac{7}{9} - \frac{2}{5}$.

Solution

$$\frac{7}{9} - \frac{2}{5} = \frac{7 \times 5}{9 \times 5} - \frac{2 \times 9}{5 \times 9}$$
$$= \frac{35}{45} - \frac{18}{45}$$
$$= \frac{17}{45}$$

A common multiple of 9 and 5 is 45, so a common denominator for $\frac{7}{9}$ and $\frac{2}{5}$ is 45. The equivalent fractions are $\frac{7}{9} = \frac{35}{45}$ and $\frac{2}{5} = \frac{18}{45}$.

A Checking

5. What common denominator can you use to add or subtract each pair of fractions?

a) $\dfrac{\blacksquare}{4}$ and $\dfrac{\blacksquare}{6}$ c) $\dfrac{\blacksquare}{5}$ and $\dfrac{\blacksquare}{7}$

b) $\dfrac{\blacksquare}{8}$ and $\dfrac{\blacksquare}{16}$ d) $\dfrac{\blacksquare}{4}$ and $\dfrac{\blacksquare}{9}$

6. Add. Show your work.

a) $\dfrac{5}{8} + \dfrac{1}{4}$ b) $\dfrac{3}{4} + \dfrac{7}{10}$

7. Subtract. Show your work.

a) $\dfrac{5}{8} - \dfrac{1}{4}$ b) $\dfrac{3}{4} - \dfrac{7}{10}$

B Practising

8. Add.

a) $\dfrac{3}{4} + \dfrac{1}{10}$ d) $\dfrac{5}{9} + \dfrac{1}{6}$

b) $\dfrac{5}{6} + \dfrac{1}{3}$ e) $\dfrac{3}{4} + \dfrac{2}{5}$

c) $\dfrac{3}{8} + \dfrac{1}{6}$ f) $\dfrac{3}{8} + \dfrac{3}{10}$

9. Subtract.

a) $\dfrac{3}{4} - \dfrac{1}{10}$ d) $\dfrac{5}{9} - \dfrac{1}{6}$

b) $\dfrac{5}{6} - \dfrac{1}{3}$ e) $\dfrac{3}{4} - \dfrac{2}{5}$

c) $\dfrac{3}{8} - \dfrac{1}{6}$ f) $\dfrac{3}{8} - \dfrac{3}{10}$

10. Add or subtract.

a) $\dfrac{1}{9} + \dfrac{1}{3}$ d) $\dfrac{1}{2} - \dfrac{1}{4}$

b) $\dfrac{4}{5} - \dfrac{3}{10}$ e) $\dfrac{3}{4} + \dfrac{5}{6}$

c) $\dfrac{3}{20} + \dfrac{2}{5}$ f) $\dfrac{6}{7} - \dfrac{2}{3}$

11. Explain why you should use equivalent fractions to add and subtract fractions when you do not have models.

12. Petra is $\dfrac{3}{5}$ of the way through a long-distance running race. Luc is $\dfrac{3}{8}$ finished. How much more of the race has Petra completed than Luc?

13. At camp, Matthew spends the morning doing painting, archery, and crafts. Painting takes up $\dfrac{1}{2}$ of the morning. Archery takes up $\dfrac{1}{3}$ of the morning.

a) What fraction of the morning does Matthew spend doing crafts?

b) How much more of the morning does Matthew spend doing crafts and painting than doing archery?

14. Copy the grid. Colour the grid so that $\dfrac{1}{18}$ more is blue than red. Use only red and blue.

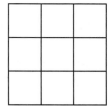

C Extending

15. Three fractions with different denominators are added. Their sum is $\dfrac{5}{8}$. What could the three fractions be? Give two possible answers.

16. When one fraction is subtracted from another fraction, the difference is zero. The fractions have different denominators. What could the fractions be? Give two possible answers.

17. Tien takes about 3 h to make a stuffed animal. Meagan takes about 4 h. If they work together, about how many stuffed animals can they make in 8 h?

18. Create a fraction puzzle like the one Yuki solved. Ask another student to solve it.

Math Game

FRACTION BINGO

In this game, you will use unit fractions to play bingo.

Number of players: 2, 3, or 4

Rules

1. Place nine cards on a table to form a square. Write a 0 on the middle card. Write fractions on the other cards. For the numerators of your fractions, choose from 1, 2, 3, 4, 5, and 6. For the denominators, choose from 2, 3, 4, 5, 6, 8, 10, 12, 15, 18, 20, and 30.

2. Roll a pair of dice. Use the numbers you roll as the denominators of two fractions. Use 1 as the numerators of the fractions.

3. If the sum or difference of your two fractions is on a card, put a counter on the number on your card.

4. Take turns rolling and calculating. Check each other's work.

5. The winner is the first player with three counters in a row horizontally, vertically, or diagonally.

Chapter Self-Test

1. a) What part of the following pattern block model shows $\frac{1}{3} + \frac{1}{6}$?

b) What is the sum of $\frac{1}{3} + \frac{1}{6}$?

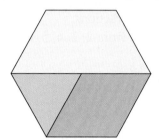

2. a) Use a model to add. Show your work.

  i) $\frac{3}{8} + \frac{1}{4}$ **ii)** $\frac{2}{5} + \frac{3}{4}$

b) Use an estimation strategy to show that your answers are reasonable.

3. Francis spent $\frac{1}{2}$ of an hour writing a story on his computer, and then played computer games for $\frac{1}{4}$ of an hour. Write an equation to describe the fraction of an hour that Francis used his computer.

4. a) Use a model to subtract. Show your work.

  i) $\frac{3}{8} - \frac{1}{4}$ **ii)** $\frac{4}{5} - \frac{3}{4}$

b) Use an estimation strategy to show that your answers are reasonable.

5. Heather is earning the money to buy a new stereo. She has earned $\frac{4}{5}$ of the money. What fraction of the amount does she still need to earn?

6. Use a model to multiply. Write each answer as a mixed number.

  a) $2 \times \frac{4}{5}$ **b)** $6 \times \frac{3}{8}$

7. Use a model to add or subtract. Show your work.

a) $2\frac{3}{4} + 3\frac{8}{9}$ **c)** $3 - 2\frac{1}{5}$

b) $4 - \frac{1}{10}$ **d)** $6 - 1\frac{3}{8}$

8. The difference between two mixed numbers is a whole number. What do you know about the two mixed numbers?

9. The difference between two mixed numbers is less than 1. What do you know about the two mixed numbers?

10. Luke made a graph to show how he spends a typical weekday.

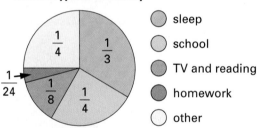

a) What fraction of the day does Luke not spend sleeping or in school?

b) How much more of the day does Luke spend in school than on homework?

c) Use the graph to make up a problem that has $\frac{7}{12}$ as the answer.

11. Explain how to add $\frac{3}{8} + \frac{2}{5}$ using equivalent fractions. Calculate the sum.

12. Subtract using equivalent fractions. Show your work.

  a) $\frac{5}{6} - \frac{4}{9}$ **b)** $\frac{7}{10} - \frac{3}{8}$

Chapter Review

Frequently Asked Questions

Q: How do you subtract fractions using grid paper and counters?

A: Outline a rectangle on grid paper to show the common denominator of the two fractions. Then use counters. For example, to subtract $\frac{3}{5} - \frac{1}{4}$:

- Outline a rectangle on grid paper with 5 rows of 4. Each row of the rectangle represents $\frac{1}{5}$, and each column represents $\frac{1}{4}$.

- Cover 3 of the 5 rows with counters to show $\frac{3}{5}$. (You cover 12 of the 20 squares since $\frac{3}{5}$ is equivalent to $\frac{12}{20}$.)

- Move counters to cover as many complete columns as possible. (This makes it easier to subtract fourths since each column represents $\frac{1}{4}$.)

- Take away 1 column of counters to model subtracting $\frac{1}{4}$. (You take counters from 5 of the 20 squares since $\frac{1}{4}$ is equivalent to $\frac{5}{20}$.)

Since 7 of the 20 squares in the rectangle still have counters, $\frac{7}{20}$ of the squares have counters. The difference is $\frac{7}{20}$.

$$\frac{3}{5} - \frac{1}{4} = \frac{12}{20} - \frac{5}{20}$$
$$= \frac{7}{20}$$

Q: How do you add mixed numbers?

A: Add the whole numbers and the fractions separately. For example, to add $2\frac{3}{4} + 4\frac{1}{2}$:

- Start by adding the whole numbers. $2 + 4 = 6$
- Then show each fraction with counters on a grid. The common denominator for $\frac{3}{4}$ and $\frac{1}{2}$ is 4, and $\frac{1}{2} = \frac{2}{4}$.

- Move counters to model adding the fractions. This total is $\frac{5}{4} = 1\frac{1}{4}$.

- Then add the whole number sum and fraction sum.

$$6 + 1\frac{1}{4} = 7\frac{1}{4}$$

Fraction Operations **339**

Q: How do you subtract a mixed number from a whole number?

A: Method 1:
Use a number line to determine the distance between the numbers.

For example, to subtract $7 - 1\frac{3}{5}$, draw an arrow from $1\frac{3}{5}$ to 7.

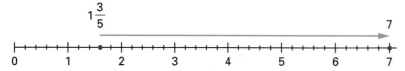

The distance from $1\frac{3}{5}$ to 2 is $\frac{2}{5}$. The distance from 2 to 7 is 5.

The total difference is $\frac{2}{5} + 5$, or $5\frac{2}{5}$.

$$7 - 1\frac{3}{5} = 5\frac{2}{5}$$

Method 2:
Use fraction strips. For example, to subtract $7 - 1\frac{3}{5}$, represent each number with fraction strips. Line up the fraction strips to compare the numbers. Then figure out the difference.

The difference is $5\frac{2}{5}$.

$$7 - 1\frac{3}{5} = 5\frac{2}{5}$$

Q: How do you add or subtract fractions using equivalent fractions?

A: Find a common multiple of the two denominators. Then write a new equation, using equivalent fractions that have the common multiple as their denominators. To add $\frac{3}{4} + \frac{3}{5}$, for example, you can use the common denominator 20, since 20 is a common multiple of 4 and 5.

$$\frac{3 \times 5}{4 \times 5} + \frac{3 \times 4}{5 \times 4} = \frac{15}{20} + \frac{12}{20}$$
$$= \frac{27}{20} \text{ or } 1\frac{7}{20}$$

Practice Questions

(9.1) **1.** Use a pattern block model to show that $\frac{1}{3} + \frac{1}{2} = \frac{5}{6}$.

(9.2) **2.** Use a model to add. Use another model or method to check your answers.

  a) $\frac{5}{8} + \frac{1}{4}$ **b)** $\frac{3}{5} + \frac{1}{2}$

(9.3) **3.** Use a model to multiply. Write each answer as a mixed number.

  a) $5 \times \frac{3}{4}$ **b)** $4 \times \frac{3}{7}$

(9.4) **4.** A short video is $\frac{1}{3}$ of an hour long. If you watch the video five times, how many hours does this take? Express your answer as a mixed number.

(9.4) **5.** Use a model to subtract. Use another model or method to check your answers.

  a) $\frac{4}{7} - \frac{1}{3}$ **c)** $\frac{5}{6} - \frac{2}{9}$

  b) $\frac{11}{12} - \frac{2}{3}$ **d)** $\frac{3}{4} - \frac{3}{5}$

(9.4) **6.** Marian has $\frac{2}{3}$ of a bag of bagels.

  a) She adds another $\frac{1}{4}$ bag of bagels. What fraction of the bag is now full of bagels?

  b) Marian has another bag of bagels that is $\frac{5}{6}$ full. What fraction describes how many more bagels are in the bag in part (a)?

(9.4) **7.** Nunavut covers about $\frac{1}{5}$ of Canada's land area. Ontario covers about $\frac{1}{9}$ of Canada's land area. What fraction describes how much more of Canada's land area is covered by Nunavut than by Ontario?

8. Use a model to add or subtract. Show your work. (9.6)

  a) $\frac{3}{10} + 2\frac{3}{5}$ **d)** $5 - \frac{5}{6}$

  b) $2\frac{1}{4} + 2\frac{1}{3}$ **e)** $6 - 2\frac{2}{7}$

  c) $4\frac{5}{9} + \frac{2}{3}$ **f)** $7 - 6\frac{7}{9}$

9. Kyle has already spent $1\frac{5}{6}$ h on a project for his technology class. If his teacher said that he should spend a total of 3 h on the project, how much longer should he work on it? Use a model to show your answer. (9.6)

10. The sum of three mixed numbers is $6\frac{2}{3}$. What is the least possible value for the greatest of the mixed numbers? Use a model or another method to explain your thinking. (9.6)

11. Decide whether each sum is between 1 and 3. Explain how you know. (9.7)

  a) $1\frac{1}{2} + 1\frac{1}{4}$ **b)** $\frac{5}{6} + \frac{1}{7}$

12. a) Estimate to decide whether each sum is greater than 1. (9.7)

  i) $\frac{2}{3} + \frac{5}{7}$ **ii)** $\frac{3}{4} + \frac{3}{7}$

  b) Calculate each sum in part (a) to find out if you estimated correctly.

13. Add or subtract using equivalent fractions. Show your work. (9.8)

  a) $\frac{3}{5} + \frac{2}{7}$ **c)** $\frac{7}{10} - \frac{2}{3}$

  b) $\frac{8}{9} + \frac{2}{3}$ **d)** $\frac{2}{3} - \frac{3}{5}$

Chapter Task

New Car Dealership

Suppose that your family has opened a car dealership in a small town. You are deciding what models and colours of vehicles to buy this year. You have surveyed visitors to the dealership about what vehicles they prefer. Your results are given below. Unfortunately, you spilled water on your results, so two of the fractions are missing.

Model	four-door family car	jeep	truck	sports car
Fraction	$\frac{1}{3}$	$\frac{1}{4}$	$\frac{1}{5}$	

Colour	silver	black	red	green	blue	beige
Fraction	$\frac{1}{4}$	$\frac{1}{10}$		$\frac{3}{10}$	$\frac{3}{20}$	$\frac{1}{20}$

? What fraction of each model/colour combination should you order?

A. What fraction of visitors prefer sports cars? Explain your calculation.

B. What fraction of visitors prefer red vehicles? Show your work.

C. You will be ordering some different models in various colours. Choose six vehicles with different model/colour combinations. Decide what fraction of your order you want to use for each combination. Justify each choice. Explain why the sum of your fractions must equal 1.

D. What is the difference between the greatest fraction and the least fraction in step C? Show your work.

E. Use addition, subtraction, or multiplication to make your own fraction problem about the dealership. Solve your problem.

Cumulative Review
Chapters 7–9

Cross-Strand Multiple Choice

(7.1) **1.** Which ordered pair represents the correct translation of point *A* that is 2 units to the left and 3 units down?

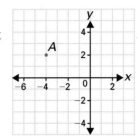

A. $(-2, -1)$ **C.** $(-2, 5)$
B. $(-6, 5)$ **D.** $(-6, -1)$

(7.3) **2.** Which figure is reflected correctly in the reflection line?

A. **C.**

B. **D.**

(7.5) **3.** Which diagram shows two triangles that are *not* congruent?

A. **C.**

D.

B.

4. Which diagram shows two triangles that are similar? (7.5)

A. **C.**

D.

B.

5. Valerie has a part-time job at a fast-food restaurant. She has $175 in her savings account. If she deposits $30 of her earnings each week, which expression describes the amount in Valerie's savings account? (8.3)

A. $30 + 175w$ **C.** $175 - 30w$
B. $175 + 30$ **D.** $175 + 30w$

6. Which algebraic expression represents a number multiplied by 2 and increased by 7? (8.3)

A. $2n + 7$ **C.** $2 \times 7 + n$
B. $n + 7 \times 2$ **D.** $7n + 2$

7. What is the value of *p* in the algebraic equation $13 = p - 27$? (8.4)

A. $p = -40$ **C.** $p = -14$
B. $p = 14$ **D.** $p = 40$

8. What is the value of $\frac{1}{5} + \frac{1}{3}$? (9.2)

A. $\frac{8}{5}$ **B.** $\frac{8}{15}$ **C.** $\frac{1}{15}$ **D.** $\frac{2}{8}$

9. What is the value of $2\frac{2}{5} - 1\frac{1}{3}$? (9.6)

A. $1\frac{1}{2}$ **B.** $1\frac{1}{15}$ **C.** $\frac{13}{15}$ **D.** $1\frac{2}{15}$

Cross-Strand Investigation

You and some friends are starting a T-shirt business called Transformational Inspirations. Each of your T-shirts will feature an original tessellation design. You have decided to create your designs in a 20 cm by 20 cm square, so that you can scan them into a computer, print them out, and apply them to the T-shirts.

10. a) Follow these steps to create a stencil for your design.
  - Cut out a 5 cm by 5 cm square from Bristol board or a similar material. Use a pencil to create a design on the square. Then cut out your design, so that you have a stencil you can trace.
  - On a piece of blank paper, draw a 20 cm by 20 cm square. Starting at the top left-hand corner of the square, trace around the edges of your stencil and then around the spaces you cut out.
  - For the rest of the first row, you can do translations, rotations, and/or reflections.
  - For the second row, repeat the same pattern you used in the first row, but translate the squares one unit to the right. (You will start with the same orientation that you ended with in the first row.)
  - Repeat this process until your design is complete. Colour each square to make your design eye-catching. (Remember that you need to colour each square the same way.)

b) Write a description of your design. (Your description is part of the copyright documentation required to make sure that no one can copy and sell your design.)

11. The cost of each blank T-shirt from the supplier is $6.00. The setup costs for your business total $200.00. Suppose that you sell your T-shirts for $14.00 each.

a) Write an algebraic expression that represents your total profit in terms of the number of T-shirts sold.

b) Write an algebraic expression that represents your total expenses in terms of the number of T-shirts sold.

c) Determine a value for t so that you "break even." (Your expression for profit, in part (a), will be equal to your expression for expenses, in part (b).) Use a calculator to help you guess and test the number of T-shirts you must sell to break even.

d) What fraction of your total expenses from part (c) is the cost of the blank T-shirts? What fraction of your total expenses are the setup costs?

CHAPTER 10

3-D Geometry

▶ GOALS

You will be able to

- build three-dimensional (3-D) shapes from nets and with cubes
- identify 3-D objects that are drawn in different ways
- recognize different views of 3-D objects
- sketch different views of 3-D objects from models and drawings

Colouring Cubes

Bright Ideas Toy Company has created a new puzzle cube. The cube is made from six large, coloured squares that can interlock along any edge. The object of the puzzle is to arrange all six squares on a flat surface so that when you form the cube, opposite faces are the same colour.

? **How can you colour the following nets for the toy company's puzzle?**

A. On grid paper, make larger copies of these nets.

a)   **b)**   **c)**

B. Visualize folding these nets into cubes. Lightly mark the faces that you predict will be opposite each other.

C. Cut out your nets. Fold the cubes to check that they fit the toy company's design. Colour the faces of the net so that when the net is a cube, opposite faces will be the same colour.

D. Construct the cubes.

E. Which net was easiest to use for matching the colours of opposite faces? What do you think made it easiest?

F. Design a different net that meets the design requirements. Use grid paper. Then construct the cube to check.

Do You Remember?

1. Sketch an example of each polygon.
   - a) isosceles triangle
   - b) equilateral triangle
   - c) quadrilateral
   - d) pentagon
   - e) hexagon
   - f) octagon

2. State whether each shape is a prism, a pyramid, or neither.

   a)   c)   e)

   b) 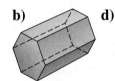 d)   f)

3. Name each prism.

   a)   b)   c)

4. Name each pyramid.

   a)   b)   c)

5. Which pyramid in question 4 is a tetrahedron?

6. Match each net with a polyhedron.

   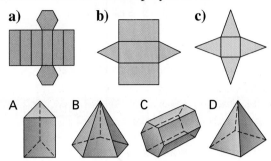

7. Write the name of each polyhedron in question 6. Choose the name from this list:
   - triangular pyramid
   - hexagonal prism
   - triangular prism
   - square-based pyramid
   - hexagonal pyramid
   - pentagonal prism
   - octagonal prism
   - pentagonal pyramid

8. What is the least number of faces that a polyhedron can have?

9. What is the least number of edges that a polyhedron can have?

10. What is the least number of vertices that a polyhedron can have?

11. Explain the difference between a polygon and a polyhedron.

12. a) Draw a polygon that is not a regular polygon.
    b) Draw the net for a prism that has the polygon in part (a) as its base.
    c) Construct the prism from the net in part (b). Place it on its base. Are the vertical faces of this prism congruent? Explain.

Building and Packing Prisms

You will need
• nets of prisms
• scissors
• tape
• a protractor

▶ **GOAL**
Build prisms from nets.

Explore the Math

Imagine that your job is to design efficient packaging. Stores and shipping companies want packages that pack into rectangular shipping cartons so that there is no wasted space between packages. They also want packages that are interesting shapes to attract buyers.

? **How can you design packages that will pack into cartons with no gaps between them?**

A. Make several congruent prisms from nets. Use nets whose bases have three, four, five, or six sides.

B. Do the prisms pack so that there are no gaps?

C. Measure the angles at each vertex where the prisms meet or almost meet. Add the angle measures.

D. Repeat steps A to C for different congruent prisms until you have used prisms with three, four, five, and six sides.

E. Summarize your findings about the types of prisms that pack without gaps.

Reflecting

1. How is packing prisms similar to tiling a plane? How is it different?

2. a) What did you notice about the touching vertices of the prisms that packed with no gaps?

   b) What did you notice when you added the angle measures? Why did this happen?

3. Which of the following properties of shapes make a difference when you are trying to pack with no gaps? Explain.

   a) the number of faces

   b) the shapes of the faces

   c) the lengths of the edges

   d) the number of vertices

   e) the sizes of the angles

Mental Imagery   PAINTING CUBES

Use nine linking cubes to make this shape.

Imagine that you painted the entire outside of this shape red.

> **You will need**
> • linking cubes

1. How many cubes would have each number of faces painted red? Explain your reasoning.

   a) 6 faces   **b)** 5 faces   **c)** 4 faces   **d)** 3 faces   **e)** 2 faces   **f)** 1 face

2. How many cubes would have some faces not painted red? Explain your reasoning.

3. Make each shape. Imagine that the entire outside of each shape is painted red. Determine each number of cubes that would have from 6 to 0 faces painted red.

   a)

   b)

Building Objects from Nets

You will need
- large grid paper
- large triangle dot paper
- a ruler
- scissors
- tape

▶ **GOAL**

Build 3-D shapes from nets.

Learn about the Math

Kaitlyn and Kwami are constructing cardboard models of the buildings in their community for a class project. Kwami uses two nets to constuct a model of a barn. He wonders if it is possible to create a single net that would work. Kaitlyn thinks that it is possible and decides to try.

? **How can you make a single net for the barn?**

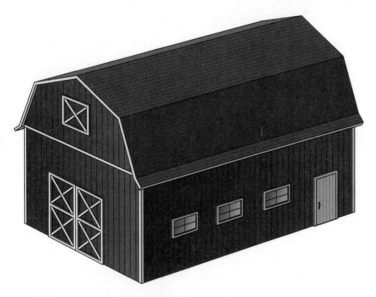

Ignore the overhang of the roof when creating your net.

A. Which two polyhedrons did Kwami use to make the top part and bottom part of the barn?

B. Draw a net for the bottom part of the barn.

C. Draw a net for the top part of the barn.

D. Combine the two nets in steps B and C to create one net for the barn.

E. Test your net to make sure it can be folded to make a model of the barn.

F. How many faces does your model have?

G. How many edges and vertices does your model have?

Reflecting

1. To build an accurate model of the barn, what measurements did you need before you started to draw your net?

2. Suppose that you wanted to build a model of the barn with the overhanging roof. How would the overhanging roof change the net?

3. How can you tell whether one net can be joined to another to make a large net for an entire structure?

4. When you folded your net for the barn, where did you start? Why?

Work with the Math

Example: Creating a net for a 3-D object

Create a net to construct a model of this bungalow.

Paul's Solution

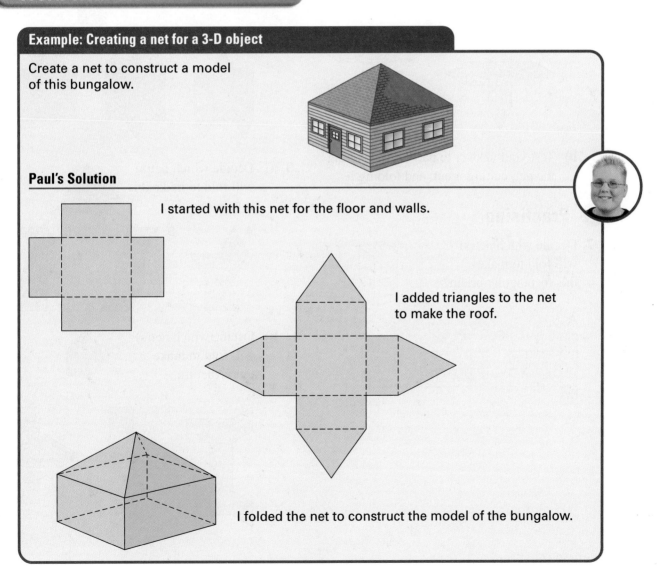

I started with this net for the floor and walls.

I added triangles to the net to make the roof.

I folded the net to construct the model of the bungalow.

A Checking

5. Examine the geometric shapes in the following model.

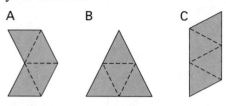

a) What polyhedrons do you see?

b) How many edges does the base of the model have?

c) How many faces does the model have?

6. a) Which net will fold to form the model in question 5?

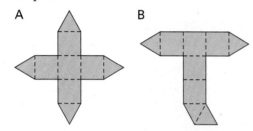

b) Test your answer to part (a) by making the net, cutting it out, and folding it.

B Practising

7. Decide which net(s) will fold to make this rectangular prism.

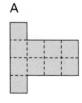

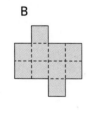

8. a) Use your visualization skills to decide which net(s) will fold to make a regular tetrahedron. You may need to make the nets, cut them out, and fold them to test your decision.

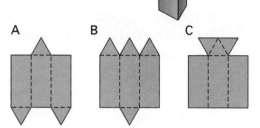

b) Decide which net(s) will fold to make this structure.

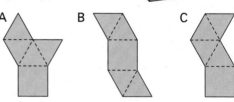

9. a) Decide which net(s) will fold to make this square pyramid.

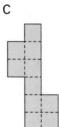

b) Decide which net(s) will fold to make this structure.

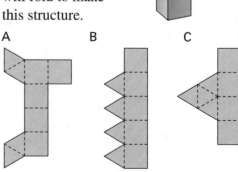

10. Examine the geometric shapes that make up the following gift box.

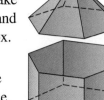

a) Name the two polyhedrons that make up the bottom part and the lid of the gift box.

b) Create a net for the lid. Assume that the lid has a bottom face.

c) Create a net for the bottom part of the gift box.

d) Use your nets from parts (b) and (c) to create a single net for the entire box, including the lid. Test your net by cutting it out and folding it.

11. Decide which net(s) will fold to make this cube structure.

A   B   C

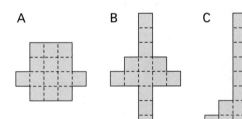

12. Examine the geometric shapes in the octahedron.

a) Name the polyhedrons that make up the top half and bottom half of the octahedron.

b) Create a net for each half of the octahedron. Assume that each half has a base face.

c) Use your nets from part (b) to create a single net for the entire octahedron. Test your net by cutting it out and folding it.

C **Extending**

13. Examine the geometric shapes in this model of a lighthouse.

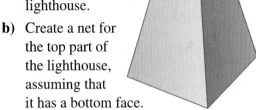

a) Name the two polyhedrons that make up the lighthouse.

b) Create a net for the top part of the lighthouse, assuming that it has a bottom face.

c) Create a net for the bottom part of the lighthouse.

d) Use your nets from parts (b) and (c) to create a single net for the entire lighthouse. Test your net by cutting it out and folding it.

14. a) Examine the following model of a house. Describe how to cut the model so that the sides and roof would fold down flat to create a net.

b) Create a net for the house. Check your net by cutting it out and folding it to create a model of the house.

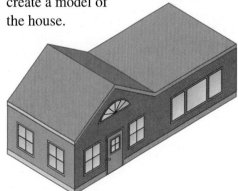

10.3 Top, Front, and Side Views of Cube Structures

You will need
• linking cubes
• a building mat
• grid paper

▶ **GOAL**

Recognize and sketch the top, back, front, and side views of cube structures.

Learn about the Math

Miguel and Bonnie are watching a construction crew build a new wing for their school. They notice that the crew follows building plans. The construction manager tells them that architects and engineers design plans with top, front, and side views to show how the parts of a building will fit together. Architects call the base of the building the *footprint* of the building.

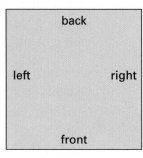

building mat

? **How can top, front, and side views represent structures?**

Miguel and Bonnie want to learn about designing building plans. They model a structure with linking cubes. They make a building mat and label their mat for the different views.

Example 1: Recognizing views of a structure

Identify the top, front, and right-side views of this cube structure.

Bonnie's Solution

I put the cube structure on the mat with the single layer of two cubes at the front.

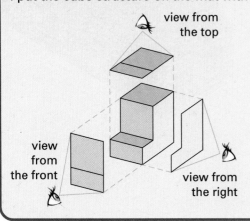

I looked directly down on the structure to see the top view. The surface I saw is a rectangular shape.

Next I brought my eye level with the structure and slowly rotated the mat to see the front view. The surface I saw is rectangular.

Then I turned the mat so that I looked directly at the right-side view. The surface I saw is like a backward L.

Miguel wonders how to show changes in depth from one layer of cubes to the next. The construction manager explains that, on a building plan, a change in depth is often shown as a thick black line.

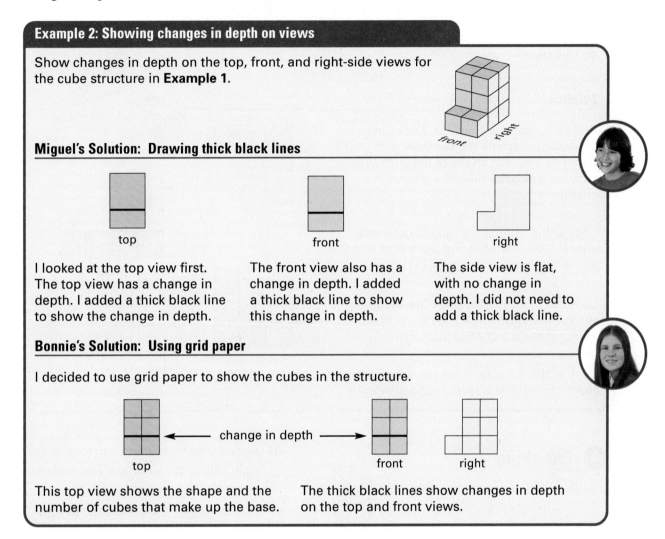

Example 2: Showing changes in depth on views

Show changes in depth on the top, front, and right-side views for the cube structure in **Example 1**.

Miguel's Solution: Drawing thick black lines

top   front   right

I looked at the top view first. The top view has a change in depth. I added a thick black line to show the change in depth.

The front view also has a change in depth. I added a thick black line to show this change in depth.

The side view is flat, with no change in depth. I did not need to add a thick black line.

Bonnie's Solution: Using grid paper

I decided to use grid paper to show the cubes in the structure.

top   ← change in depth →   front   right

This top view shows the shape and the number of cubes that make up the base.

The thick black lines show changes in depth on the top and front views.

Reflecting

1. What relationship do you see between the width of the top view and the width of the front view?

2. What relationship do you see between the depth shown on the front view and the width of the right-side view?

3. What relationship do you see between the height of the front view and the height of the right-side view?

4. How is it possible for two buildings to have the same front view but different side views?

Work with the Math

Example 3: Drawing views on grid paper

Draw top, front, and side views of the following structure on grid paper.

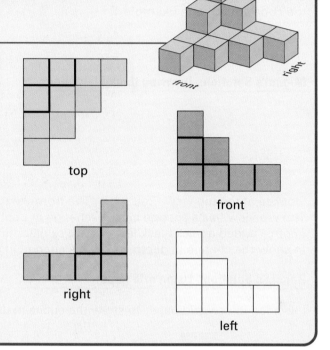

Solution

Build the model with cubes. Look down at the structure to see the top view. Use grid paper to draw the shape of the surfaces. Draw a thick black line where there is a change in depth.

Look at the front of the model. Draw the shape of the surfaces and use a thick black line where there is a change in depth.

Look at the right side. Draw the shape of the surfaces and use a thick black line where there is a change in depth.

Look at the left side. Draw the shape of the surfaces. There is no change in depth.

Ⓐ Checking

5. a) Build this structure using linking cubes.

b) Place the structure on your building mat in the position shown. Visualize what the structure would look like from the top, front, and left side.

c) State whether each diagram is the top, front, or left-side view.

A   B   C

d) Rotate your building mat to look at the right side of the structure. Use grid paper to draw the right-side view.

e) Which view(s) show a change in depth?

Ⓑ Practising

6. a) Build this structure with linking cubes and place it on your building mat.

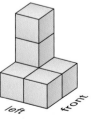

b) Rotate your building mat to see the top, front, and left-side views. Draw the three views on grid paper. Draw a thick black line where there is a change in depth.

7. a) Build this structure with linking cubes and place it on your building mat.

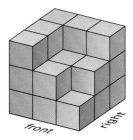

b) Rotate your mat to see the top, front, right-side, and left-side views. Draw the four views on grid paper. Draw a thick black line where there is a change in depth.

8. a) Use these top, front, and side views to build a linking-cube structure.

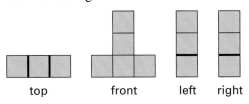

top   front   left   right

b) Draw the back view.

9. a) Use 27 linking cubes to build a solid structure that has the following top, front, and left-side views.

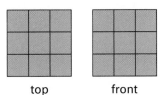

top   front   left

b) Use a smaller number of linking cubes to build another structure with top, front, and left-side views that are all the same.

c) What is the minimum number of linking cubes you can use to build a structure with top, front, and left-side views that are all the same? How do you know?

10. a) Use linking cubes to build this structure.

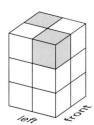

b) Attach a cube to the green face.

c) Draw the top, front, left-side, and right-side views of the new structure. Draw a thick black line where there is a change in depth.

11. Draw the top, front, and left-side views of this structure, with the green cubes removed. Draw a thick black line where there is a change in depth.

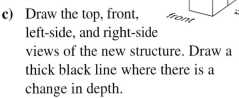

ⓒ Extending

12. a) Use these top, front, and side views to build a linking-cube structure.

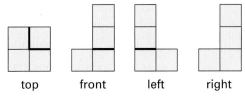

top   front   left   right

b) Draw the back view.

13. a) Use these top, back, and side views to build a linking-cube structure.

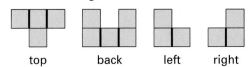

top   back   left   right

b) Draw the front view.

14. Build two linking-cube structures that have the same front views, but different top and right-side views. Draw the front, top, and right-side views.

Mid-Chapter Review

Frequently Asked Questions

Q: How can you predict whether congruent prisms will fill a space without any gaps?

A: If the base of the prism can tile an area, then the prism will fill the space with no gaps between the prisms.

Q: When and how can you combine two different nets to make a single net for a shape?

A: If the two nets are for two parts of the shape, they can be joined where the parts meet to create a single net. The two parts of the shape have to join along a common edge. Faces that are hidden in the shape are removed from the nets before they are joined.

Q: What are the top, front, and side views of a cube structure?

A: The top, front, and side views allow you to see the faces of a cube structure as if you were looking at it "straight on." To draw the top, front, and side views, you draw the shape of the surfaces you see and show any change in depth with a thick black line.

For example, the top, front, left-side, and right-side views of a cube structure are shown below.

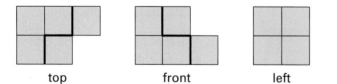

| top | front | left | right |

Q: Do three views of a cube structure always provide enough information to build the cube structure?

A: No. Consider, for example, a linking-cube structure with a missing cube, as seen from the left side. The top, front, and right-side views do not give you enough information to build the cube structure. You also need the left-side view.

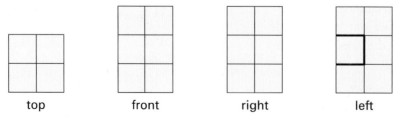

| top | front | right | left |

Practice Questions

(10.2) **1.** Name the polyhedron that each net represents. You may need to make each net, cut it out, and fold it to find out.

a) 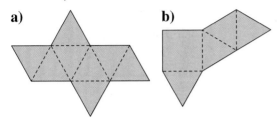 **b)**

(10.2) **2.** State whether each net will fold to make a box with a lid. If the net will not fold to make a box with a lid, explain why.

a) **c)**

b) **d)**

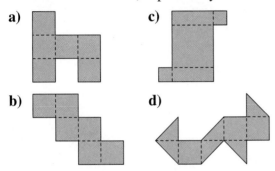

(10.2) **3.** The following dollhouse is under construction and needs walls.

a) Create a net that will fold to make the walls and floor for the dollhouse.

b) Is it possible to create a single net for the entire dollhouse? Explain.

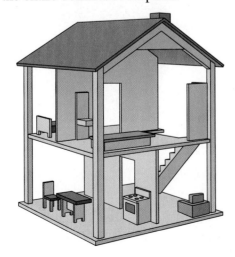

4. Decide which net(s) will fold to make this cube structure. (10.2)

A   B   C

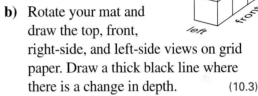

5. a) Build this structure with linking cubes and place it on your building mat.

b) Rotate your mat and draw the top, front, right-side, and left-side views on grid paper. Draw a thick black line where there is a change in depth. (10.3)

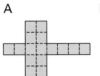

6. a) Use linking cubes to build this structure.

b) Attach a cube to the yellow face.

c) Draw the top, front, left-side, and right-side views of the new structure. Draw a thick black line where there is a change in depth. (10.3)

7. Draw the top, front, and right-side views of this structure, with the yellow cubes removed. Draw a thick black line where there is a change in depth. (10.3)

8. a) Use these top, front, and right-side views to build a linking-cube structure.

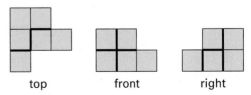

top   front   right

b) Draw the left-side view.

c) Draw the back view. (10.3)

Top, Front, and Side Views of 3-D Objects

You will need
- a building mat
- triangle dot paper
- large grid paper
- scissors
- tape
- coloured pencils

▶ **GOAL**

Recognize and sketch the top, front, and side views of 3-D objects.

Learn about the Math

You can use the skills you learned for drawing top, front, and side views of cube structures to help you draw top, front, and side views of 3-D objects.

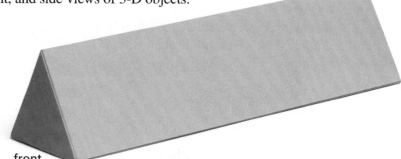

? **How can you sketch views of 3-D objects?**

A. Create a net to make a model of a triangular chocolate bar package. Construct the model.

B. Place your model on your building mat. Look directly down on the model to see the top view. Draw the top view. Show any changes in depth with a thick black line. Label your drawing to show that it is the top view.

C. Bring your eyes level with the front of the model so that you are looking directly at the front view. Draw the front view. Label your drawing to show that it is the front view.

D. Repeat step C for the back view, the right-side view, and the left-side view.

E. Are the front and back views the same? Are the left-side and right-side views the same?

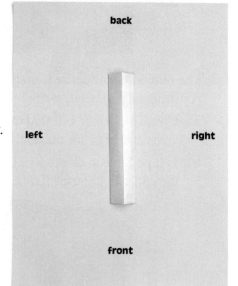

Reflecting

1. a) Could a different face of your model have been the front view? Explain.

b) Could a face of your model have been the top view? Explain.

2. How could your views be different if you positioned your prism differently on the mat?

3. Why is it important to keep the object you are viewing in the same position on the building mat to draw different views?

Example 1: Drawing views of an object

Draw top, front, left-side, and back views of a step stool shaped like stairs.

Romona's Solution

When I drew views of cube structures, I used a building mat to keep the structure in one position. Here I'll keep the stairs in the same position while I draw the top, front, and side views.

top

I looked straight down at the top of the stairs. I noticed the change in depth at the lower step. I drew the shape of the steps and showed the change in depth by drawing a thick black line.

front

Next I moved to the front and brought my eyes level with the stairs. I noticed the two rectangular surfaces and a change in depth. I drew the shape of the surfaces. I showed the change in depth by drawing a thick black line.

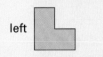
left

I moved to the left side and brought my eyes level with the stairs. I noticed only one surface, with no changes in depth. I drew the shape.

back

Finally I drew the back view.

Example 2: Drawing views of a polyhedron

Draw top, front, and left-side views of this polyhedron.

Fawn's Solution

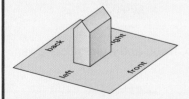

top

front

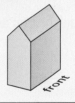
left

I needed to keep the polyhedron in the same position on my building mat to draw the views.

I looked straight down at the top of the polyhedron and sketched the view.

I moved to look directly at the front of the polyhedron. I drew the front view.

I moved to look directly at the left side of the polyhedron. I drew this side view.

Ⓐ Checking

4. Identify the polyhedron that has these views.

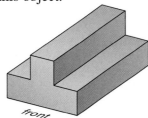

top   front   left

5. Draw the top, front, right-side, and left-side views of this object.

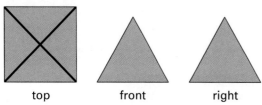

front

Ⓑ Practising

6. Identify the polyhedron that has these views.

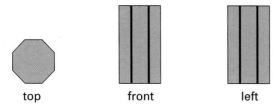

top   front   right

7. Identify the polyhedron that has these views.

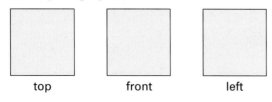

top   front   left

8. Identify the polyhedron that has these views.

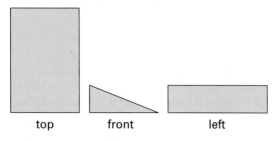

top   front   left

9. a) Identify the polyhedron that has this top view.

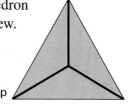

top

b) Draw the front and left-side views.

10. a) Identify two polyhedrons that have this side view.

b) Sketch the top view of each polyhedron you named in part (a).

side

11. a) Visualize folding this net into a 3-D shape. Draw and colour the top view and the front view.

b) Explain how you decided which view would be the top view.

12. Draw the top, front, and left-side views of this box.

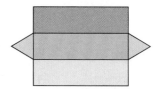

front

13. Draw the top, front, and right-side views of this box.

front

14. a) Draw the top, front, and left-side views of this block structure.

b) What polyhedrons make up the block structure?

front

15. Draw the top view of this Mayan temple.

16. a) Decide which view of this mailbox will be the front view. Draw the top, front, and left-side views.

b) Identify as many polyhedrons as you can that have the same three views.

C Extending

17. Sketch a different view of each object.

a)

b)

c)

d)

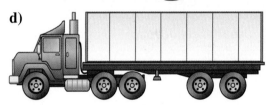

18. Find an object that has different views. Draw the top, front, left-side, and right-side views.

19. Describe a situation in everyday life when you would use top, front, and side views. Explain how these views would be useful.

20. Sketch the top view, front view, and a side view of each object.

a) a can

b) a pylon

c) a computer mouse

Isometric Drawings of Cube Structures

You will need
- linking cubes
- a ruler
- triangle dot paper
- coloured pencils

▶ **GOAL**

Make realistic drawings of cube structures on triangle dot paper.

Learn about the Math

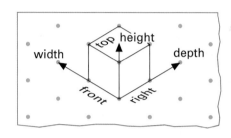

isometric drawing

a 3-D view of an object in which
- vertical edges are drawn vertically
- width and depth are drawn diagonally
- equal lengths of the object are equal on the drawing

Miguel built a cube structure. He wants to fax an **isometric drawing** of his structure to Heather.

? How can Miguel draw the cube structure?

Example 1: Making an isometric drawing

Use triangle dot paper to make an isometric drawing.

Miguel's Solution

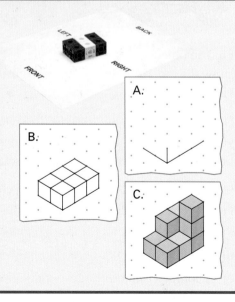

I started by using the bottom layer of the structure. I placed the structure in the position for drawing.

A. First I drew the vertical part of the front cube and the bottom of the front and right side on triangle dot paper. I used one space to represent the height, two spaces for the width, and three spaces for the depth.

B. Then I completed the bottom layer. I used one space to represent the height, width, and depth of each cube.

C. I added the other layers and erased hidden lines to finish my drawing.

Reflecting

1. Explain how Miguel placed the bottom layer to show a view that he could use to make an isometric drawing.

2. Describe a different way Miguel could have placed the bottom layer to make an isometric drawing.

3. Could Miguel have made an isometric drawing of the structure without first breaking up the layers? Explain.

Work with the Math

Example 2: Drawing a cube structure

Make an isometric drawing of this cube structure.
Shade your drawing to make it look 3-D.

Solution

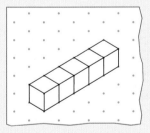

Place the cube structure to prepare for making an isometric drawing. Draw one row of the bottom layer.

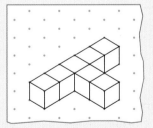

Complete the bottom layer, erasing hidden lines.

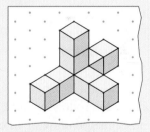

Finish the drawing, erasing the remaining hidden lines and shading.

Ⓐ Checking

4. Build this cube structure. Make an isometric drawing to represent your cube structure.

Ⓑ Practising

5. a) Build this cube structure.

b) Turn it so that you can see the left side.

c) Make an isometric drawing to show what you see.

6. a) Build this cube structure.

b) Make an isometric drawing of the left-side view.

7. Build the following cube structures. Make an isometric drawing of each structure.

a)

b)

c)

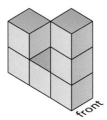

d)

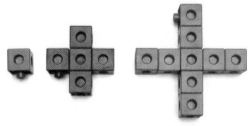

8. a) Build these cube structures. Make an isometric drawing of each structure.

b) Make an isometric drawing of the cube structure you get by layering the cube structures in part (a) with the orange layer on the bottom and the green layer on the top.

c) Imagine adding another layer to the bottom of the cube structure in part (b), or build this layer. Make an isometric drawing to represent the new cube structure.

9. Explain how you know each of the following could be an isometric drawing of the same cube structure.

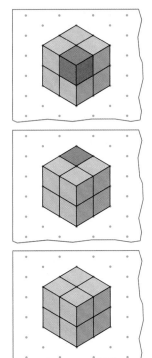

10. The top, front, and right-side views are shown for different cube structures. Build each cube structure and represent it using an isometric drawing.

a)

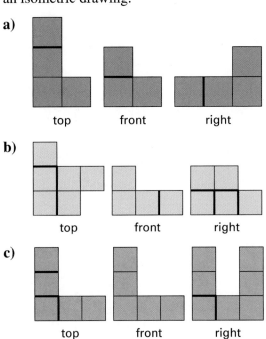

top   front   right

b)

top   front   right

c)

top   front   right

11. For this cube structure, which view is incorrect? Explain.

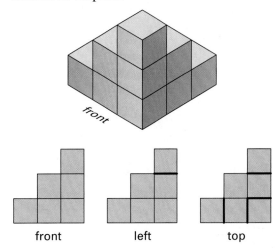

front

front   left   top

⊙ Extending

12. Use the cube structures in question 7.

a) Turn each cube structure a quarter turn clockwise. Make an isometric drawing of each.

b) Repeat part (a) for a half turn clockwise.

13. Sometimes different linking cube structures can be represented by the same isometric drawing.

a) Use different numbers of cubes to build two structures that can be represented by the same isometric drawing for at least one view.

b) Are the structures different for at least one view? Explain.

c) Draw the top, front, left-side, and right-side views for one of your structures.

d) Ask someone to use your drawings to build the cube structure.

e) Explain whether your drawings provided enough information to rebuild your cube structure accurately.

You will need
• a ruler
• triangle dot paper
• 3-D models

10.6 Isometric Drawings of 3-D Objects

▶ **GOAL**

Make realistic drawings of 3-D objects on triangle dot paper.

Learn about the Math

Romona is working on a drawing of the Arc de Triomphe for a geography project.

? **How can you make a realistic drawing of a 3-D object?**

A. Draw the front face of the Arc de Triomphe so that the height is vertical and the width is diagonal.

B. Join the bottom right corner to a grid point to show the depth.

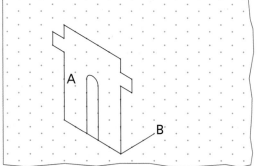

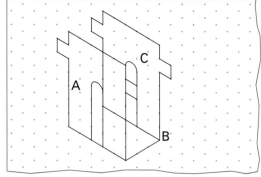

C. Draw a copy of the front face from step A so that it appears behind the front sketch. Use the grid point from step B as the back right corner.

D. Join vertex points on the front face to matching points on the back face.

E. Erase any lines or arcs that would be blocked from view by a solid surface.

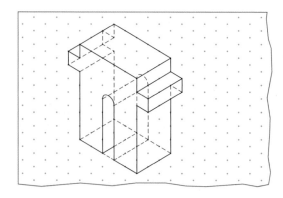

Reflecting

1. Why must the back face of the sketch be a copy of the front face?

2. How does joining matching points on the front and back faces make the sketch look 3-D?

3. How is making an isometric drawing of a 3-D object like making an isometric drawing of a cube structure? How is it different?

Work with the Math

Example 1: Sketching a realistic diagram

Sketch a realistic diagram of a house, using triangle dot paper and shading.

Paul's Solution

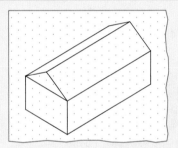

First I drew the outline of a house.

Then I added details (such as doors and windows) and used shading to make my diagram more realistic.

Example 2: Using views to make an isometric drawing

Use triangle dot paper and the following top, front, and right-side views to make a realistic drawing of the structure.

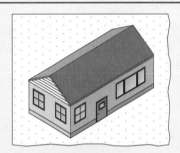

top   front   right

Bonnie's Solution

I marked the vertices of the front and drew the shape.

I used the width of the top and front, and extended the parallel edges. I finished the back edges.

I shaded parallel faces to show depth.

A Checking

4. Use triangle dot paper to make an isometric drawing of this rectangular prism. Look at a 3-D model if you need help.

B Practising

5. Make an isometric drawing of each prism on triangle dot paper. Look at 3-D models if you need help.

a)

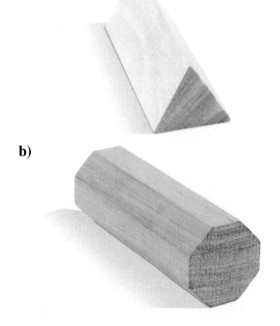

b)

6. Make an isometric drawing of each object. Use triangle dot paper.

a)

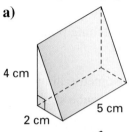

4 cm

5 cm

2 cm

b)

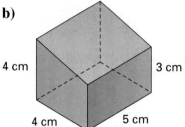

4 cm

3 cm

4 cm

5 cm

c)

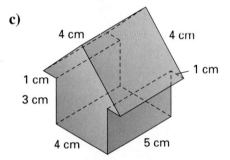

4 cm

4 cm

1 cm

1 cm

3 cm

4 cm

5 cm

7. Make an isometric drawing of each object.

a)

b)

7. c)

d)

8. Make an isometric drawing of each view of the recipe box on triangle dot paper.

a)

b)

9. Choose an object in a picture in this lesson. Make an isometric drawing of the object from a different view.

10. a) Choose an object in the classroom. Make an isometric drawing of it.

b) Make an isometric drawing of the same object from a different view.

c) Explain how your drawings in parts (a) and (b) are the same and how they are different.

11. a) Make an isometric drawing of a prism that has a pentagonal base.

b) Explain how you chose the view for your drawing in part (a).

12. Explain why an isometric drawing looks 3-D.

13. Make an isometric drawing of the object whose top, front, and side views are shown below.

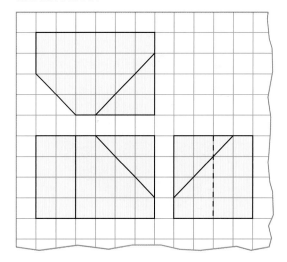

Communicating about Views

▶ **GOAL**

Use mathematical language to describe views of 3-D objects.

Communicate about the Math

Dan is building 3-D objects to display in a math fair. He is creating a brochure about how to build each object.

Dan asked Rosa to read his description and comment on the steps he wrote for building the object.

? **How can Dan improve his description?**

Dan's Brochure

Description of the shape:
The shape is a pentagonal prism. ◀————

Rosa's Questions

What type of pentagon is its base?

Instructions for building the shape:

Step 1: Cut out enough 2 cm squares to make the side faces of the prism. ◀———

How many squares are needed?

Step 2: Cut out the shape for the ends. ◀———

Is the prism open at the back? How large are these shapes for the ends?

Step 3: Tape the pieces together to make the prism. ◀———

Why not show a net for the object?

Reflecting

1. a) Would providing a net improve Dan's instructions for building the prism? Explain.

b) Would showing top, front, and side views improve the instructions? Explain.

2. What parts of the Communication Checklist did Dan cover well? Explain your answer.

3. What parts of the Communication Checklist did Dan not cover well? How can he improve his description in his brochure?

> **Communication Checklist**
>
> ☑ Did you explain your thinking?
>
> ☑ Did you give all the important details?
>
> ☑ Did you use sketches to make your thinking clear?
>
> ☑ Did you use proper math language and terminology?

Work with the Math

Example: Describing how to draw a cube on triangle dot paper

Describe how to draw a cube on triangle dot paper, and use shading to show depth.

Miguel's Solution

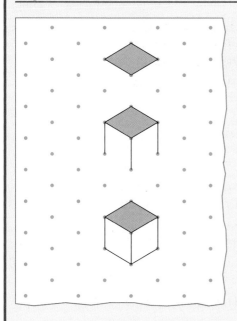

I started by drawing a rhombus on triangle dot paper to represent the top square face. Then I shaded the rhombus.

Then I drew three of the edges of the cube that are perpendicular to this face of the cube. These three edges are parallel on the cube and in the drawing. Each vertical edge is attached to one of the vertices that is on the bottom of the cube.

Next I drew two more edges of the cube that are parallel to the two closest edges of the top face. I drew these edges diagonally to connect them to the ends of the three edges I had already drawn.

The finished shape looks like a cube. Shading the top face is one way to show depth.

A Checking

4. Use Rosa's and your own ideas and write a second draft of Dan's description for his brochure.

B Practising

5. a) Write a first draft for a paragraph. Choose one of the following topics:
  - how to create a net for a 3-D object made up of six equilateral triangles

  or

  - how to make realistic 3-D sketches with shading to show depth

b) Use the Communication Checklist to help you decide
  - what is good about your writing
  - what you can improve

c) Write a second draft.

6. a) Write a first draft for a paragraph that explains how to build a structure using the top, front, and side views shown below.

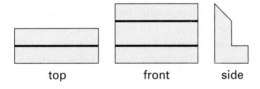

| top | front | side |

b) Use the Communication Checklist to revise your paragraph. Write a second draft.

7. a) Write a paragraph that explains how to use these top, front, and left-side views to build a linking-cube structure.

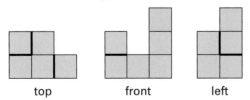

| top | front | left |

b) Use the Communication Checklist to help you decide
  - what is good about your writing
  - what you can improve

c) Check your paragraph by building the structure with linking cubes. Make any changes you think would make the instructions clearer.

8. a) Write a paragraph that explains how to draw the right-side view of your cube structure in question 7.

b) Check your paragraph by drawing the right-side view of the cube structure. Make any changes that you think would help.

9. a) Build a cube structure with up to 10 linking cubes.

b) Write a paragraph that explains how to make an isometric drawing of your cube structure on triangle dot paper.

c) Check your paragraph by making the isometric drawing. Use the Communication Checklist to revise your paragraph.

10. a) Choose an object in your classroom. Use the Communication Checklist to help you write a paragraph that explains one of the following:
  - how to draw the front, top, and right-side views of the object

  or

  - how to create an isometric drawing of the object

b) Trade with a partner. Follow the steps.

11. Suppose that you buy a desk in an assemble-it-yourself kit. Why might you find different views of the assembled desk helpful?

Number of players: 2 to 4

Each group uses 52 blank file cards to create a deck with 13 sets of four cards. Each set must have

- a *name card* with the name of a 3-D shape
- a *shape card* with a sketch of the 3-D shape
- a *view card* with the top, front, and right-side views
- a *net card* with a net that could be folded to make the 3-D shape

Fishing for Solids is similar to the game Go Fish. Players collect sets of four cards: a name card, a shape card, a view card, and a net card. After playing, groups can trade decks for more games.

Rules

1. Deal eight cards to each player. Place the remaining cards face down in a stack.

2. If possible, the first player makes any sets in her or his hand, and places the cards face up for the other players to check.

3. Then the first player chooses another player and describes a card that she or he needs to make a set. A player must have at least one card in a set before asking for another card in that set. To describe a card, the player

   - names the solid
   - states whether the card is a name, shape, view, or net card
   - describes any art on the card so that the other player can identify it

You will need
- 52 file cards

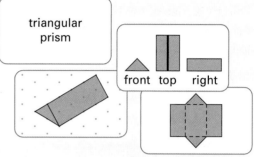

A set of four cards

4. If the other player does not have the card, the first player draws a card from the stack. The turn ends. Play continues clockwise to the next player.

5. If the other player does have the card, it must be given to the first player. The first player then asks any player for any other card in the same set. If the other player does not have the card, the first player draws a card from the stack. The turn ends.

6. If a player has no cards left after giving a card to another player, he or she draws a card from the stack.

7. If a player can make a set after drawing a card from the stack, she or he must wait for the next round before showing the set.

8. The game ends when all 13 sets are collected. The player with the most sets is the winner.

Curious Math

PERSPECTIVE DRAWING IN ART

Albrecht Dürer was a German artist and printmaker (1471–1528). Before Dürer's time, drawings looked 2-D. The art did not give the feeling that some objects were farther away than others.

Dürer was one of the first Renaissance artists to use perspective. He introduced the term "net." He used nets to help him visualize 3-D objects and show perspective.

In 1525, Dürer presented the first examples of polyhedral nets. Dürer's drawing of the net of an icosahedron is shown.

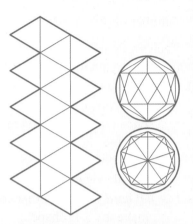

Melancholia I

Kaufmann Haggadah

1. How many different polyhedrons can you see in Dürer's woodcut *Melancholia I*?

2. Compare Dürer's *Melancholia I* with the initial panel from the *Kaufmann Haggadah*, which was created in the late 1300s. Which piece of art gives a better impression of depth? Which appears more two-dimensional? How can you tell? Are there any measurements that you can make to support your conclusions?

Chapter Self-Test

1. a) Name the two polyhedrons that make up this perfume bottle.

b) Create a net for the bottle.

2. Which structure matches these top, back, front, and right-side views?

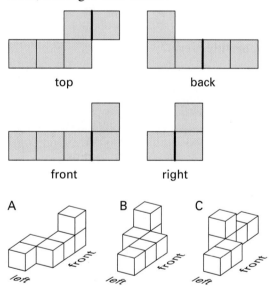

top

back

front

right

3. a) Use these top, front, and left-side views to build a linking-cube structure.

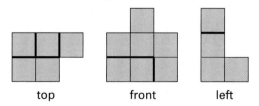

top

front

left

b) Draw the right-side view of the structure in part (a).

4. Draw the top, front, and side views of each structure. (Ignore the ropes.)

a)

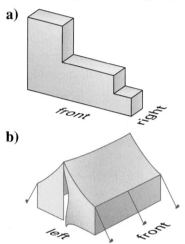

b)

5. Build this cube structure. Make an isometric drawing of it on triangle dot paper.

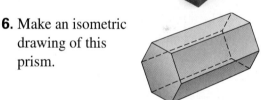

6. Make an isometric drawing of this prism.

7. Write a paragraph that explains how to draw this net and how to use it to build a 3-D shape.

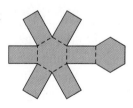

8. Make an isometric drawing of this desk on triangle dot paper.

Chapter Review

Frequently Asked Questions

Q: What are the top, front, and side views of polyhedrons and 3-D objects?

A: The top, front, and side views of polyhedrons and 3-D objects present views that you see by looking straight at the objects.

For example, the top, front, and right-side views of a desk are shown below.

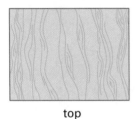 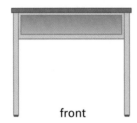 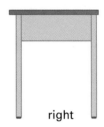

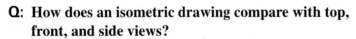

top   front   right

Q: Can different objects have the same front, top, or side views?

A: Yes, different objects can have the same view. For example, these two polyhedrons have the same front and left-side views.

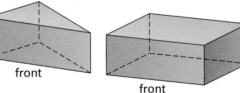

front   front

The top view gives information about the width and depth of an object. The front view gives information about the height and width. Side views give information about the depth and height. Some details may be hidden from any of the views.

Q: What is an isometric drawing?

A: An isometric drawing is a 3-D view of an object in which vertical edges of the object are drawn vertically and width and depth are drawn diagonally. Isometric drawings can be made by joining dots on triangle dot paper.

Q: How does an isometric drawing compare with top, front, and side views?

A: An isometric drawing is a single representation of a 3-D object. The drawing is 2-D, but it appears to be 3-D. Top, front, and side views are separate drawings of a 3-D object. Each view is 2-D, and it appears to be 2-D.

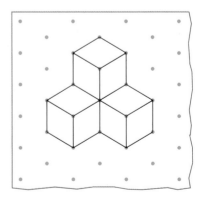

Practice Questions

(10.1) **1. a)** Construct a net for a package that will pack so that there are no gaps.

   b) Without building models, how can you determine whether the package in part (a) will pack so that there are no gaps?

(10.2) **2.** Decide which net(s) will fold to make this structure with a pentagonal pyramid on top of a pentagonal prism.

A   B   C

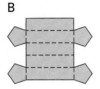

(10.3) **3.** Suppose you are given the right-side view of a cube structure.

   a) Which of the following is a possible left-side view?

right

   A   B   C

   b) Build the shape to check.

(10.4) **4.** Draw the top, front, and right-side views of each structure.

   a)   **b)**

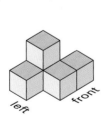 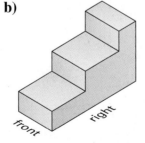

left   front   front   right

5. Identify the polyhedrons with the given views. (10.4)

   a)

   top   front   right

   b)

   top   front   right

6. a) Which 3-D shape matches all three views? (10.4)

   top   front   left

   A   B   C   D

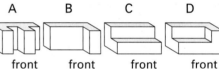

   front   front   front   front

   b) Build the shape to check.

7. Attach a cube to each yellow face. Use triangle dot paper to make an isometric drawing of the new figure. (10.5)

   a)   **b)**

 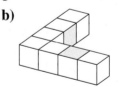

8. Make an isometric drawing of each prism on triangle dot paper. (10.6)

   a)   **b)**

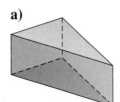 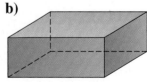

Gift Box Contest

Suppose that Sweet Tooth Candy Company has created a new line of chocolates and needs a fancy gift box that will attract attention. The company is holding a contest at your school to find the best design. The student who creates the best design will win a computer.

? **What will your gift box look like?**

A. Design and construct a unique gift box that will be appealing and attractive to customers.

B. Your contest entry must include
- a description of your gift box and a convincing argument about why your design should be chosen
- a net that can be used to make a model of your gift box
- the top, front, and side views of your gift box that are suitable to be included in an advertisement
- a realistic sketch that can be used in an advertisement

Task Checklist

- ☑ Is your description clear?
- ☑ Did you use correct math language?
- ☑ Did you include all the required diagrams and sketches?
- ☑ Did you draw your diagrams accurately?
- ☑ Did you include a carefully constructed model?

CHAPTER 11

Surface Area and Volume

▶ GOALS

You will be able to

- develop a formula and calculate the surface area of a rectangular prism
- develop a formula and calculate the volume of a rectangular prism
- solve problems involving surface area and volume of rectangular prisms
- describe the relationship between the dimensions and volume of a rectangular prism

Getting Started

You will need
- centimetre linking cubes
- a shoe box
- centimetre grid paper

Filling Space

Simon, Indira, and Kwami are wondering how to win the DVD player. They each decide to use a shoe box and centimetre linking cubes to make a model.

? **How can you estimate how many foam cubes the play box holds?**

Guess how many foam cubes the play box holds, and win **A DVD PLAYER!**

Simon says, "Fill the box with cubes packed together."

Indira says, "Fill the bottom of the box with cubes. This gives the area of the first layer."

Kwami says, "Use cubes to make a corner that shows the length, width, and height of the box."

A. Predict what each student will do next. Explain why.

B. How could you use each student's idea to determine the number of foam cubes the play box holds?

C. How are the students' ideas alike? How are they different?

D. How is estimating the number of foam cubes different from guessing?

Do You Remember?

1. a) Use centimetre grid paper to draw a net for this rectangular prism.

b) Calculate the area of each face.

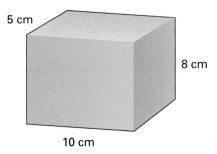

5 cm
8 cm
10 cm

2. Which unit would you use (centimetres, square centimetres, or cubic centimetres) to measure each of the following?

a) the length of ribbon around a photo frame

b) the wrapping paper covering a gift

c) the distance around a pattern block

d) the amount of space in a shoe box

e) the amount of soil in an ant farm

3. a) Draw three rectangles, each with an area of 36 cm^2.

b) Determine the perimeter of each rectangle.

4. The **capacity** of a 1 cm cube is 1 mL. How many 1 cm ice cubes would you need to melt to fill a 1.5 L jar?

1 mL

1 cm^3

5. A centimetre linking cube is smaller than a regular linking cube, as shown. What is the volume of each cube (in cm^3)?

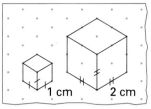

1 cm   2 cm

6. Count the centimetre linking cubes to calculate the volume of each prism.

a)

b)

7. Which dimensions match each item?

A. 7 cm × 7 cm × 20 cm

B. 12 cm × 6 cm × 3 cm

C. 7 cm × 4 cm × 10 cm

D. 17 cm × 6 cm × 25 cm

a)

c)

b)

d)

Surface Area of a Rectangular Prism

You will need
- a calculator
- centimetre square dot paper
- triangle dot paper
- a ruler

▶ **GOAL**

Develop a formula to calculate the surface area of a rectangular prism.

Learn about the Math

Kwami and Jody are wrapping small boxes of wedding cake for souvenirs at a wedding. Kwami says, "We'll need enough wrapping paper to cover the surface, plus about 10% to allow for overlaps and folding."

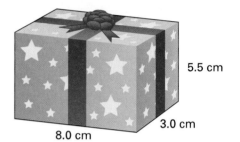

5.5 cm

3.0 cm

8.0 cm

? **How much paper do Kwami and Jody need to wrap each box?**

Example 1: Determining surface area by adding the areas

Use a net to determine the amount of wrapping paper that Kwami and Jody need to wrap each box.

Kwami's Solution

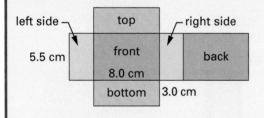

left side — top — right side

5.5 cm

front

8.0 cm

back

bottom 3.0 cm

I imagined unwrapping the box and laying the paper flat. I drew the net of the box, and labelled all the faces.

Each face is a rectangle.

I calculated the area of each face using the formula $A = l \times w$.

Area of front $= 8.0 \text{ cm} \times 5.5 \text{ cm}$ $= 44.0 \text{ cm}^2$	Area of right side $= 3.0 \text{ cm} \times 5.5 \text{ cm}$ $= 16.5 \text{ cm}^2$	Area of top $= 8.0 \text{ cm} \times 3.0 \text{ cm}$ $= 24.0 \text{ cm}^2$
Area of back $= 8.0 \text{ cm} \times 5.5 \text{ cm}$ $= 44.0 \text{ cm}^2$	Area of left side $= 3.0 \text{ cm} \times 5.5 \text{ cm}$ $= 16.5 \text{ cm}^2$	Area of bottom $= 8.0 \text{ cm} \times 3.0 \text{ cm}$ $= 24.0 \text{ cm}^2$

To find the total surface area, I added these areas.

Total Surface Area = front + back + right side + left side + top + bottom
$$= 44.0 \text{ cm}^2 + 44.0 \text{ cm}^2 + 16.5 \text{ cm}^2 + 16.5 \text{ cm}^2 + 24.0 \text{ cm}^2 + 24.0 \text{ cm}^2$$
$$= 169.0 \text{ cm}^2$$

The total surface area is 169.0 cm².

10% of 169.0 cm² is 16.9 cm².

The total amount of wrapping paper that we need is 169.0 cm² + 16.9 cm² = 185.9 cm².

Example 2: Determining surface area by doubling congruent areas

Determine the amount of wrapping paper that Kwami and Jody need to wrap each box.

Jody's Solution

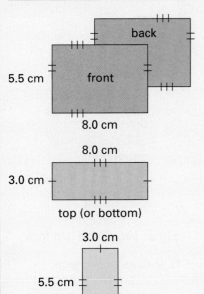

5.5 cm

back

front

8.0 cm

8.0 cm

3.0 cm

top (or bottom)

3.0 cm

5.5 cm

left side (or right side)

I noticed that the front face and the back face have the same area. Also, the top face and the bottom face have the same area. The same is true for the left side and the right side.

I just need to find the area of one of each rectangle, and double it.

Area of front and back = 2 × (8.0 cm × 5.5 cm)
= 2 × 44.0 cm²
= 88.0 cm²

Area of top and bottom = 2 × (8.0 cm × 3.0 cm)
= 2 × 24.0 cm²
= 48.0 cm²

Area of both sides = 2 × (5.5 cm × 3.0 cm)
= 2 × 16.5 cm²
= 33.0 cm²

Total Surface Area = 88.0 cm² + 48.0 cm² + 33.0 cm²
= 169.0 cm²

I added 10% more to allow for overlaps and folding.

The total is 185.9 cm², but I think I should round it so the answer is reasonable.

We need about 190 cm² of wrapping paper for each box.

16.9 cm²
+ 169.0 cm²
185.9 cm²

Reflecting

1. How are Kwami's and Jody's solutions alike? How are they different?

2. Why was Jody able to do half the area calculations that Kwami had to do?

3. After Kwami was done, he noticed that he could add the area of the front, one side, and the top, and then double the sum to get the same answer. Check if he is correct.

4. Write a formula to calculate the surface area of a rectangular prism.

Example 3: Calculating surface area using a formula

Calculate the surface area of this rectangular box.

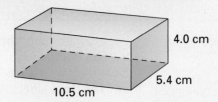

4.0 cm

5.4 cm

10.5 cm

Solution A

Sketch the different faces.

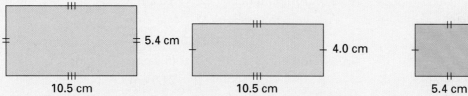

5.4 cm

4.0 cm

4.0 cm

10.5 cm

10.5 cm

5.4 cm

Bottom or top is the 1st face.   Front or back is the 2nd face.   Either side is the 3rd face.

Surface Area = (2 × Area of 1st face) + (2 × Area of 2nd face) + (2 × Area of 3rd face)
= (2 × 10.5 cm × 5.4 cm) + (2 × 10.5 cm × 4.0 cm) + (2 × 5.4 cm × 4.0 cm)
= 113.4 cm^2 + 84.0 cm^2 + 43.2 cm^2
= 240.6 cm^2

Solution B

length = 10.5 cm, width = 5.4 cm, height = 4.0 cm

Area of bottom or top	Area of front or back	Area of either side
= length × width	= length × height	= width × height
= 10.5 cm × 5.4 cm	= 10.5 cm × 4.0 cm	= 5.4 cm × 4.0 cm
= 56.7 cm^2	= 42.0 cm^2	= 21.6 cm^2

Surface Area = 2 × (sum of the 3 different faces)
= 2 × (56.7 cm^2 + 42.0 cm^2 + 21.6 cm^2)
= 2 × 120.3 cm^2
= 240.6 cm^2

Ⓐ Checking

5. a) Draw a net for this rectangular prism on centimetre square dot paper. Then calculate the surface area.

b) Use a formula to calculate the surface area of the prism.

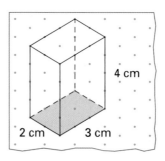

4 cm

2 cm   3 cm

B Practising

6. State the length, width, and height of each prism. Then calculate each surface area.

a)   **b)**   **c)**

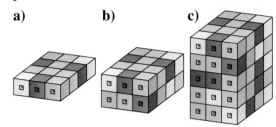

7. a) Sketch a rectangular prism that is 3 cm by 5 cm by 6 cm.

   b) What is the surface area?

8. Determine the surface area of each cube.

a)   **b)**

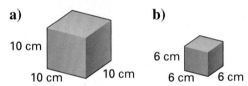

9. a) List the dimensions of all the rectangular prisms you could make with twelve centimetre linking cubes.

   b) Which prism has the greatest surface area?

   c) Which prism has the least surface area?

10. Julie built a small scene inside an open box for a history project. She is wrapping the box to take it to school.

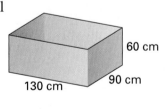

   a) Exactly how much wrapping paper is needed to cover the outside of the box, including the top?

   b) How much wrapping paper will Julie need if she needs an extra 10% for overlaps and folds?

11. Karen is painting the floor and four walls of a basement. The basement is 12.0 m long, 4.0 m wide, and 2.5 m high. Paint comes in 4 L cans. One litre of paint covers 10 m².

   a) What area does Karen need to paint?

   b) How many cans of paint does Karen need?

12. A sports company packages its basketballs in boxes, as shown. The boxes are shipped in wooden crates. Each crate holds 24 boxes.

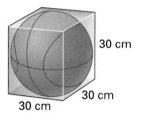

   a) Sketch all the possible shapes of crates that could be used. The dimensions must be whole numbers of centimetres.

   b) Calculate the surface area of each crate you sketched in part (a).

   c) Which crate is made from the least amount of wood?

13. Jordan cuts a cube into small pieces. Is the total surface area of all the pieces less than, greater than, or equal to the surface area of the original cube? Explain your thinking with words, diagrams, and calculations.

C Extending

14. a) Sketch any rectangular prism that has a surface area of 24 cm².

   b) Double the lengths of the sides to draw a new rectangular prism. How does the surface area of your new prism compare with the surface area of your original prism?

   c) Sketch another new rectangular prism with sides that are half the length of the original prism's sides. How does the surface area of this new prism compare with the surface area of your original prism?

Volume of a Rectangular Prism

You will need
- centimetre grid paper
- centimetre linking cubes
- a calculator
- a ruler
- triangle dot paper

▶ **GOAL**

Develop a formula to calculate the volume of a rectangular prism.

Learn about the Math

James says, "I have to pack dice in boxes like this. Each edge of each die is 1 cm long. If I calculate the volume of a box, I'll know how many dice it can hold."

height = 4 cm

length = 5 cm

width = 3 cm

? **How many centimetre linking cubes can you put in the box?**

A. Model James's problem. Use centimetre grid paper to represent the base of the box. Use centimetre linking cubes to represent the dice. What is the area of the base?

B. Stack the cubes to the height of the box. How many layers of cubes do you have?

C. Use your answers in steps A and B to calculate the volume of the box.

D. Repeat steps A to C using a different face as the base of the box.

Reflecting

1. How many choices do you have when selecting a face to use as the base of the box?

2. How does your choice for the base affect the height of the box?

3. How does your choice for the base affect your calculation of the volume of the box?

4. Write a formula that describes how to calculate the volume of a rectangular box, no matter which face you choose as the base.

> **Communication Tip**
>
> Units of volume always have a small, raised 3 written after them: 12 m³. The raised 3 indicates that three dimensions are involved: length, width, and height.

Work with the Math

Calculate the volume of a rectangular box that measures 3.0 m by 4.5 m by 2.0 m.

Indira's Solution

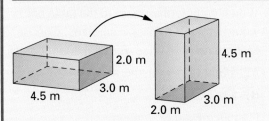

First I sketched the box. To make the calculations easier, I used the face that is 2.0 m by 3.0 m as the base.

Area of base = length × width
= 2.0 m × 3.0 m
= 6.0 m²

I multiplied the length by the width to find the area of the base.

Volume = Area of base × height
= 6.0 m² × 4.5 m
= 27.0 m³

I multiplied the area of the base by the height to find the volume.

The volume of the box is 27.0 m³.

What are some possible dimensions of a rectangular box with a volume of 36 cm³?

Simon's Solution

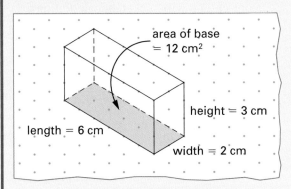

I modelled the box using 36 centimetre linking cubes.

I put 12 centimetre linking cubes in the first layer.

I stacked two more layers to complete the model.

Then I drew my model on triangle dot paper.

Volume = length × width × height
36 cm³ = 6 cm × 2 cm × 3 cm
36 cm³ = 4 cm × 3 cm × 3 cm

I wrote the dimensions for my diagram. Then I calculated a different way to form a base of 12 cm². I think there are others. What are they?

A Checking

5. Which coloured face would you use for the base if you wanted to calculate the volume of each prism? Why?

a)

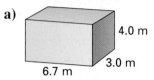

4.0 m
3.0 m
6.7 m

b)

4.3 cm
2.0 cm
5.0 cm

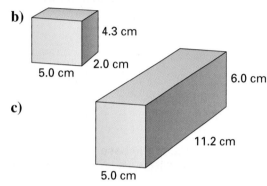

6.0 cm
11.2 cm
5.0 cm

c)

6. Calculate the volume of each prism in question 5.

B Practising

7. Calculate the volume of each prism.

a) 2 cm
12 cm   3 cm

d) 1.5 m
8.0 m   1.0 m

b) 6.5 cm
20.0 cm   10.0 cm

e) 4.5 m
7.5 m   2.5 m

c) 8.5 cm
12.0 cm   10.0 cm

f) 4.0 m
3.2 m   3.0 m

8. a) Sketch a rectangular prism that has a volume of 500 cm³ and sides that are whole numbers.

b) Sketch a rectangular prism that has a volume of 500 cm³ and some sides that are *not* whole numbers.

9. Sketch a rectangular prism with each set of dimensions. Calculate the volume of the prism.

a) $l = 8$ cm, $w = 8$ cm, $h = 8$ cm

b) $l = 0.5$ m, $w = 0.5$ m, $h = 2.0$ m

c) $l = 3.5$ km, $w = 2.0$ km, $h = 3.0$ km

10. Which rectangular prism in the table below would you use to pack each item? Choose each prism only once. Explain your choices.

A. a box of pencils

B. a basketball

C. earrings

D. one CD

E. ten CDs

	Length (cm)	Width (cm)	Height (cm)
a)	14	12	1
b)	6	6	1
c)	40	40	40
d)	14	12	10
e)	19.0	3.5	1.5

11. Anthony needs to buy a box of nails for his carpentry project. The hardware store sells these two boxes of nails for the same price. Which box should Anthony buy? Explain your choice with a sketch, calculations, and words.

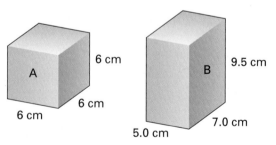

6 cm
A
6 cm
6 cm

9.5 cm
B
7.0 cm
5.0 cm

12. List two sets of dimensions (length, width, and height) that will result in a box with each volume.

a) 24 cm³   **b)** 27 cm³   **c)** 51.2 cm³

13. Copy and complete the table by filling in the missing dimensions of the rectangular prisms.

	Length (cm)	Width (cm)	Height (cm)	Volume (cm³)
a)	6	6	8	
b)	4.5	5.0		216.0
c)	3		3	27
d)	2.5	8.4		52.5
e)		7	7	343

14. This pool is 10 m long by 6 m wide. The pool is 1.5 m deep.

a) If a volume of 1 m³ holds 1000 L, how many litres of water can the pool hold?

b) How many litres of water are in the pool when the pool is 90% full?

15. Regan folded this net into an open box. Calculate the volume of the box. Explain what you did.

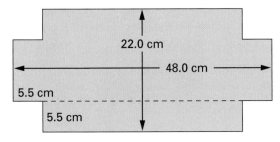

16. Brand A tissues come in a box that is 22.5 cm long, 12.0 cm wide, and 6.8 cm high. It sells for 79¢. Brand B tissues come in a box that is 13 cm long, 13 cm wide, and 8 cm high. It sells for $1.99. Which brand is the better buy? Explain your answer.

17. Humidifier W is 37.0 cm long, 24.0 cm wide, and 23.5 cm high. Humidifier X is 50 cm long, 40 cm wide, and 11 cm high. Humidifier Y is 40 cm long, 35 cm wide, and 20 cm high. Which humidifier holds the most water? Explain your answer.

● Extending

18. In each case, rectangle A has the same shape as rectangle B, but it is a different size. The rectangles form the sides shown in each prism. Determine the unknown side length of B, and calculate the volume of each prism.

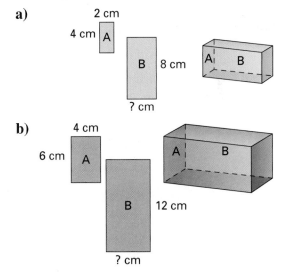

19. A calculator is tightly wrapped in protective padding. It is sold in a box that is 14.5 cm wide, 24.0 cm long, and 6.0 cm high. The calculator itself is 8.5 cm wide, 18 cm long, and 2.5 cm high. What is the volume of the padding?

Mid-Chapter Review

Frequently Asked Questions

Q: How do you calculate the surface area of a rectangular prism?

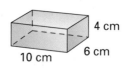

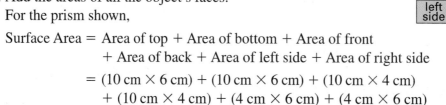

A1: Add the areas of all the object's faces.
For the prism shown,

Surface Area = Area of top + Area of bottom + Area of front
+ Area of back + Area of left side + Area of right side

= (10 cm × 6 cm) + (10 cm × 6 cm) + (10 cm × 4 cm)
+ (10 cm × 4 cm) + (4 cm × 6 cm) + (4 cm × 6 cm)

= 60 cm² + 60 cm² + 40 cm² + 40 cm² + 24 cm² + 24 cm²

= 248 cm²

A2: Notice that the front and the back are congruent, the left side and the right side are congruent, and the top and the bottom are congruent. Calculate the areas of the front or back, left side or right side, and top or bottom. Double each area, and then add.

Surface Area = (2 × 60 cm²) + (2 × 40 cm²) + (2 × 24 cm²)

= 120 cm² + 80 cm² + 48 cm²

= 248 cm²

A3: Calculate the areas of the front or back, left side or right side, and top or bottom. Add the three areas, and then double.

Surface Area = 2 × (60 cm² + 40 cm² + 24 cm²)

= 2 × 124 cm²

= 248 cm²

Q: How do you calculate the volume of a rectangular prism?

A: Choose one face to use as the base. For the prism shown, the darker blue face is a good choice because the dimensions are whole numbers and are easier to multiply.

Multiply the length by the width to calculate the area of the base. Multiply the area of the base by the height to calculate the volume of the prism.

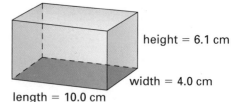

Volume = Area of base × height

= (10.0 cm × 4.0 cm) × 6.1 cm

= 244.0 cm³

Practice Questions

(11.1) **1.** Calculate the surface area of each item. Sketch the item, and label its dimensions.

Item	Length	Width	Height
a) tissue box	22 cm	7 cm	10 cm
b) cereal box	16.3 cm	5.0 cm	27.5 cm
c) candy box	17.5 cm	2.5 cm	6.5 cm
d) CD case	140 mm	10 mm	125 mm

(11.1) **2.** Meagan needs to paint the outside of a box for an art project. The box is 18 cm by 5 cm by 2 cm. What surface area does she need to paint?

(11.1) **3.** Explain how you would determine the surface area of an open-top rectangular box. Draw a net to support your explanation.

(11.1) **4.** A cube is cut in half. Explain why the total surface area of the two halves is more than the surface area of the original cube.

(11.1) **5.** Can you use the same number of linking cubes to create rectangular prisms with different surface areas? Sketch some prisms made from 16 cubes to support your answer.

(11.2) **6.** Calculate the volume of this aquarium.

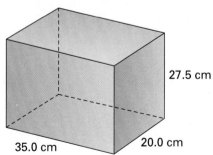

35.0 cm 20.0 cm 27.5 cm

7. Calculate the volume of each item. (11.2)

a) a printer cartridge box
13.5 cm by 3.5 cm by 11.5 cm

b) a magazine box
24.0 cm by 10.0 cm by 30.5 cm

c) a guitar amplifier road case

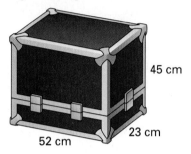

45 cm 23 cm 52 cm

8. Mercury Courier charges $0.10/cm³ to deliver a package. How much will the courier service charge to deliver each of these packages? (11.2)

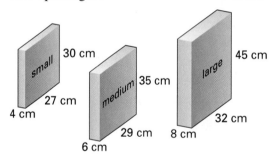

small 30 cm 27 cm 4 cm
medium 35 cm 29 cm 6 cm
large 45 cm 32 cm 8 cm

9. What is the capacity of the aquarium in question 6? (*Hint*: 1 cm³ = 1 mL) (11.2)

10. a) Which rectangular prism has a greater volume?

b) Which rectangular prism has a greater surface area? (11.2)

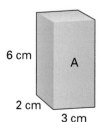
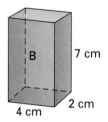

6 cm A 2 cm 3 cm
B 7 cm 4 cm 2 cm

11.3 Solve Problems by Guessing and Testing

▶ **GOAL**

Use guess and test and a table to solve measurement problems.

You will need
• a calculator

Learn about the Math

Jody is designing a snack box. The box must be 10 cm high and have a volume of 450 cm^3. Jody wants to use the least amount of cardboard possible.

Jody says, "I want the surface area to be as small as possible so I don't waste the cardboard."

? **How can Jody determine the best dimensions the box should have?**

1 Understand the Problem

Jody knows that the volume of the box must be 450 cm^3 and the box must be 10 cm high. She knows that she has to use the least amount of cardboard to make the surface area as small as possible.

2 Make a Plan

The height is 10 cm. Jody reasons that the area of the base must multiply with the 10 cm height to produce a volume of 450 cm^3. The area of the base must be 45 cm^2.

Jody decides to use a calculator to find combinations of numbers that multiply to 45 for the length and width. She will then use the combination that gives the least surface area.

3 Carry Out the Plan

Jody organizes her guesses in a table. She uses a calculator and rounds her results when necessary. She compares her results to find the set of measurements that gives the least surface area.

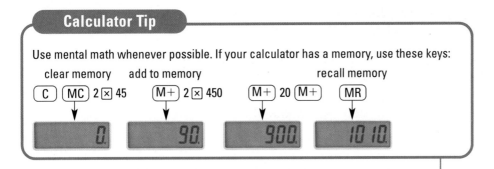

Length (cm)	Width (cm)	Height (cm)	Surface Area $= 2 \times (l \times w) + 2 \times (w \times h) + 2 \times (l \times h)$
1.0	45.0	10.0	$= 2 \times (1.0 \text{ cm} \times 45.0 \text{ cm}) + 2 \times (45.0 \text{ cm} \times 10.0 \text{ cm}) + 2 \times (1.0 \text{ cm} \times 10.0 \text{ cm})$ $= 90.0 \text{ cm}^2 + 900.0 \text{ cm}^2 + 20.0 \text{ cm}^2$ $= 1010.0 \text{ cm}^2$
2.0	22.5	10.0	580.0 cm^2
3.0	15.0	10.0	450.0 cm^2
4.0	11.25	10.0	395.0 cm^2
5.0	9.0	10.0	370.0 cm^2
6.0	7.5	10.0	360.0 cm^2
7.0	6.43	10.0	358.6 cm^2
8.0	5.625	10.0	362.5 cm^2

Area of base $= l \times w$

$45 \text{ cm}^2 = 7 \text{ cm} \times ? \text{ cm}$

$45 \div 7 = $ *6.4285714*

Round to 6.43.

Jody finds that a length of 7 cm and a width of about 6.4 cm result in a box with the least surface area.

4 **Look Back**

Jody looks at the column for surface area. She notices that the numbers decrease until you reach the box with length 8.0 cm. She thinks the numbers will increase again because she already knows the result for the box with length 9.0 cm. Also, Jody notices that the box with length 7.0 cm has an almost equal length and width.

Reflecting

1. How did the guess-and-test strategy help Jody find a solution to the problem?

2. Suppose that Jody did not have a calculator. How would this have affected her solution to the problem?

3. Jody found a good solution, but not the best possible solution. How does her reasoning in step 4 (Look Back) provide a clue to improving her solution?

4. What other strategy could Jody have used to solve the problem?

Work with the Math

Example: Determining the dimensions of a box using guess and test

Simon is designing a rectangular box that will hold 42 dice in a single layer. The dice are 1 cm cubes. What dimensions will require the least amount of cardboard?

Simon's Solution

1 Understand the Problem

I have to find the set of dimensions that gives the least surface area. The dimensions have to be whole numbers, since the dice are 1 cm cubes. The height of the box must be 1 cm, because the dice have to be arranged in a single layer.

2 Make a Plan

Since I know the height, I only need to calculate the area of the base. I'll try all the possible length and width combinations that give the area I need.

3 Carry Out the Plan

Since the height is 1 cm, the area of the base must be 42 cm². Since the length and the width are whole numbers, they must be factors of 42. I'll try all the factors of 42.

l (cm)	*w* (cm)	*h* (cm)	$SA = 2 \times (l \times w) + 2 \times (w \times h) + 2 \times (l \times h)$
42	1	1	$= 2 \times (42 \text{ cm} \times 1 \text{ cm}) + 2 \times (1 \text{ cm} \times 1 \text{ cm}) + 2 \times (42 \text{ cm} \times 1 \text{ cm})$ $= 84 \text{ cm}^2 + 2 \text{ cm}^2 + 84 \text{ cm}^2$ $= 170 \text{ cm}^2$
21	2	1	130 cm²
14	3	1	118 cm²
7	6	1	110 cm²

A box that is 7 cm by 6 cm by 1 cm will have the least surface area, 110 cm².

4 Look Back

Simon checks his work. He makes sure that he has used all the factors of 42 and that his calculations are correct. Simon concludes that his answer is correct.

A Checking

5. a) Determine two sets of dimensions for a box with a volume of 1500 cm³. Use a calculator.

b) Calculate the surface area for each set of dimensions.

c) Explain which set of dimensions would be better if you wanted to use less cardboard to make a box with a volume of 1500 cm³.

B Practising

6. a) Determine two sets of dimensions for a box with a volume of 3000 cm³.

   b) Calculate the surface area for each set of dimensions.

   c) Explain which set of dimensions would be better if you wanted to use less material to make the box.

7. Someone spilled juice on these plans for a playground.

   a) List two different sets of dimensions (length and width) that the playground might be.

   b) Which set of dimensions results in a playground with a smaller area?

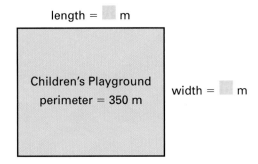

length = ▓ m

Children's Playground
perimeter = 350 m

width = ▓ m

8. Winnie is using this diagram to make a wooden picture frame. Some of the writing has been smudged.

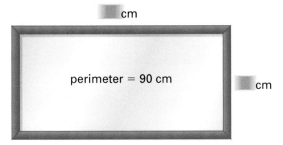

▓ cm

perimeter = 90 cm

▓ cm

   a) List two different sets of dimensions (length and width) that the picture frame could be.

   b) Which set of dimensions results in a frame with a greater area?

9. An Egyptian mummy stands upright in a case that is 3.5 m high and has a volume of 14 m³. The width of the case is half the length. What are the length and width of the case?

10. a) A box has a volume of 570 cm³ and a surface area of 478 cm². What are the length, width, and height of the box?

   b) A box has a volume of 546 cm³ and a surface area of 422 cm². What are the length, width, and height of the box?

11. a) A box has a volume of 308 cm³. Its height is 4 cm greater than its length. Its length is 3 cm greater than its width. What is the width of the box?

   b) A box has a volume of 195 cm³. Its height is 8 cm greater than its length. Its length is 2 cm greater than its width. What is the width of the box?

12. A CD jewel case is 12.5 cm by 1.0 cm by 13.5 cm. You want to design a storage box that will hold 24 CDs. Use your calculator to determine two different sets of dimensions for the CD storage box. Use words, diagrams, and calculations to explain which set of dimensions you would use.

13. Expand the work you did in question 6. Use a calculator to find the best solution for using the least amount of material.

Relating the Dimensions of a Rectangular Prism to Its Volume

You will need
- centimetre linking cubes
- triangle dot paper

▶ **GOAL**

Discover how changing the sides of a rectangular prism affects its volume; sketch a rectangular prism given its volume.

Learn about the Math

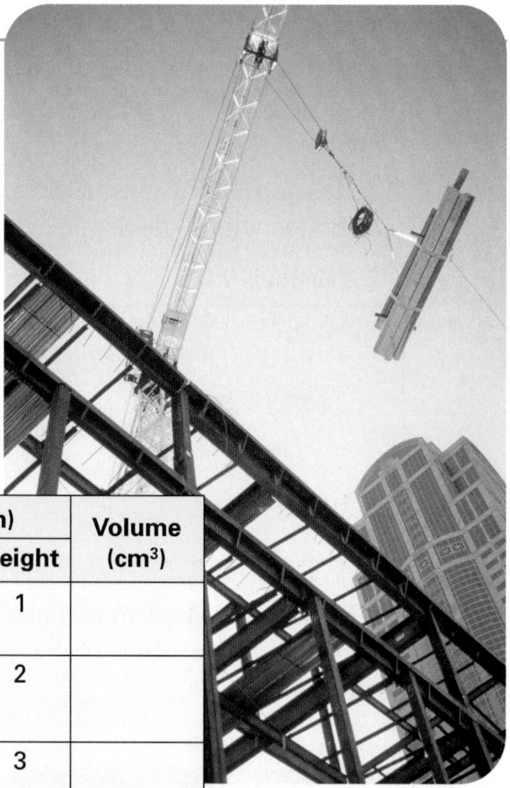

A new building has a square base and will have 16 rooms on each floor.

? **How many rooms could be in the finished building?**

A. Use centimetre linking cubes to model a rectangular prism that is one layer high. Calculate the volume, and record it in a copy of this table.

Model	Dimensions (cm)			Volume (cm³)
	Length	Width	Height	
	4	4	1	
	4	4	2	
	4	4	3	
	4			

B. Increase the height by one layer at a time. Calculate each volume, and record it in your table.

C. How does the volume change as the height increases?

D. How many rooms will there be if the finished building has 4 storeys? 8 storeys?

Reflecting

1. Why did the volume increase each time you added a layer?

2. Find two volumes in your table, where one is double the other. Compare the heights. What do you notice?

3. Find two volumes in your table, where one is triple the other. Compare the heights. What do you notice?

Work with the Math

Example: Doubling other dimensions in a rectangular prism

Does the volume double if you double one of the dimensions? Use diagrams to support your answer.

James's Solution

I drew diagrams on triangle dot paper and recorded the measurements in a table.

Dimensions (cm)			Volume (cm³)
Length	Width	Height	
2	1	1	2
2	2	1	4
2	4	1	8
2	8	1	16
2	16	1	32

First I doubled the width.

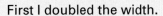

If you double the width, you double the volume.

Dimensions (cm)			Volume (cm³)
Length	Width	Height	
1	2	2	4
2	2	2	8
4	2	2	16
8	2	2	32
16	2	2	64

Then I doubled the length.

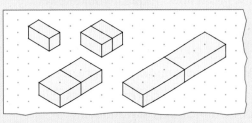

If you double the length, you double the volume.

Ⓐ Checking

4. a) Sketch a rectangular prism that has a volume of 200 m³.

b) The height of the prism is doubled. The other dimensions remain the same. What is the volume now?

c) The height of the original prism is tripled. The other dimensions remain the same. What is the volume now?

d) The length of the original prism is doubled. The other dimensions remain the same. What is the volume now?

B Practising

5. The volume of a rectangular prism is 200 cm³. The height of the prism is 10 cm.

    a) The height of the prism changes, causing its volume to become 400 cm³. The other dimensions remain the same. What is the new height of the prism? Use sketches to support your answer.

    b) The height of the original prism changes, causing its volume to become 600 cm³. The other dimensions remain the same. What is the height of the prism now? Explain your answer.

    c) The height of the original prism changes again, causing its volume to become 100 cm³. The other dimensions remain the same. What is the height of the prism now? Explain your answer.

6. A rectangular prism is 5 m high by 3 m long by 6 m wide.

    a) What is the volume of the prism?

    b) The height of the prism doubles. The other dimensions stay the same. What is the volume of the prism now?

    c) The length of the original prism doubles, and the other dimensions stay the same. What is the volume now?

7. A box is 7.5 cm high, 10.2 cm long, and 2.0 cm wide.

    a) What is the volume of the box?

    b) What is the volume of the box if its height triples?

    c) What is the volume of the box if its length is multiplied by 4?

8. A toy company sells model cars, trucks, and airplanes in boxes. The company's standard box is 9.0 cm high, 14.5 cm long, and 8.0 cm wide.

    a) What is the volume of the standard box?

    b) Models of transport trucks are sold in boxes that are twice as long as the standard box. The other dimensions are the same. What is the volume of a model truck box?

    c) Models of jet airplanes are sold in boxes that are half as high and twice as wide as the standard box. The length is the same. What is the volume of a model airplane box?

9. A rectangular prism is 2.5 cm high, 3.4 cm long, and 5.6 cm wide.

    a) What is the volume of the prism?

    b) The height of the prism changes, causing its volume to become 142.8 cm³. The length and the width stay the same. What is the new height?

    c) The width of the original prism changes, causing its volume to become 95.2 cm³. The length and the height stay the same. What is its width now?

C Extending

10. A box of chocolates is 5 cm high, 7 cm long, and 12 cm wide.

    a) What is the volume of the box?

    b) The length of the box is doubled, and the width is halved. Has the volume changed? Explain your answer.

    c) The height and length of the original box are both doubled. The width is divided by 4. Has the volume changed? Explain your answer.

Number of players: any number

Rules

1. Place your counter on START.

2. When it is your turn, move one space. (More than one player may be on a space.)

3. Spin to get the height (in centimetres) of the prism you chose.

4. Calculate the volume of the prism using the given area of the base.

5. Keep a running total of the volumes and a tally for the number of turns.

6. The winner is the player who is on a space that touches FINISH after exactly 11 turns and has the greatest total volume.

You will need

- Turn Up the Volume! gameboard
- a calculator
- a spinner (or die)
- different counters

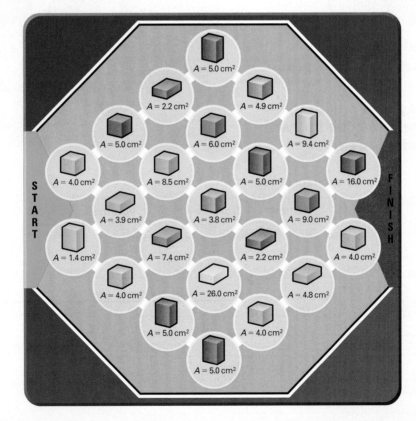

Exploring the Surface Area and Volume of Prisms

▶ **GOAL**

Investigate relationships between surface area and volume of cubes and other rectangular prisms.

Explore the Math

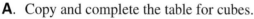

For a science fair, Jody and her group are designing a small oven. The oven will be in the shape of a rectangular prism. Jody knows that they need to combine the least possible surface area with the greatest possible volume to design the most heat efficient oven.

Jody says, "I wonder if there's a cube that has the same number value for both its surface area and its volume."

Kwami says, "Would this cube be the prism with the least surface area for its volume?"

? **Can you find a prism with the same volume as a cube but with a smaller surface area?**

A. Copy and complete the table for cubes.

Length (cm)	Width (cm)	Height (cm)	Surface Area (cm²)	Volume (cm³)
1	1	1		
2	2	2		
3	3	3		
4	4	4		
5	5	5		
6	6	6		
7	7	7		
8	8	8		
9	9	9		
10	10	10		

B. Which cube has the same number value for its surface area and its volume?

C. Sketch rectangular prisms that are not cubes but have the same volume as the cube you named in step B. Sketch as many different prisms, with whole number dimensions, as you can.

D. Calculate the surface area of each prism you sketched in step C.

Reflecting

1. What are the dimensions of the cube whose surface area and volume have the same number value?

2. When does the volume of a cube have a greater value than the surface area of the cube? Describe the pattern in the table.

3. In step C, you examined rectangular prisms that have the same volume as one cube. Describe how the surface areas of these prisms compare with the surface area of the cube. What conclusions can you make, based on your results?

4. Of all the rectangular prisms you examined, which one has the least volume compared with its surface area? Use your answer to explain what shape Jody's oven should have.

Mental Math

CHOOSING EASILY MULTIPLIED PAIRS

You can often multiply three or more numbers by choosing pairs of easily multiplied numbers.

To multiply $2 \times 8 \times 4.5$, multiply 2 and 4.5 first instead of multiplying 2 and 8.

$2 \times 8 \times 4.5 \qquad 8 \times 9 = 72$

$\underbrace{\qquad\qquad}_{9}$

1. How would the steps differ if you multiplied 8 and 4.5 first?

2. Explain how to multiply any even number by 0.5 in your head.

3. Multiply.

a) $4 \times 9 \times 2.5$ d) $3.5 \times 9 \times 2$ g) $2.5 \times 2.5 \times 4$ j) $20 \times 8 \times 50$

b) $32 \times 8 \times 0.5$ e) $9 \times 5 \times 1.2$ h) $7.5 \times 8 \times 2$ k) $40 \times 1.17 \times 2.5$

c) $0.5 \times 9 \times 8$ f) $24 \times 0.5 \times 10$ i) $87 \times 20 \times 0.5$ l) $1.1 \times 0.5 \times 18$

TO THE MAX

Get paper, a ruler, and scissors.

Cut four congruent squares from the corners.

You can now fold the paper into an open box.

1. Cut ten pieces of paper that each measure 16 cm by 12 cm.

2. Cut out four congruent squares from the corners of each piece of paper. Use the side lengths in the following table, plus five more side lengths, to create ten different nets.

Side of square corner (cm)	Volume of box (cm³)
1.0	
1.5	
2.0	
2.5	
3.0	

Copy and extend the table to record the ten side lengths.

3. Fold each net to form an open box.

4. Calculate the volume of each box, and record the volume in your table.

5. What corner size makes the box with the greatest volume?

1. Calculate the surface area of each rectangular prism.

a)

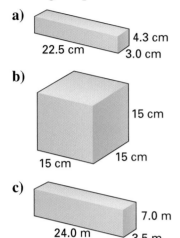

4.3 cm
22.5 cm
3.0 cm

b)

15 cm

15 cm

15 cm

c)

7.0 m
24.0 m
3.5 m

2. a) Determine the volume of this prism.

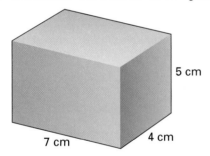

5 cm

7 cm   4 cm

b) Triple the width, length, and height of the prism. What is the volume now?

3. These three backpacks each cost $20. Which backpack is the best buy? Explain your thinking.

28 cm   C

14 cm   12 cm

26 cm   A

32 cm   B

20 cm   10 cm

16 cm   8 cm

4. Copy and complete the table by calculating the missing dimensions. Sketch each prism.

	Height	Length	Width	Volume
a)	cm	6 cm	1 cm	30 cm³
b)	5 m	m	2 m	50 m³

5. Icarus Airlines does not allow luggage that is more than 22 704 cm³ in volume to be carried on board an airplane.

a) Suppose that you work as a security guard for this airline. Will you allow this suitcase on the plane? Why or why not?

60 cm

35 cm

15 cm

b) Calculate the volume of this suitcase.

c) How much fabric did the manufacturer use to make this suitcase?

6. A rectangular prism has a volume of 1716 cm³. Its three dimensions are consecutive whole numbers. What is its surface area?

7. The sides of a cube are doubled in length. How many times larger is the surface area of the cube?

Chapter Review

Frequently Asked Questions

Q: How does the volume of a rectangular prism change if one of its dimensions changes?

A: The volume increases or decreases by the same factor that the dimension increases or decreases. For example, suppose that a rectangular prism has a volume of 10 cm³. If its length is doubled, its volume will double to 20 cm³. If its length is halved, its volume will be half, or 5 cm³.

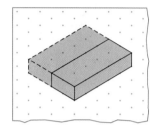

Q: Do the surface area and volume of a cube increase at the same rate when the dimensions of the cube are increased?

A: The surface area and volume increase at different rates.

Suppose that the length, width, and height of a cube increase in the sequence 1 cm, 2 cm, 3 cm, 4 cm, 5 cm, 6 cm, 7 cm, ….

The surface area of the cube will increase in the sequence 6 cm², 24 cm², 54 cm², 96 cm², 150 cm², 216 cm², 294 cm², ….

The volume of the cube will increase in the sequence 1 cm³, 8 cm³, 27 cm³, 64 cm³, 125 cm³, 216 cm³, 343 cm³, ….

At first, the value of the surface area is greater than the value of the volume. After six terms, however, the value of the volume becomes greater than the value of the surface area.

Practice Questions

(11.1) **1. a)** Determine the surface area of this prism.

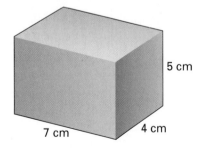

5 cm

7 cm   4 cm

b) Double the width, length, and height of the prism. What is the surface area now?

2. Calculate the surface area of each rectangular prism.   (11.1)

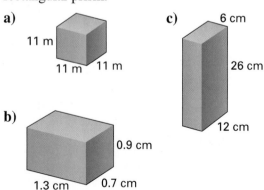

a) 11 m
11 m   11 m

b) 0.9 cm
1.3 cm   0.7 cm

c) 6 cm
26 cm
12 cm

(11.2) **3.** Calculate the volume of each rectangular prism.

a)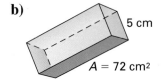
7 cm
$A = 49$ cm²

c)
8 cm
$A = 48$ cm²

b)
5 cm
$A = 72$ cm²

(11.2) **4.** Use this net to determine the surface area and volume of the folded-up box.
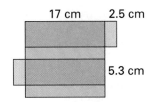
17 cm   2.5 cm
5.3 cm

(11.2) **5.** Jeanette is comparing two full boxes of buttons at a store. Both boxes cost $2.99. Explain which box is the better buy.

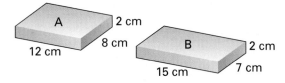
A
2 cm
8 cm
12 cm
B
2 cm
15 cm
7 cm

(11.3) **6.** Mr. Armour is designing a rectangular cargo car for a railway train. The car must have a volume of 400 m³ and be at least 20 m long.

a) Determine two possible sets of dimensions the car could have.

b) Which set of dimensions has the least surface area?

7. Three different rectangular prisms have the same volume, 24 m³. Will they have the same surface area? Explain your thinking using numbers, diagrams, and words. (11.4)

8. The volume of a rectangular box is 50 m³.

a) The height of the box is doubled. The other dimensions remain the same. What is the volume now?

b) The width of the original box is tripled. The other dimensions remain the same. What is the volume now?

c) The length of the original box is doubled. The other dimensions remain the same. What is the volume now?

d) The height and width of the original box are doubled. The length remains the same. What is the volume now?

e) The height and length of the original box are doubled, and the width is halved. What is the volume now?　　(11.4)

9. A cube is 2 cm by 2 cm by 2 cm. Its height is doubled once, then again and again.

a) Copy and complete this table.

Area of base (cm²)	Height (cm)	Volume (cm³)	Surface Area (cm²)
4	2		
4	4		
4	8		
4	16		

b) What happens to the volume of a cube as its height doubles?

c) What happens to the surface area of a cube as its height doubles?　　(11.5)

Adventurepark Statue

The Adventurepark planning committee wants to put a large statue of the park mascot beside the park gate. The committee is holding a competition to choose a design for the statue. You need to create a design, build a model (using cubes) that is based on your design, and write a proposal that will persuade the committee to choose your design.

Task Checklist

☑ Did you show all your steps?

☑ Did you explain your thinking?

☑ Did you draw and label your diagrams neatly and accurately?

☑ Did you use appropriate math vocabulary?

? **What information about the size and cost of your design can you give the committee?**

A. Design a statue for Adventurepark. Use 50 linking cubes, in different colours, to build a model based on your design.

B. Calculate the surface area and volume of your model.

C. Suppose that 1 cm on your model represents 1.5 m on the statue.

    a) Exterior paint comes in 3.78 L cans that cost $24 each. The paint in one can will cover 18 m². How much will the paint for the statue cost?

    b) How many litres of each colour of paint will be needed to paint the statue?

    c) The statue will be hollow. How many cubic metres of sand will be needed to fill it?

D. Write a proposal that will persuade the Adventurepark committee to choose your design. Support your proposal with diagrams and calculations, as well as your model.

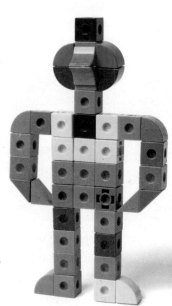

Probability

Lucky 7

Tynessa has a 20-sided die that shows the numbers 1 to 20. She also has two spinners, a special 10-by-10 dartboard, and a standard deck of 52 playing cards.

Tynessa's favourite number is 7. To get a 7, she could use any of the following methods:

- Roll the die, and read the number on the top face.
- Spin spinner A.
- Spin spinner B.
- Throw a dart at the 10-by-10 dartboard.
- Remove the 12 face cards from the deck, and draw any card without looking. (Treat each ace as a 1.)

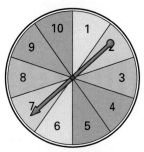

spinner A

? **Which method is most likely to give Tynessa a 7?**

A. Copy the table. Use fractions to write the probability of getting a 7 using each method. Record the fractions in the "Fraction form" column.

Method	Probability of getting a 7		
	Fraction form	Decimal form	Percent form
20-sided die			
spinner A			
spinner B			
10-by-10 dartboard			
deck of cards (no J, Q, or K)			

B. Complete the table with the decimal form and percent form.

C. Which method gives Tynessa the greatest probability of getting a 7?

D. Which way of writing the probability (fraction, decimal, or percent form) do you think is best for comparing the methods? Explain.

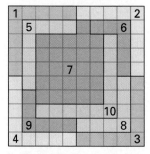

spinner B

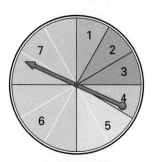

10-by-10 dartboard

Do You Remember?

1. Kyle rolled two six-sided dice, with the numbers 1 to 6, and added the values. This bar graph shows his results.

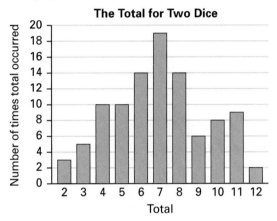

The Total for Two Dice

a) Which total occurred most often?

b) Which total occurred least often?

c) Which totals occurred the same number of times?

d) Which total occurred six times?

e) How often did the total 10 occur?

2. Copy and complete the table. Write the fractions in **lowest terms**.

	Fraction	Decimal	Percent
a)	$\frac{7}{10}$		
b)			10%
c)	$\frac{9}{100}$		
d)		0.88	
e)			55%
f)			34%
g)	$\frac{3}{5}$		
h)		0.52	

3. Tell how to use a calculator to express a fraction as a decimal; for example, $\frac{4}{5} = 0.8$.

4. What are the **possible outcomes** when you toss a coin?

5. A paper bag holds three marbles: one green, one red, and one blue.

   a) Suppose that you remove one marble at **random** from the bag. What are the possible outcomes?

   b) Suppose that you put the marble from part (a) back in the bag. Then you remove a marble from the bag again. What are the possible two-marble outcomes?

   c) Suppose that you put the marble back, then remove a marble again. What are the possible three-marble outcomes?

6. Suppose that you remove three marbles as in question 5, but you don't put the marble back each time. Copy and complete the **tree diagram** to determine the number of possible outcomes.

1st marble	2nd marble	3rd marble	Outcome
G	R → B	B	G-R-B
	B →	R	G-B-R

7. Toss two coins 20 times and record your results. Use your results to write these fractions.

   a) the fraction for the number of times out of 20 that both coins show Heads

   b) the fraction for the number of times out of 20 that one coin shows Heads and one coin shows Tails

Exploring Probability

▶ **GOAL**

Determine probability from an experiment.

Explore the Math

Last year, Omar played 20 baseball games. He was up to bat three times per game. On average, he gets a hit once in these three times per game.

? **What is the probability that Omar will get two hits in one game?**

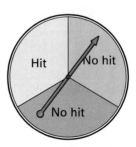

A. Do an experiment with a spinner like this. Each trial of three spins represents Omar's three at-bats per game. Copy the following table, and record the results of each at-bat for 20 trials.

Trial number	First at-bat	Second at-bat	Third at-bat
1			
2			
3			

B. Copy the following table. One possible outcome for a trial is no hit, no hit, hit. (You can write this as N-N-H.) Copy the following table. List all the possible outcomes for your experiment in the "Outcome" column.

Outcome	Frequency	Experimental probability based on your data		
		Fraction	Decimal	Percent

event

a set of one or more possible outcomes for a probability experiment

C. In your experiment, the **event** you were investigating was Omar getting two hits in a game. A **favourable outcome** is H-H-N. Circle the other favourable outcomes in your table. Record the frequency for each outcome from your experiment in the "Frequency" column.

favourable outcome

the result that you are investigating in a probability experiment

D. Use your results to calculate the probability that Omar will get two hits in a game. **Experimental probability** is calculated using the ratio $\dfrac{\text{number of trials in which the investigated event was observed}}{\text{total number of trials in experiment}}$. Record the experimental probability for each outcome as a fraction, a decimal, and a percent in your table.

experimental probability

a measure of the likelihood of an event, based on data from an experiment

E. Repeat the experiment. Record the results in a new table like the one in step A. Then complete another table like the one in steps B to D for the 20 trials in your second experiment.

F. Copy the following table. Record the possible outcomes and the results for your first and second experiments in your table. Compare the experimental probabilities from your two experiments.

| Outcome | Probability based on your data | | | Probability based on class data |
	First experiment	Second experiment	Combined	

G. Combine your data from both experiments. Compare the experimental probability from the combined results with the experimental probabilities calculated in the individual experiments.

H. Combine your results with your classmates' results. Complete your table.

Reflecting

1. a) Will the probabilities from an experiment with 20 trials be exactly the same if you repeat the experiment? Explain.

   b) Will the probabilities from an experiment with 100 trials be exactly the same if you repeat the experiment? Explain.

2. Suppose that you conducted an experiment with 1000 trials, and then repeat it. How would the experimental probability from the first experiment compare with the experimental probability from the second?

3. Suppose that you are designing an experiment to determine the probability of an event. How should you design your experiment so that the experimental probabilities will be as close as possible to each other every time the experiment is repeated?

12.2 Calculating Probability

▶ **GOAL**

Identify and state the theoretical probability of favourable outcomes.

Learn about the Math

Colin and Kaitlyn are buying tickets for the pizza lottery. The tickets are numbered from 1 to 100. Colin buys all the tickets that are multiples of 10. Kaitlyn buys all the tickets that are factors of 32.

Enter the draw for a free pizza for lunch!
100 tickets sold every week!
1 winner every week!

? **What is the probability of each student winning the lottery?**

Probability is a number that shows how likely it is that an event will happen. It can be expressed as a decimal from 0 (will never happen) to 1 (is certain to happen). **Theoretical probability** is calculated using the ratio $\dfrac{\text{number of favourable outcomes for an event}}{\text{total number of possible outcomes}}$.

theoretical probability

a measure of the likelihood of an event, based on calculations

Communication Tip

The probability of an event is often written as $P(X)$, where X is a description of the event; for example, if $P(H)$ represents the probability of tossing a coin and getting Heads, $P(H) = \dfrac{1}{2}$, or 0.5, or 50%.

Example 1: Calculating probability

Calculate the probability that Colin will win. Express the number in different ways.

Colin's Solution

$P(\text{Colin winning}) = \dfrac{\text{number of tickets Colin bought}}{\text{total number of tickets sold}}$

$= \dfrac{10}{100}$

$= \dfrac{1}{10}$ or 0.1 or 10%

I bought the 10 tickets with numbers that are multiples of 10: #10, #20, #30, #40, #50, #60, #70, #80, #90, and #100.

Example 2: Calculating probability

Calculate the probability that Kaitlyn will win. Express the number in different ways.

Kaitlyn's Solution

$P(\text{Kaitlyn winning}) = \dfrac{\text{number of tickets Kaitlyn bought}}{\text{total number of tickets sold}}$

$= \dfrac{6}{100}$ or 0.06 or 6%

I bought the 6 tickets with numbers that are factors of 32: #1, #2, #4, #8, #16, and #32.

Reflecting

1. Explain the difference between a possible outcome and a favourable outcome.

2. How does the probability of drawing ticket #24 compare with the probability of drawing ticket #25?

3. Suppose that Colin had won the pizza lottery two weeks in a row. Does this affect his probability of winning the next week? Explain.

Work with the Math

Example 3: Tossing three coins

Determine the probability of getting Tails twice and Heads once, in any order, when tossing three coins. Express the number in fraction form.

Rana's Solution

HHH, THH, HTH, HHT, ⟨TTH⟩, ⟨THT⟩ ⟨HTT⟩ TTT

$P(\text{two Tails and one Heads}) = \dfrac{3}{8}$

I listed all the possible outcomes for tossing three coins. There are eight possible outcomes.

I circled the three outcomes that show two Tails and one Heads.

Example 4: Drawing cards from a standard deck

Calculate the probability of drawing an ace from a standard deck of 52 playing cards. Express the number in different ways.

Tynessa's Solution

P(drawing an ace)

$= \dfrac{4}{52}$

or about 0.08 or 8%

There are 52 cards in a standard deck, and 4 of these cards are aces.

I used a calculator to write the fraction as a decimal.

4 ÷ 52 = `0.076923`

I rounded the result to two decimal places.

A Checking

4. Express each probability in different ways.

a) P(yellow)

b) P(green)

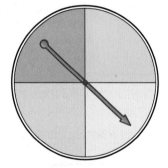

c) P(one Heads and one Tails)

d) P(Q♠)

B Practising

5. What is each probability?

a) P(4)

b) P(♥)

c) P(any 4 or any heart)

d) P(a vowel)

6. Suppose that you toss three pennies.

a) What is the probability of all three pennies landing heads up: $P(H, H, H)$?

b) Suppose the three pennies land heads up and you toss a fourth penny. What is the probability of it landing heads up? Explain.

7. Tien chooses one ball without looking. What is each probability?

a) $P(\text{black})$

b) $P(10)$

c) $P(\text{an odd number})$

d) $P(\text{an even number})$

e) $P(\text{solid red, yellow, or green})$

f) $P(\text{a number less than 20})$

8. A paper bag contains a $5 bill, a $10 bill, a $20 bill, a $50 bill, and a $100 bill. The probability of drawing the $100 bill from the bag is $\frac{1}{5}$.

Explain why the probability of drawing the $1 coin from this bag of coins is not $\frac{1}{5}$.

9. Describe an event with each probability.

a) $\frac{1}{2}$ **c)** $\frac{1}{31}$ **e)** $\frac{1}{1}$

b) $\frac{1}{7}$ **d)** $\frac{1}{365}$ **f)** $\frac{0}{4}$

10. For the following spinner, what is each probability?

a) $P(\text{multiple of 3})$

b) $P(\text{factor of 12})$

c) $P(\text{prime number})$

d) $P(3, 5, \text{or } 8)$

e) $P(\text{a number less than 12})$

11. Heidi's group uses the spinner from question 10. They spin it 600 times. How many times would you expect the arrow to land on each number below? Explain your thinking.

a) 7

b) an even number

c) a number less than 4

d) a number greater than 12

12. A regular die shows the numbers from 1 to 6.

a) Calculate $P(\text{even number})$.

b) Calculate two other probabilities for the die.

C Extending

13. Anthony draws a block from this bag, checks its colour, and puts it back. He continues drawing one block at a time and putting it back until he draws the red block.

a) What is the probability that Anthony will draw the red block for the first time on his second try?

b) Suppose that Anthony does not replace each block after he draws it. Will the probability that he draws a red block on his second try change? Explain.

Mid-Chapter Review

Frequently Asked Questions

Q: What is probability?

A: Probability is a number between 0 and 1 that tells the likelihood of something happening.

You can express probability as a fraction, a decimal, or a percent. For example, when a coin is tossed, the theoretical probability of Tails is one-half. It is written $P(T) = \frac{1}{2}$ or 0.5 or 50%.

Q: What is the difference between a possible outcome, a favourable outcome, and an event?

A: • A possible outcome is any single result that can happen.

• An event is a set of one or more possible outcomes that can happen.

• A favourable outcome is the outcome in which you are interested.

For example, the possible outcomes of rolling a die numbered from 1 to 6 are 1, 2, 3, 4, 5, and 6. An event might be rolling an odd number. For this event, 1, 3, and 5 are the favourable outcomes.

Q: What is the difference between experimental probability and theoretical probability?

A: • The experimental probability of an event is a measure of the likelihood of the event, based on data from an experiment. It is calculated using the ratio

$$\frac{\text{number of trials in which the investigated event was observed}}{\text{total number of trials in experiment}}.$$

For example, suppose Ralph draws a card from a standard deck 30 times. If he draws a king 4 times, then the experimental probability of drawing a king is $\frac{4}{30}$ or $\frac{2}{15}$, which is about 0.13 or 13%.

• The theoretical probability of an event is a measure of the likelihood of an event, calculated using the ratio

$$\frac{\text{number of favourable outcomes for an event}}{\text{number of possible outcomes}}.$$ For example, the theoretical probability of drawing a king from a deck of 52 cards is $\frac{\text{number of kings in deck}}{\text{number of cards in deck}} = \frac{4}{52}$ or $\frac{1}{13}$, which is about 0.08 or 8%.

Practice Questions

(12.1) **1.** Raj is conducting an experiment by spinning this spinner three times for each trial. List all the possible outcomes for a trial.

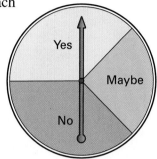

(12.1) **2.** When Andrea tossed a coin 10 times, she got four Heads. Write the fraction, decimal, and percent for the experimental probability of tossing Tails.

(12.2) **3.** Heather rolled a regular six-sided die 16 times. Her experimental probabilities for three events are given below:

A. even: $\dfrac{9}{16}$

B. less than 4: $\dfrac{8}{16}$

C. 6: $\dfrac{1}{16}$

a) Which result matches the theoretical probability?

b) Which result is close to the theoretical probability?

c) Which result is not at all close to the theoretical probability?

(12.2) **4.** Determine the probability of landing on each colour.

a) blue
b) red
c) yellow
d) green

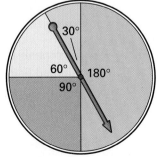

5. Zach made dice using these solids. He marked each face with a different number. Does each solid have an equal probability for every outcome? Explain. (12.2)

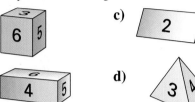

6. Calculate the probability of each event. Use the notation $P(\ ) =$. (12.2)

a) Joe tosses two coins and gets two Heads.

b) Rana draws one card from a standard deck of 52 cards. She draws a heart.

c) Elizabeth rolls a 10-sided die with the numbers 1 to 10. The face shows an odd number.

7. Kyle and Winnie reported the following results for their experiments. Whose results are most likely to happen again? Why? (12.2)

a) Each tossed a coin eight times. Kyle tossed four Heads and four Tails. Winnie tossed eight Heads and zero Tails.

b) Each drew two cards from a standard deck. Kyle drew two hearts. Winnie drew a heart and a club.

c) Each rolled a regular six-sided die 12 times. Kyle rolled a 6 two times. Winnie rolled a 6 four times.

Suppose that you have two bags. Each bag contains some red marbles and black marbles. If you draw a red marble from the bag, you win.

Here are two situations to investigate:

Situation A

Marbles	Bag 1	Bag 2
red	5	3
black	6	4

Situation B

Marbles	Bag 1	Bag 2
red	6	9
black	3	5

1. In both situations, you have a better probability of winning with bag 1. Explain why.

2. Suppose you place the marbles from bag 1 of both situations into a single bag. Calculate the probability of winning using this bag.

3. Suppose you place the marbles from bag 2 of both situations into a single bag. Calculate the probability of winning using this bag.

4. How is the probability of winning using combined bags different from the probability of winning using separate bags?

Number of players: Any number of people can play. If you play as a class, form two or three teams and take turns.

You will need
• two dice

Rules

1. Roll two dice, and add the numbers. The sum is your score. You can roll as many times as you want and continue adding to your score until you reach 100—as long as you do *not* roll a 1. You can stop whenever you want and bank your points.

2. If you roll a 1 on either die, your turn is over and you lose all your points from *this turn*. If you roll two 1s, you lose *all* your points.

3. The first person (or team) to reach 100 points wins.

For example, Kaitlyn rolls a 2 and a 3 (2 + 3 = 5 points). Then she rolls a 6 and a 4 (5 + 10 = 15 points), followed by a 5 and a 6 (15 + 11 = 26 points). Then she stops. Kaitlyn has 26 points.

On her second turn, Kaitlyn rolls a 3 and a 5 (8 points), and then a 3 and a 1. Her score for this turn is 0. She still has 26 points from her first turn.

On her third turn, Kaitlyn rolls a 5 and a 4 (9 points), and then a 1 and a 1. Kaitlyn's total score returns to 0.

12.3 Solve Problems Using Organized Lists

▶ **GOAL**
Use organized lists to determine all possible outcomes.

Learn about the Math

"Geoffrey, I have exactly five coins in my hand, worth a total of 50¢. There are no pennies. If you guess what the coins are, I'll give them to you. If you guess wrong, you give me 50¢."

"Let me think about it, Samantha."

? **What is the probability of Geoffrey guessing correctly and winning?**

1 Understand the Problem

The following information is given in the problem:

- Samantha has exactly five coins.
- The coins have a value of 50¢.
- They can be any combination of nickels, dimes, and quarters.
- There are no pennies.
- There may be more than one of some types of coins.
- There may be none of some types of coins.

2 Make a Plan

Geoffrey decides to write all the possible combinations in an organized list. He will make a table and work systematically so that he will not miss or repeat any combination.

3 Carry Out the Plan

Quarters	Dimes	Nickels	Total value	Summary of possible combinations
5	0	0	$1.25	with five quarters, one combination
4	1	0	$1.10	with four quarters, two combinations
4	0	1	$1.05	
3	2	0	$0.95	with three quarters, three combinations
3	1	1	$0.90	
3	0	2	$0.85	
2	3	0	$0.80	with two quarters, four combinations
2	2	1	$0.75	
2	1	2	$0.70	
2	0	3	$0.65	
1	4	0	$0.65	with one quarter, five combinations
1	3	1	$0.60	
1	2	2	$0.55	
1	1	3	$0.50 *	
1	0	4	$0.45	
0	5	0	$0.50 *	with no quarters, six combinations
0	4	1	$0.45	
0	3	2	$0.40	
0	2	3	$0.35	
0	1	4	$0.30	
0	0	5	$0.25	

I have one chance out of two of winning. My probability of guessing correctly is $\frac{1}{2}$. I'll take Samantha's challenge.

* Only two combinations add up to 50¢. Either Samantha has one quarter, one dime, and three nickels, or she has five dimes.

4 Look Back

Geoffrey sees a pattern in the numbers, so he is sure that he listed all the possible combinations.

Reflecting

1. Describe the number patterns in Geoffrey's list.

2. What other strategies could Geoffrey have used to determine all the possible combinations?

3. How did an organized list help Geoffrey determine the probability of guessing correctly?

4. Do you think Samantha's challenge is fair? Explain your answer.

Example: Making an organized list to determine all possible outcomes

Tynessa cannot remember the combination for her bicycle lock. She knows that the numbers are 1, 2, or 3 and they add up to 7. What are the possible combinations? What is the probability that the first of the possible combinations that Tynessa tries will open the lock?

Tynessa's Solution

1 Understand the Problem

The numbers in the correct combination add up to 7.
I need to find all the possible combinations with a sum of 7.

2 Make a Plan

I'll make an organized list of all the possible combinations, starting with a first number of 1, then of 2, and then of 3.

3 Carry Out the Plan

First number	1	1	1	1	1	1	1	1	1	2	2	2	2	2	2	2	2	2	3	3	3	3	3	3	3	3	3
Second number	1	1	1	2	2	2	3	3	3	1	1	1	2	2	2	3	3	3	1	1	1	2	2	2	3	3	3
Third number	1	2	3	1	2	3	1	2	3	1	2	3	1	2	3	1	2	3	1	2	3	1	2	3	1	2	3
Sum	3	4	5	4	5	6	5	6	⑦	4	5	6	5	6	⑦	6	⑦	8	5	6	⑦	6	⑦	8	⑦	8	9

Six of these combinations have a sum of 7. The probability that the first of these combinations I try will open the lock is $\frac{1}{6}$.

4 Look Back

I see patterns in my list, so I'm sure that I've found all the combinations. The first number, which is 1, appears nine times. Then the number 2 appears nine times. Then the number 3 appears nine times.

A Checking

5. Suppose that you have two coins in your pocket. The coins can be pennies, nickels, dimes, or quarters.

   a) How many different combinations are possible?

   b) How many of these combinations add up to less than 20¢?

   c) What is the probability that the two coins in your pocket add up to less than 20¢?

6. Meagan has three dogs: Fido, Spot, and Rover. The sum of the dogs' ages is 15. Rover is the oldest, and Spot is the youngest. List all the different combinations of ages the dogs could be.

B Practising

7. In a baseball tournament, teams get 5 points for a win, 3 points for a tie, and 1 point for a loss. Nathan's team has 29 points.

   a) Use an organized list to show how many different combinations of wins, ties, and losses Nathan's team could have.

   b) What is the probability that Nathan's team had more losses than ties?

8. Tien has six coins in her pocket. She has at least one quarter, one dime, one nickel, and one penny. Use an organized list to show how many different ways Tien might have more than 60¢.

9. The digits in the number 743 have two properties:

   • Their sum is 14.

   • They are in decreasing order.

   List all the other three-digit numbers that have the same two properties.

10. Samantha threw three darts and hit the dartboard each time.

   a) Use an organized list to show all the possible scores Samantha might have.

   b) What is the likelihood that her score will be less than 30?

11. Matthew has three brothers: Alex, Mark, and Luke. The sum of their three ages is 12. List all the different combinations of ages that Matthew's brothers could be.

12. In chess, each player starts with eight pawns, two knights, two bishops, two rooks, one queen, and one king. A pawn is worth 1 point. A knight and a bishop are each worth 3 points. A rook is worth 5 points. Suppose that the pieces you capture in a game are worth 10 points. List all the combinations of pieces you could have.

13. The coins in Yan's pocket are worth 27¢. Yan could have any combination of pennies, nickels, dimes, or quarters.

   a) How many different combinations of coins are possible?

   b) Suppose that Yan has one nickel. What is the probability that you could guess the combination of coins he has in his pocket?

14. Make up a problem that can be solved by using an organized list. Include a complete solution.

12.4 Using Tree Diagrams to Calculate Probability

▶ **GOAL**

Use tree diagrams to determine all possible outcomes.

Learn about the Math

Mei has three sisters and no brothers.

? **What are the chances that a family with four children will have all girls?**

Example 1: Using a tree diagram to analyze outcomes

Determine the probability that a family with four children will have all girls.

Rana's Solution

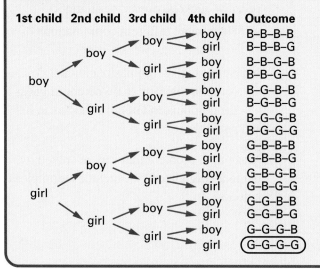

I listed all the possibilities in a tree diagram. Each child will be either a boy or a girl.

There are 16 possible outcomes. In only one of the outcomes are all the children girls.

$P(4 \text{ girls}) = \frac{1}{16}$

Example 2: Using a tree diagram to analyze outcomes

Determine the probability that a family with four children will have two girls and two boys.

Omar's Solution

I used Rana's tree diagram. There are 16 possible outcomes.
In six of these outcomes, there are two girls and two boys.

P(2 girls and 2 boys) $= \frac{6}{16}$ or $\frac{3}{8}$

Reflecting

1. a) How does a tree diagram help you list all the possible outcomes?

b) How does a tree diagram help you calculate the probability of an event?

c) Explain why a tree diagram is an organized list.

2. a) How many outcomes are possible for a family of five children? How does this compare with the outcomes for a family of four children?

b) How many outcomes are possible for a family of six children? Explain what you did.

3. How can you use the number of branches at each stage of a tree diagram to calculate the total number of branches in the diagram? Explain why this makes sense.

Work with the Math

Example 3: Determining the probability of travelling to the end of a maze

A small robot is put at the entrance to this maze. The robot stops when it comes to a dead end. At each "gate" in the maze, the robot randomly turns left or right. What is the probability of the robot reaching the exit?

Solution

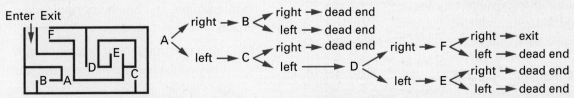

Put a letter at each gate and draw a tree diagram to try each possibility.

The tree diagram shows that seven routes are possible. Only one of these routes leads to the exit.

P(reaching the exit) $= \frac{1}{7}$

Example 4: Using a tree diagram to determine all the possible outcomes

The Belleville Bills and the Arnprior Allstars are in a playoff series. The first team to win two games will win the series. The two teams have an equal probability of winning any game. What is the probability of the series lasting three games?

Solution

Draw a tree diagram to represent the possible results of the series.

Game 1	Game 2	Game 3	Games played
Team A wins	Team A wins	not needed	2
	Team B wins	Team A wins	3
		Team B wins	3

Game 1	Game 2	Game 3	Games played
Team B wins	Team A wins	Team A wins	3
		Team B wins	3
	Team B wins	not needed	2

The tree diagram shows that there are six possible outcomes.
- Team A wins games 1 and 2, so game 3 is not played.
- Team A wins games 1 and 3.
- Team B wins games 2 and 3.
- Team A wins games 2 and 3.
- Team B wins games 1 and 3.
- Team B wins games 1 and 2, so game 3 is not played.

In four of these outcomes, the series lasts three games.

P(series lasts 3 games) $= \frac{4}{6}$ or $\frac{2}{3}$

A Checking

4. Ioanna sells irises and lilies in bunches of four. If she randomly selects the flowers, determine the probability that a bunch will have one iris and three lilies.

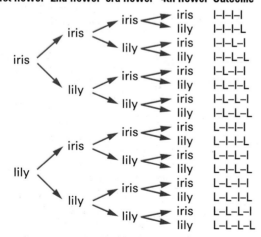

1st flower	2nd flower	3rd flower	4th flower	Outcome

5. Draw a tree diagram to show all the possible outcomes of tossing a coin three times.

B Practising

6. Use a tree diagram to determine the probability of tossing a total of 7 using two regular dice.

7. Braydon wants a triple-scoop ice cream cone. In his freezer, he has strawberry, maple walnut, and blueberry ripple ice cream.

a) Use a tree diagram to determine how many possible combinations Braydon can choose.

b) What is the probability that at least one scoop will be strawberry?

8. Omar has three T-shirts: one red, one green, and one yellow. He has two pairs of shorts: one red and one black. Use a tree diagram to answer the following questions.

a) How many different outfits can Omar put together?

b) What is the probability of Omar's outfits including a red T-shirt or red shorts?

9. Takumi tosses these coins, one after the other.

a) In how many different ways can they land?

b) What is the probability of at least three coins landing on Heads?

10. The combination for a lock uses three of the digits 1, 2, 3, and 4.

a) How many three-digit combinations can the lock have? Explain your answer.

b) How many three-digit combinations can the lock have, if the same digit cannot be used more than once?

11. a) Rosa has three blouses, three skirts, and three jackets. How many different outfits can she put together?

b) Asha has three blouses, two skirts, and four jackets. Can she put together more or fewer outfits than Rosa? Explain.

12. George is the third of five children. He has two older sisters (S) and two younger brothers (B). This arrangement can be written as SSBBB.

a) Use a tree diagram to determine the probability of a family with five children beginning with two sisters.

b) Is the probability of a family of five beginning with two sisters the same as the probability of a family of five ending with two sisters? Explain.

c) What is the probability of the sixth child being a boy?

13. Leonard places a small robot at the entrance to this maze. Each choice of direction is random. The robot stops when it comes to a dead end. What is the probability of the robot reaching the exit?

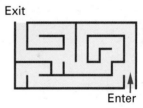

ⒸExtending

14. Marian wants to buy a double scoop of ice cream. The ice cream shop has 23 flavours, which can be served in a regular cone, a sugar cone, or a dish. How many different combinations can Marian order?

Applying Probabilities

You will need
• two six-sided dice
• a calculator

▶ **GOAL**

Calculate and compare probabilities.

Learn about the Math

Tynessa and Kaitlyn are creating a new game about tossing two dice.

Tynessa says, "Let's make it simple. All you have to do is roll the dice and look at the numbers."

Kaitlyn says, "Four results are possible.
- Alike: The numbers are not the same, but both numbers are either even or odd.
- Consecutive: The second die is either one greater or one less than the first die.
- Matching: The numbers are the same.
- Different: The numbers are not alike, consecutive, or matching."

Tynessa says, "Our rules should say that the player who gets the least likely result wins."

? **Which result is least likely?**

A. Use a tree diagram or an organized list to determine all the possible outcomes.

B. Calculate the probability of each result.

C. List the results in order from most probable to least probable.

D. Write the rules that determine a winner, based on your list in step C.

E. Try Tynessa and Kaitlyn's game. Roll the dice at least 20 times. Record your results in a table like this one.

Result	Tally
alike	
consecutive	
matching	
different	

F. How do your experimental probabilities compare with the theoretical probabilities you calculated in step B?

Reflecting

1. The probability of each result can be written as a fraction, a decimal, or a percent. Which form makes comparing the probabilities easiest? Explain.

2. You could base the rules of the game on either the theoretical probabilities or the experimental probabilities. Which would you use? Why?

Work with the Math

Example: Calculating batting averages

A player's batting average is the probability that the player will get a hit, based on previous data. It is calculated using the ratio $\frac{\text{number of hits}}{\text{number of times at bat}}$ and is expressed as a number to three decimal places. (Zero is not recorded in the ones place of a batting average.)

Player	Number of hits	Number of times at bat
Heather	25	75
Indu	32	100
Meagan	18	50

Use the data in the table to calculate each player's probability of getting a hit. List the batting averages from least to greatest.

Colin's Solution

For Heather, $P(\text{hit}) = \dfrac{25}{75}$
$= \dfrac{1}{3}$
or about 0.333

For Indu, $P(\text{hit}) = \dfrac{32}{100}$
or 0.320

For Meagan, $P(\text{hit}) = \dfrac{18}{50}$
or 0.360

The batting averages from least to greatest are Indu .320, Heather .333, and Meagan .360.

A Checking

3. Which player has the greatest probability of getting a hit?
   - Raj's batting average is .575.
   - Bella gets a hit three out of every five times at bat.
   - Connie has a 58% chance of getting a hit.
   - For Derek, $P(\text{hit}) = \frac{2}{3}$.

B Practising

4. Which player has the least probability of getting a hit?
   - Jasleen has a batting average of .360.
   - Claudia gets two hits for every five times at bat.
   - Maxine has a 40% chance of getting a hit.
   - The likelihood of Ernie getting a hit is $\frac{1}{4}$.

5. For which of these spinners is black as equally likely to occur as at least one other colour? Explain.

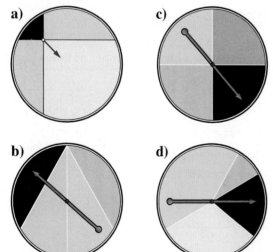

a)   c)

b)   d)

6. Explain which event is more likely:
   - rolling two dice and getting a total of 11

   or
   - tossing a coin and getting three Heads in a row

7. a) What is the probability of winning each amount each time you fish: $10, $5, $0.10?

100 fish in the pond
A winner every time!
Two fish are worth $10 each.
Eight fish are worth $5 each.
Every other fish is worth 10¢.
Only $1 per try.

   b) What is the total value of the 100 fish?

   c) All 100 fish are caught. How much profit did the operator of the fish pond make?

   d) What is the probability that you would win more than you pay?

8. To play the birthday wheel, you pay $1. Then you spin the wheel. If your birthday month comes up, you win $10. What is the probability that you will win on one spin?

The Birthday Wheel

(November, June, January, August, March, October, May, April, February, July, December, September)

9. Suppose that you want to get either four consecutive cards from the same suit or three cards with the same value. Your hand is shown here. You draw a card from the remaining 42 cards in the standard deck. Which probability is greater? Why?

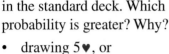

- drawing 5♥, or
- drawing 8♥ or 8♠

10. Bag A has 3 black marbles and 5 white marbles. Bag B has 8 black marbles and 10 white marbles. Bag C has 17 black marbles and 20 white marbles.

   a) Which bag gives the greatest probability of drawing a black marble?

   b) Which bag gives the greatest probability of drawing a white marble?

   c) If all the marbles are placed in one bag, what is the probability of drawing a white marble?

C Extending

11. In a lottery, one four-digit number is chosen at random. The digit 0 is not used. Suppose that you buy a ticket.

   a) What are your chances of winning?

   b) What are your chances of losing?

12. In basketball, each shot at the basket will either go in or not go in. Explain why the probability of getting a basket is not $\frac{1}{2}$, even though only two outcomes are possible.

13. Mei and Janice are playing a game with three dice. The nets of the dice are shown below. Which two dice should Mei choose to be most likely to roll the highest score?

Die A		
2	2	2
	5	
	5	
	5	

Die B		
1	3	1
	3	
	5	
	5	

Die C		
1	4	1
	4	
	6	
	6	

Mental Math

EXPRESSING A FRACTION AS A PERCENT

When the denominator of a fraction is a factor of 100, you can use mental math to change it to a percent.

$$\frac{3}{4} = \frac{75}{100} \text{ or } 75\%$$

$\times 25$ (top)
$\times 25$ (bottom)

1. Express each fraction as a percent.

   a) $\frac{1}{4}$

   b) $\frac{2}{4}$

   c) $\frac{1}{5}$

   d) $\frac{3}{5}$

   e) $\frac{1}{20}$

   f) $\frac{17}{20}$

   g) $\frac{7}{25}$

   h) $\frac{16}{25}$

   i) $\frac{3}{50}$

   j) $\frac{27}{50}$

1. Katya writes each letter in the word "PROBABILITY" on a different card and turns over the cards. Then she mixes the cards and draws one card at random. What is each theoretical probability?

   a) P(B)

   b) P(a vowel)

   c) P(B or I)

   d) P(K)

2. What is the difference between theoretical probability and experimental probability? Give an example.

3. a) What is the probability of guessing someone's birth month?

   b) Express the probability as a number rounded to two decimal places.

4. In Bingo, the numbers 1 to 75 can be drawn. Each Bingo card has 24 numbers and one free space. Suppose that you have this Bingo card.

B	I	N	G	O
2	16	32	46	61
4	18	33	48	62
7	20	FREE	56	67
8	23	40	57	71
10	27	42	59	75

   a) How many outcomes are possible?

   b) How many outcomes are favourable?

   c) What is the probability that the number called is on your Bingo card?

5. Suki has $2 coins, $1 coins, and quarters in her wallet. She owes her brother $2.50. Use an organized list to show all the possible combinations of coins that she could use to get exactly $2.50.

6. Alice is going to the drugstore, the dry cleaners, and the bakery.

   a) Use a tree diagram to show how many possible orders she could go to all three stores.

   b) Alice chooses her route randomly. What is the probability that she will go to the bakery first?

7. Jim has the following spinner, two regular dice, and three coins. Which outcome is most probable?

   A. landing on red with the spinner

   B. rolling a 9 with the dice

   C. tossing two Heads and one Tails with the coins

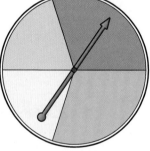

8. A spinner has four colours: red, blue, yellow, and green. The probability of landing on red is 25%. The probability of landing on blue is 12%. The probability of landing on yellow is 50%.

   a) What is the probability of landing on green?

   b) If you spin the spinner 75 times, how many times can you expect to land on blue?

Chapter Review

Frequently Asked Questions

Q: How can you use an organized list to calculate probability?

A: An organized list shows all the possible outcomes. This makes the favourable outcomes easier to see. For example, in a school checkers tournament, a win is worth 5 points, a tie is worth 2 points, and a loss is worth 1 point. Helene has 10 points.

Combination for 10	Number of wins	Number of ties	Number of losses
5 + 5	2	0	0
5 + 2 + 2 + 1	1	2	1
5 + 2 + 1 + 1 + 1	1	1	3
5 + 1 + 1 + 1 + 1+1	1	0	5
2 + 2 + 2 + 2 + 2	0	5	0
2 + 2 + 2 + 2 + 1 + 1	0	4	2
2 + 2 + 2 + 1 + 1 + 1 + 1	0	3	4
2+2+1+1+1+1+1+1	0	2	6
2+1+1+1+1+1+1+1+1	0	1	8
1+1+1+1+1+1+1+1+1+1	0	0	10

What is the probability that she had more wins than ties?

The table shows that there are ten ways to earn 10 points. Only one way has more wins than ties.

$$P(\text{more wins than ties}) = \frac{1}{10} \text{ or } 10\%$$

Q: How do you use a tree diagram to calculate probability?

A: Suppose that your school is in a volleyball tournament and every team has an equal chance of winning. The winner must win two out of three games. What is the probability that your school will win in the first two games?

The tree diagram shows that six outcomes are possible and equally likely. In one of these outcomes, your school wins the first two games.

$$P(\text{win first 2 games}) = \frac{1}{6}$$

Note: Theoretical probability is not very reliable in sports because there are many factors that affect who wins. Some of these factors are

- the latest performance records of the teams
- the confidence of each team over its opponents
- the combinations of the different players
- the number of supporters of each team at the game

Game 1	Game 2	Game 3	Games played
win	win	not needed	2
		win	3
	loss	loss	3
loss	win	win	3
		loss	3
	loss	not needed	2

Practice Questions

(12.2) **1.** Travis tossed three coins 16 times. Here are the experimental probabilities for three outcomes:

A. H, T, H: $\dfrac{2}{16}$

B. H, H, H: $\dfrac{3}{16}$

C. two H's and one T in any order: $\dfrac{11}{16}$

   a) Which outcome matches the theoretical probability exactly?

   b) Which outcome is close to the theoretical probability?

   c) Which outcome is not at all close to the theoretical probability?

(12.2) **2.** Ali draws one card at random from a standard deck of playing cards. What is each probability?

   a) *P*(black card)   **b)** *P*(6)

(12.3) **3.** On Saturdays, Elliott practises the piano, walks his dog, does his homework, and cleans his room. Use an organized list to show all the possible orders in which Elliott can do these tasks.

(12.3) **4.** Only Dave, Tony, Colin, and Joel are competing in the 400 m hurdles. They all have an equal chance of winning.

   a) Use an organized list to show all the outcomes for the first three hurdlers crossing the finish line.

   b) What is the probability that Joel will finish in the top three?

5. For her meal, Heidi can choose
   • a hamburger or a hot dog
   • juice, water, or lemonade

   Use a tree diagram to show all the possible combinations for Heidi's meal.   (12.4)

6. Alan has two jackets (tan and blue), three caps (red, white, and blue), and three pairs of running shoes (white, blue, and black). He always wears a jacket, a cap, and running shoes.   (12.4)

   a) Use a tree diagram to show how many different combinations Alan can wear.

   b) In how many combinations will Alan be wearing something blue?

   c) In how many combinations will Alan be wearing two things of the same colour?

7. Franz rolls a single die, numbered 1 to 6, two times. Use a tree diagram to determine the probability of rolling two consecutive numbers in increasing order.   (12.4)

8. Determine the probability of a mother's first three children all being boys.   (12.4)

9. Which player has the best batting average?   (12.5)
   • Chanelle's batting average is .380.
   • Tim hit the ball 12 times in his last 30 times at bat.
   • Francine's probability of getting a hit is $\dfrac{1}{3}$.
   • Ted hits the ball 35% of the time.

10. In a school football league, the Panthers won 6 of their last 10 games. The Hawks won 8 of their last 14 games. The Pythons won 7 of their last 12 games. Which team has the greatest theoretical probability of winning the championship?   (12.5)

Chapter Task

Rock, Paper, Scissors

Rock, Paper, Scissors is an old game.
The rules are as follows:
- Rock flattens scissors.
- Scissors cut paper.
- Paper covers rock.

? **Does using a strategy affect the probability of winning?**

A. Calculate your theoretical probability of winning if each choice is equally likely.

B. With a partner, play 20 rounds of Rock, Paper, Scissors. Do not use any strategy for your choices. Create a tally chart to keep track of the number of times you win, lose, or draw.

C. What is your experimental probability of winning?

D. Play another 20 rounds. This time, one of you plays as usual and the other uses a strategy. For example, one of you could
- always pick paper, *or*
- use a pattern, such as paper, paper, paper, rock, *or*
- always pick the last choice your partner made

E. Record your results in another tally chart. What is your experimental probability of winning now?

F. How do your results compare with the results of other students in your class? Explain.

G. Would you use a strategy to play Rock, Paper, Scissors? Justify your decision.

> ### Task Checklist
>
> ☑ Did you explain your thinking?
>
> ☑ Did you express the probabilities using math language?
>
> ☑ Did you give enough detail in your answers?
>
> ☑ Is your conclusion justified by the data and your analysis?

Cross-Strand Multiple Choice

(10.2) **1.** Which polyhedron does this net form when folded?

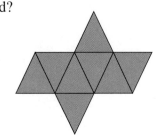

A.

C.

B.

D.

(10.3) **2.** What is the top view of the cube structure?

A.

C.

B.

D.

3. Which top view matches this polyhedron? (10.4)

A. **B.** **C.** **D.**

4. Examine the building in the photo. Which polyhedrons make up this building? (10.4)

A. rectangular prism and cone

B. rectangular prism and rectangular pyramid

C. cylinder and rectangular pyramid

D. rectangular prisms and square-based prism

5. What is the surface area of this rectangular prism? (11.1)

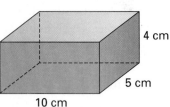

A. 200 cm^3 **C.** 45 cm^2

B. 90 cm^3 **D.** 220 cm^2

(11.2) **6.** Which rectangular prism has a volume of 196.8 cm³?

A.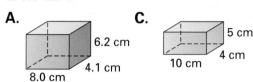
6.2 cm
4.1 cm
8.0 cm

C. 5 cm
4 cm
10 cm

B.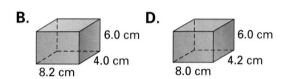
6.0 cm
4.0 cm
8.2 cm

D. 6.0 cm
4.2 cm
8.0 cm

(12.1) **7.** Which statement is an example of experimental probability?

A. The probability of picking a face card from a deck of cards is $\frac{3}{13}$.

B. A coin is tossed ten times to determine the probability of Heads coming up.

C. When tossing a coin, the probability of Heads coming up is $\frac{1}{2}$.

D. The probability of rolling a 4 with one die is $\frac{1}{6}$.

8. Kyle owns two pairs of basketball shorts (red and black) and three pairs of basketball shoes (blue, yellow, and green). How many different combinations of shorts and shoes can he wear? (12.4)

A. 6 **B.** 5 **C.** 8 **D.** 12

9. Which event is more likely to occur? (12.5)

A. Tossing a coin and getting three Heads in a row.

B. Drawing an ace from a standard deck of playing cards.

C. Rolling two dice and getting a total of 7.

D. Drawing a white marble from a bag that has 1 white marble and 9 black marbles.

Cross-Strand Investigation

10. Super Juice has created a new fruit punch, which is a blend of strawberry, cherry, and kiwi juices. The fruit punch will be packaged and sold in a small box, like a Tetra Pak™, that holds 300 mL. The company has asked you to design the box.

a) Decide what shape you want to use for your box.

b) Draw a side view, top view, and front view to show the prisms that make up your box.

c) Draw a three-dimensional sketch of your box, with the measurements of the different sides.

d) Draw a net of your box.

e) Cut out the net, and create an attractive cover for your box.

f) Calculate the surface area of your box, so the company will know how much material is required. Show your work.

g) Calculate the volume of your box. Remember that it must hold 300 mL of punch. Show your work.

Review of Essential Skills from Grade 6

Chapter 1: Factors and Exponents

Place Value

Every digit in a number has a different place value. The location of the digit determines its place value.

The number 5692 is shown in the place value table below. Notice, for example, that the place value of the 9 is the tens column.

Thousands	Hundreds	Tens	Ones
5	6	9	2

1. Copy and complete the table.

		Ten thousands		Hundreds	Tens	
1	4	3	6	9	2	7

Factors

Ms. Zimm has 24 girls in a physical education class. She uses factors of 24 to group them for different exercises. A **factor** is a number that is used to make a product.

Here is an organized list for finding all the *factors* of 24.

Try 1: $1 \times 24 = 24$ 1 and 24 are factors of 24.
Try 2: $2 \times 12 = 24$ 2 and 12 are factors of 24.
Try 3: $3 \times 8 = 24$ 3 and 8 are factors of 24.
Try 4: $4 \times 6 = 24$ 4 and 6 are factors of 24.
Try 5: $5 \times ? = 24$ 5 is not a factor of 24.
Try 6:   6 is already known to be a factor of 24.

The factors of 24 are 1, 2, 3, 4, 6, 8, 12, and 24.

2. Find all the factors of each number.
   a) 28   **b)** 32   **c)** 56   **d)** 29   **e)** 39   **f)** 51

3. Which numbers have 8 as a factor?
   a) 30   **b)** 40   **c)** 48   **d)** 54   **e)** 87   **f)** 104

4. a) What number do you try first when finding all the factors of 27?
   b) What pair of factors do you get?

c) What number do you try next?

d) Do you get a factor?

e) Continue until you have all the factors of 27.

Prime and Composite Numbers

A number with only two different factors, 1 and itself, is a **prime number**.

$5 = 5 \times 1$

A number with more than two factors is a **composite number**.

$8 = 4 \times 2$ $\qquad\qquad$ $8 = 8 \times 1$

The number 1 is neither prime nor composite. $1 = 1 \times 1$

5. Copy and complete the table.

Number	Factors	Number of factors	Prime or composite
1	1		
2	1, 2		
3	1, 3		
4	1, 2, 4		
5			
6			
...			
19			
20			

Operations with Whole Numbers

"The contest winner must first correctly answer a skill-testing question."

Find the value of $48 \div 6 + 2 \times 5 - 4$.

There are rules for the order of operations.
- Perform any operations within brackets.
- Multiply and divide in order from left to right.
- Then add and subtract in order from left to right.

$48 \div 6 + 2 \times 5 - 4$
$= 8 + 2 \times 5 - 4$
$= 8 + 10 - 4$
$= 14$

6. Evaluate.

a) $4 \times 24 \div 6 - 10 + 5$ $\qquad$ **d)** $9 \times 12 \times 12 \div 12$ $\qquad$ **g)** $6 \times 5 - 4 \div 2$

b) $5 + 8 \div 2 - 3$ $\qquad$ **e)** $12 + 5 \times 3 - 36 \div 2$ $\qquad$ **h)** $9 - 5 + 4 \times 8 - 15 \div 3$

c) $47 - 9 \times 2 \div 6$ $\qquad$ **f)** $7 + 6 \times 3$ $\qquad$ **i)** $27 \div 3 \times 8 - 15 \times 4 + 2$

7. Copy each statement, and use $<$, $>$, or $=$ to make it true.

a) $8 + 7 \times 5 \ \blacksquare\ 8 \times 7 + 5$ $\qquad$ **d)** $23 - 12 \div 6 \ \blacksquare\ 2 + 6 \times 3$

b) $7 + 2 \times 4 \ \blacksquare\ 5 \times 2 + 6$ $\qquad$ **e)** $5 \times 5 \div 5 \ \blacksquare\ 5 \div 5 \times 5$

c) $5 \times 9 \div 3 \ \blacksquare\ 8 + 4 \times 8$ $\qquad$ **f)** $36 \div 12 \div 3 \ \blacksquare\ 24 \div 12 \div 2$

Chapter 2: Ratio, Rate, and Percent

Comparing and Ordering Decimal Numbers

The results for the school 100 m race are shown below.

John	14.30 s	Sam	12.48 s
Henrietta	14.14 s	Christa	13.95 s
Alfred	13.98 s		

Henrietta's time is less than John's time.

14.14 < 14.30

1 tenth is less than 3 tenths.

Alfred's time is greater than Christa's time.

13.98 > 13.95

8 hundredths is greater than 5 hundredths.

1. Copy and complete.

a)

b)

2. Draw a number line 10 cm long.

a) Mark your number line to show 6.0, 6.1, …, 7.0.

b) Use a ruler to mark accurately the point halfway between 6.3 and 6.4.

c) What is the number you marked in part (b)?

Multiplying and Dividing by Powers of 10

A calculator was used to multiply and divide by powers of 10. What patterns are shown below?

$2.4 \times 100 = 240$	$2.4 \div 1000 = 0.0024$
$2.4 \times 10 = 24$	$2.4 \div 100 = 0.024$
$2.4 \times 1 = 2.4$	$2.4 \div 10 = 0.24$
$2.4 \times 0.1 = 0.24$	$2.4 \div 1 = 2.4$
$2.4 \times 0.01 = 0.024$	

3. Evaluate.

a) 45×10
   45×0.1
   45×0.01
   45×0.001

b) 100×32.15
   0.1×32.15
   0.01×321.5
   0.01×3.215

c) $40.32 \div 100$
   $4.032 \div 100$
   $403.2 \div 10$
   $40.32 \div 1$

4. Evaluate.

a) 7.3×10

b) $33.65 \div 10\ 000$

c) $10\ 000 \times 87.56$

d) $100\ 000 \times 85.48$

e) 4.315×0.01

f) $0.7462 \div 100$

g) $425.7 \div 10$

h) 9.8×1000

i) 76.93×0.001

Ratios

A **ratio** is a comparison of two numbers. For example, the front gear on Rob's bicycle has 46 teeth and the back gear has 18 teeth. The gear ratio of a bicycle compares the number of teeth on the front gear to the number of teeth on the back gear.

The gear ratio can be written in three different ways:

$$46 \text{ to } 18 \qquad\qquad 46:18 \qquad\qquad \frac{46}{18}$$

The numbers 46 and 18 are the terms of the ratio. The units of the measurements are not written because they are the same.

5. One of the gear ratios possible for a five-speed bike is 40:17. How many teeth are on the front gear? How do you know?

6. Express each ratio in three different ways.
   a) 4 out of 9 bicycles are 15-speed bicycles
   b) 5 out of 9 bicycles are not 15-speed bicycles

7. There are about 26 students for every teacher in a school. What is the student-teacher ratio?

Meaning of Percent

In a survey, 54 out of 100 Grade 5 students could name the ten provinces of Canada. The result can be shown as a fraction or as a decimal number.

$$\frac{54}{100} \quad \text{``54 hundredths''} \qquad 0.54 \quad \text{``54 hundredths''}$$

To write this result as a percent, replace "hundredths" with "%." The symbol % is read as "percent."

This means that 54% of the students asked could name the ten provinces of Canada.

8. Write each ratio as a percent.
   a) $\frac{83}{100}$ **b)** $\frac{1}{100}$ **c)** 50:100   **d)** 9 out of 100   **e)** 100 out of 100

9. Write each decimal number as a percent.
   a) 0.45   **b)** 0.76   **c)** 0.40   **d)** 0.08   **e)** 0.80   **f)** 0.8

10. Estimate the percent of each shape that is shaded.
   a) 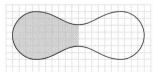 **b)** 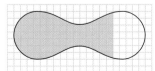 **c)**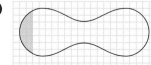

Fractions and Decimal Numbers as Percents

About $\frac{3}{4}$ of a circle is shaded. What percent of the circle is shaded?

First express the fraction as a decimal number. To do this, divide the numerator by the denominator.

$$\frac{3}{4} \qquad\qquad \begin{array}{r} 0.75 \\ 4\overline{)3.00} \end{array} \qquad\qquad \frac{3}{4} = 0.75 \text{ or } 75\%$$

About 75% of the circle is shaded.

11. Write each decimal number as a percent.

    a) 0.08   **b)** 0.7   **c)** 0.45   **d)** 0.37   **e)** 0.008   **f)** 0.5543

12. Write each fraction as a decimal number and as a percent.

    a) $\frac{1}{10}$ **b)** $\frac{4}{5}$ **c)** $\frac{1}{2}$ **d)** $\frac{6}{8}$ **e)** $\frac{12}{16}$ **f)** $\frac{5}{8}$

13. There are 16 boys and 9 girls in a class.

    a) What percent of the students are girls?   **b)** What percent of the students are boys?

Percents as Decimal Numbers and Fractions

12.5% of an iceberg is above the surface of the ocean.

To write the percent as a decimal number or a fraction, replace "%" with "hundredths."

$$12.5\% = 12.5 \text{ hundredths} \qquad\qquad 12.5\% = 12.5 \text{ hundredths}$$

$$= 0.125 \qquad\qquad\qquad\qquad = \frac{12.5}{100}$$

$$= \frac{125}{1000}$$

$$= \frac{1}{8}$$

About 0.125, or $\frac{1}{8}$, of an iceberg is above the surface of the ocean.

14. Copy and complete the table.

Fraction	Decimal number	Percent
	0.01	
$\frac{1}{10}$		
		5%
$\frac{3}{8}$		
	0.2	
		25%

Chapter 3: Data Management

Information on a Graph

Depending on the type of graph, some or all of the following information must be included: a main title, an *x*-axis title, a *y*-axis title, a scale or interval, and a legend.

1. List the information that is missing from this graph.

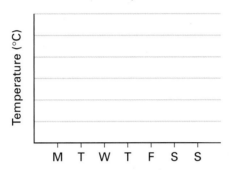

Temperatures for a Week in London, Ontario

2. List the points given on the scatter plot shown.

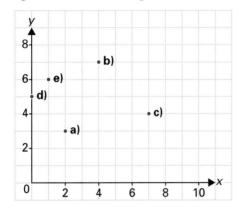

3. Look at the circle graph shown.

a) What is the largest part of Kevin's day?

b) What does he spend the least amount of time doing?

c) How much more time does he spend in school than playing basketball?

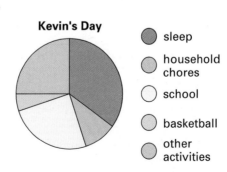

Kevin's Day

- sleep
- household chores
- school
- basketball
- other activities

Range

The range of a set of data is the difference between the maximum value and the minimum value.

For example, Karen examined the range of the prices of gasoline over a two-week period. She noticed that the highest price was $0.89 per litre and the lowest price was $0.71 per litre. The range of the prices was $0.89 − $0.71 = $0.18.

4. Canadian Joanne Malar swam the 200 m individual medley in the following times. State the range of the swimming times.

Year	1991	1993	1995	1997	1998
Swimming time (s)	139.14	134.35	135.66	136.17	133.26

Mean, Median, and Mode

Canadian diver Alexandre Despatie received the following scores for a 10 m platform dive.

Judge	France	Spain	Canada	Italy	United States
Score	8.5	9.0	9.5	10.0	9.5

To find the mean of a set of data, calculate the sum of all the pieces of data and divide by the number of pieces of data.

$(8.5 + 9.0 + 9.5 + 10.0 + 9.5) \div 5$
$= 46.5 \div 5$
$= 9.3$

The mean of Despatie's scores is 9.3.

To find the median of a set of data, order the data from smallest to largest and then identify the piece of data that is in the middle.

8.5, 9.0, **9.5**, 9.5, 10.0

The median of Despatie's scores is 9.5.

To find the mode of a set of data, identify the piece of data that occurs most often.

The score that appears most often is 9.5, so the mode is 9.5.

5. Despatie received these scores for another 10 m platform dive.

Judge	Canada	Japan	South Africa	Russia	China
Score	9.5	9.0	8.5	9.0	8.5

a) What is the mean of Despatie's scores?

b) What is the median of Despatie's scores?

c) What is the mode of Despatie's scores?

Chapter 4: Patterns and Relationships

A calculator can be a helpful tool when exploring number patterns. You may use a calculator in the following activities.

1	2	3	4	5	6	7	8	9	10
11	12	13	14	15	16	17	18	19	20
21	22	23	24	25	26	27	28	29	30
31	32	33	34	35	36	37	38	39	40
41	42	43	44	45	46	47	48	49	50
51	52	53	54	55	56	57	58	59	60
61	62	63	64	65	66	67	68	69	70
71	72	73	74	75	76	77	78	79	80
81	82	83	84	85	86	87	88	89	90
91	92	93	94	95	96	97	98	99	100

Addition Patterns in a Hundreds Chart

Look at the hundreds chart at the right.

1. Put a square box around any four numbers in the hundreds chart. For example, you could put a square box around 32, 33, 42, and 43.

   a) Add the numbers in each diagonal of your box.

   b) What do you notice about each sum in part (a)?

   c) Repeat parts (a) and (b) for other square boxes around groups of four numbers. What do you notice?

2. Now place a square box around any group of nine numbers in the hundreds chart.

   a) Find the sum of the numbers in each diagonal. What do you notice?

   b) How does your pattern in part (a) compare with any pattern you saw in question 1?

   c) What other patterns do you see?

   d) Find the sum of the numbers in the centre column and the centre row. What do you notice? Is there a pattern?

   e) Repeat parts (a) to (d) for other square boxes around groups of nine numbers in the hundreds chart. Do you need to modify your patterns based on other discoveries?

3. Predict what addition patterns you might discover if you drew a square box around any group of 16 numbers in the hundreds chart.

   a) Check your predictions by adding.

   b) Try other 4-by-4 squares in the hundreds chart to verify your predictions.

4. Suppose that you turned a square so it was in the shape of a diamond and placed it around any five numbers in the hundreds chart.

    a) Predict what addition patterns you might find.

    b) Check your predictions by adding numbers in the hundreds chart.

5. Now place a rectangle around any six numbers in the hundreds chart so that the rectangle covers three numbers in one row and three numbers in the row directly underneath it.

    a) Find all the patterns that you can.

    b) Compare your patterns with the patterns of other students in your class. What do you notice?

6. Create a pattern of your own based on the numbers in the hundreds chart. Show your pattern to other students in your class, and verify to them that your pattern works in all cases.

Multiplication Patterns

7. Put a square box around any four numbers in the hundreds chart. For example, you could put a square box around 32, 33, 42, and 43.

    a) Multiply the numbers in each diagonal of your box. Then subtract the smaller number from the larger number.

    b) What do you notice about your difference in part (a)?

    c) Repeat parts (a) and (b) for other square boxes around groups of four numbers. What do you notice?

8. Now place a square box around any group of nine numbers in the hundreds chart.

    a) Multiply the numbers in the opposite corners of your box. Find the difference between the two products.

    b) Multiply the three numbers in each of the diagonals of your box. Find the difference of the two products.

    c) Divide your answer in part (b) by your answer in part (a). What do you notice?

    d) Repeat parts (a) to (c) for other square boxes around groups of nine numbers in the hundreds chart. Do you need to modify your patterns based on other discoveries?

9. Create a pattern of your own based on the numbers in the hundreds chart. Show your pattern to other students in your class, and verify to them that your pattern works in all cases.

Chapter 5: 2-D Measurement

Converting between Units

To convert between units, you can multiply or divide by powers of 10.

To convert 5 km to metres, you need to know that 1 km = 1000 m.

$$5 \text{ km} = 5 \times 1000 \text{ m}$$
$$= 5000 \text{ m}$$

To convert 12 cm to metres, you need to know that 100 cm = 1 m.

$$12 \text{ cm} = 12 \div 100 \text{ or } 0.12 \text{ m}$$

1. Copy and complete the table. The first row is done for you. What patterns can help you complete the table?

Millimetres (mm)	Centimetres (cm)	Decimetres (dm)	Metres (m)	Kilometres (km)
4 800	480	48	4.8	0.0048
90 000				
		2		
			8000	
				3
	76 400			

Perimeter

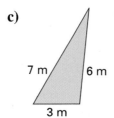

4 cm

7 cm

10 cm

16 cm

What is the distance around this figure?

Perimeter = 7 cm + 16 cm + 10 cm + 4 cm + 3 cm + 12 cm
= 52 cm

2. Find the perimeter of each shape.

a)
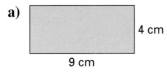
4 cm
9 cm

c)
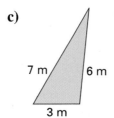
7 m   6 m
3 m

b)

12 cm

d)
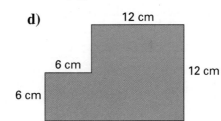
12 cm
6 cm
6 cm   12 cm
6 cm

Area of a Rectangle

The amount of surface that a region occupies is the **area** of the region. Area is measured in square units.

Find the area of this rectangle.

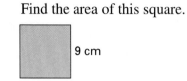

3 m

8 m

Find the area of this square.

9 cm

Area = length × width
= 8 m × 3 m
= 24 m²

Area = length × width
= 9 cm × 9 cm
= 81 cm²

3. Find the area of each shape.

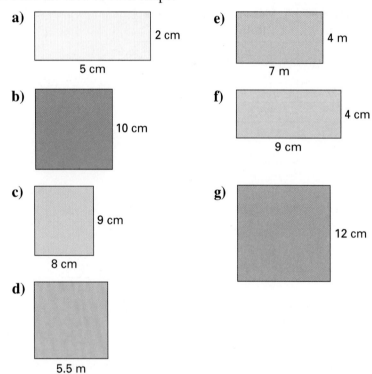

a)

2 cm

5 cm

b)

10 cm

c)

9 cm

8 cm

d)

5.5 m

e)

4 m

7 m

f)

4 cm

9 cm

g)

12 cm

Area of a Parallelogram

Finding the area of a parallelogram is similar to finding the area of a rectangle.

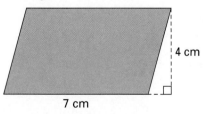

4 cm

7 cm

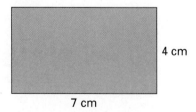

4 cm

7 cm

The area of the rectangle and the area of the parallelogram are the same.
The area of the parallelogram is found as follows:

Area = base length × height
   = 7 cm × 4 cm
   = 28 cm²

4. Find the area of each parallelogram.

a)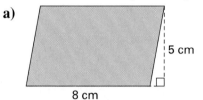
5 cm
8 cm

b)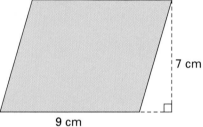
7 cm
9 cm

c)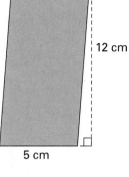
12 cm
5 cm

d)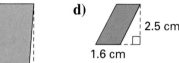
2.5 cm
1.6 cm

Area of a Triangle

The area of a triangle is half the area of a parallelogram.

The area of this triangle is found as follows:

Area = 0.5 × base length × height
   = 0.5 × 3 cm × 4 cm
   = 6 cm²

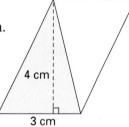
4 cm
3 cm

5. Find the area of each triangle.

a)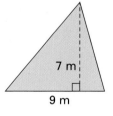
7 m
9 m

b)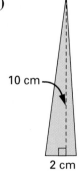
10 cm
2 cm

c)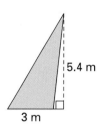
5.4 m
3 m

Chapter 6: Addition and Subtraction of Integers

Positive and Negative Numbers

$+3$ is a **positive number**.
$+1, +2, +3, +4, \ldots$ are all positive numbers.
These numbers can also be written as 1, 2, 3, 4, … without the $+$ sign.

-3 is a **negative number**.
$-1, -2, -3, -4, \ldots$ are all negative numbers.
When writing a negative number, the negative ($-$) sign is always shown.

1. Copy the word. Next to it, write a word with the opposite meaning.

a) increase   **c)** minus   **e)** down   **g)** rise   **i)** below

b) backwards   **d)** win   **f)** gain   **h)** spend   **j)** positive

2. What temperature is shown on each thermometer?

a)   **c)**   **e)**

b)   **d)**   **f)**

3. Which number in each pair of numbers would be to the right on a number line? Use $<$ or $>$ to make a true statement.

a) 5 or 7   **b)** 10 or 0   **c)** 19 or 9   **d)** 23 or 28

4. Arrange the numbers in each group in order from least to greatest.

a) 12, 3, 9, 0, 18, 5   **b)** 15, 7, 108, 15, 6, 7   **c)** 13, 10, 7, 4, 109, 88

5. Evaluate the following expressions. Evaluate as many of them as you can using mental math.

a) $12 - 5$

b) $5 + 9$

c) $4 + 3 + 8$

d) $7 + 0 + 9 + 4$

e) $5 + 9 - 2$

f) $8 + 6 - 12 + 1$

g) $12 - 5 - 7$

h) $15 - 8 - 3$

i) $9 - 6 - 8$

j) $7 - 3 - 2$

k) $10 - 5 + 3 - 7 + 9 - 2$

l) $3 + 9 - 6 - 2 - 3 + 1$

m) $5 - 3 + 6 - 2 + 4 + 7 - 13$

n) $12 - 7 - 3 - 2 + 8 - 5 + 1$

o) $11 - 9 + 7 + 1 - 5 - 3 + 2$

p) $4 + 5 - 6 + 10 - 11$

q) $12 - 5 + 9 - 12 - 3 + 10 - 9$

r) $3 + 14 - 12 - 4 + 8 - 7 + 10$

Chapter 7: 2-D Geometry

Polygon

A polygon is a two-dimensional closed shape that is formed by line segments.

not a polygon
(not a closed shape)

not a polygon
(not formed with
straight lines)

a polygon

1. Name the shape, and state whether or not it is a polygon.

a)

b)

c)

d)

e)

f)

Angles

An angle is the amount of space between two lines that intersect. An angle is measured in degrees. A protractor can be used to measure an angle. For example, $\angle ABC$ is 45°.

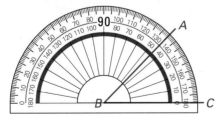

2. Use a protractor to measure each angle.

a)

b)

c)

d)

3. Use a protractor to draw each angle.

   a) 40°   **b)** 75°   **c)** 110°   **d)** 145°

Types of Angles

Angles of different measurements have different names.

Angle measurement	Name of angle
less than 90°	acute
90°	right
between 90° and 180°	obtuse
180°	straight
between 180° and 360°	reflex

4. What is the name of each angle?

   a) 40°   **b)** 90°   **c)** 220° **d)** 180° **e)** 14° **f)** 95°

Congruent and Similar

Two shapes are congruent if they have the same size and shape. Two shapes are similar if they have the same shape but are not the same size.

5. Are the following shapes congruent or similar?

a)    **c)**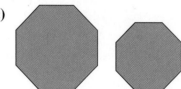

b)   **d)**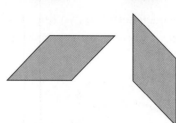

Slides, Flips, and Turns

In previous years, you worked with geometric patterns and discovered how their properties can help you solve problems.

6. Would you slide, flip, or turn the coloured shape to match the tracing? How do you know?

a)    **b)** 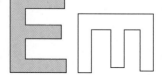   **c)**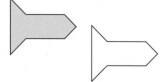

7. Copy and complete the other half of each figure.

a)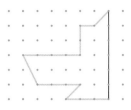

b)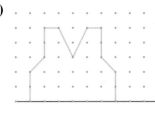

c)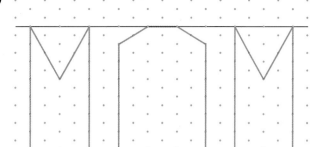

8. Copy each diagram onto grid paper. Draw the slide or flip image indicated.

a)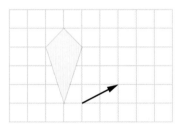

b)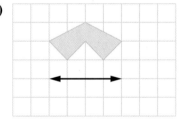

c)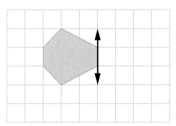

d)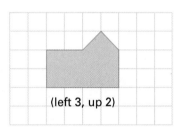
(left 3, up 2)

e)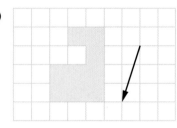

f)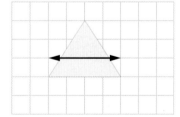

Chapter 8: Variables, Expressions, and Equations

Symbols

Symbols can be used to show information in a more compact form. Symbols can be used to represent numbers. For example, some companies use letters to represent their phone numbers to help you remember their phone numbers.

Look at the keypad on any phone.

1. a) What phone number is represented by 555-SAVE?

  b) What type of company might use this number?

2. a) What phone number is represented by 555-TIME?

  b) What type of company might use this number?

3. Use the phone number 555-3325.

  a) Create a word that could be used to remind you of this number.

  b) What type of company might use this number? Give reasons for your answer.

4. a) Think of your own company. What does it sell? Who are its customers?

  b) What phone number might you want so that you could use it in an advertisement?

  c) Create an advertisement that uses this phone number.

Bar Codes

In many stores, products are labelled with bar codes like the one on the back of this student textbook. The label is a number of bars, both thin and thick, that are quickly read by a bar code reader.

5. What do you think a bar code represents?

6. Why do you think stores and companies use bar codes instead of numbers?

7. Do you think each store or company has a different set of bar codes? Why?

8. Look at the bar code on the back of this textbook. How do you think each number is represented by each bar? Compare your answer with other students' answers.

Chapter 9: Fraction Operations

Fractions

A fraction is a part of a whole. The top part of a fraction is called the numerator. The numerator represents how many parts of the whole there are. The bottom of a fraction is called the denominator. The denominator represents how many equal parts the whole is divided into.

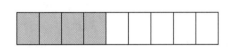

The numerator indicates how many sections of the rectangle are coloured in.

$$\frac{4}{9}$$

The denominator indicates how many equal sections the whole rectangle is divided into.

1. Write the fraction that represents each description.

a) a circle divided into 6 sections, with 5 sections coloured in

b) a fraction strip divided into 5 sections, with 2 sections coloured in

c) a 4-by-4 grid, with 7 squares coloured in

d) a pizza cut into 10 slices, with 3 slices eaten

Proper Fractions

A proper fraction is a fraction with a numerator that is smaller than the denominator. For example, $\frac{1}{3}$ is a proper fraction.

Improper Fractions

An improper fraction is a fraction with a numerator that is larger than the denominator. For example, $\frac{11}{4}$ is an improper fraction.

Mixed Numbers

A mixed number is a combination of a whole number and a proper fraction. For example, $2\frac{1}{4}$ is a mixed number.

To convert a mixed number to an improper fraction, multiply the denominator by the whole number. Add the result to the numerator to get the numerator of the improper fraction. The denominator stays the same.

Convert $3\frac{1}{2}$ to an improper fraction: $3\frac{1}{2} = \frac{2(3) + 1}{2}$

$$= \frac{7}{2}$$

2. Convert each mixed number to an improper fraction.

a) $4\frac{1}{3}$ **b)** $7\frac{2}{9}$ **c)** $1\frac{5}{6}$ **d)** $8\frac{3}{5}$

To convert an improper fraction to a mixed number, divide the numerator by the denominator. Your whole number answer becomes the whole number in the mixed number. The remainder becomes the numerator of the fraction, and the denominator stays the same.

Convert $\frac{11}{4}$ to a mixed number: $\frac{11}{4} = 2\frac{3}{4}$

3. Convert each improper fraction to a mixed number.

a) $\frac{7}{3}$ **b)** $\frac{16}{9}$ **c)** $\frac{31}{5}$ **d)** $\frac{13}{6}$

Comparing and Ordering Fractions

Fractions are easily compared if they have the same denominator. If they do not have the same denominator, follow these steps:
- First find the LCM of the denominators. This number is called the lowest common denominator (LCD).
- Then write equivalent fractions using the LCD.

For $\frac{2}{3}$ and $\frac{4}{5}$, the LCM of 3 and 5 is 15.

$\frac{10}{15} < \frac{12}{15}$ therefore $\frac{2}{3} < \frac{4}{5}$

4. Which is more?

a) $\frac{1}{2}$ dollar or $\frac{1}{4}$ dollar   **c)** $\frac{1}{5}$ of a pizza or $\frac{1}{6}$ of a pizza

b) $\frac{1}{3}$ off or $\frac{1}{4}$ off the regular price   **d)** $1\frac{3}{4}$ apples or $1\frac{7}{8}$ apples

5. Copy and complete the following diagram to compare $\frac{3}{4}$ and $\frac{4}{5}$. Which is greater?

Chapter 10: 3-D Geometry

Polyhedrons

A polyhedron is a three-dimensional shape that is formed by polygons.

Prisms and pyramids are two types of polyhedrons.

Prisms are like stretched polygons.   Pyramids come to a point.

A prism is named according to the polygon that is its base.   A pyramid is also named according to the polygon that is its base.

          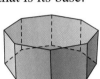

An octagon forms the base of this prism, so this prism is called an octagonal prism.   A hexagon forms the base of this pyramid, so this pyramid is called a hexagonal pyramid.

1. Name each polyhedron.

a)   c)   e)

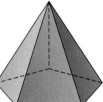   

b)   d)   f)

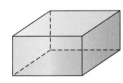   

A face is a flat surface (polygon) of a polyhedron.

An edge is formed where two of the polygons that make up a polyhedron meet.

A vertex is formed where two or more edges meet. (The plural of vertex is vertices).

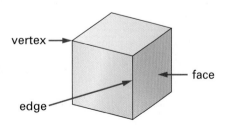

2. State how many faces, edges, and vertices are in each polyhedron from question 1.

3. State the polygons that form each polyhedron below. The first one is done for you.

Polyhedron	Polygons
	2 triangles 3 rectangles

Nets

A net is a 2-D representation of what a 3-D shape would look like if it were taken apart and laid out flat.

3-D hexagonal pyramid

net of hexagonal pyramid

4. Draw a net for each polyhedron.

a)

b)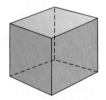

c)

Chapter 11: Surface Area and Volume

Volume is the amount of space that is occupied by a 3-D object. Volume is measured in units cubed. Below are some of the units that are used to measure volume.

$$1 \text{ cm}^3 = 1 \text{ mL}$$
$$1000 \text{ cm}^3 = 1 \text{ L}$$
$$1 \text{ m}^3 = 1000 \text{ L}$$

1. A 1 cm³ block is shown. Estimate how many 1 cm³ blocks are needed to fill each object below.

   a) a pop can

   b) your shoe

   c) a box that is the same size as this student textbook

The volume of any rectangular solid can be found as follows:
   Volume = length × width × height

2. Find the volume of each rectangular solid.

   a) length 6 cm, width 5 cm, height 2 cm

   b) dimensions 2.5 m by 5.5 m by 6.0 m

   c) dimensions 2.5 cm by 2.5 cm by 4.0 cm

   d) length 1.4 m, width 2.3 m, height 5.4 m

   e) dimensions 0.3 cm by 0.4 cm by 0.9 cm

3. A rectangular mould for making bricks is 90 mm by 90 mm by 57 mm. What is the volume of one brick that is made from this mould?

4. A wall is 5 m long, 1 m high, and 19 cm thick. There are 100 blocks in a cubic metre. How many blocks are in the wall?

5. The Czukars built a swimming pool in their yard. The pool is 10 m long, 5 m wide, and 2 m deep.

   a) What is the maximum volume of the pool?

   b) The walls and floor of the pool are 15 cm thick. How many cubic metres of soil were removed to allow for the pool, including the walls and floor?

   c) The depth of the water is only 1.7 m. What volume of water was in the pool the first time it was filled?

Chapter 12: Probability

Possible Outcomes

Possible outcomes are the different events that have a possibility of occurring.

1. For each spinner, how many different possible outcomes are there?

a)

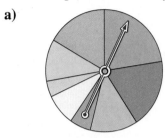

c)

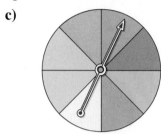

b)

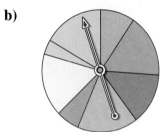

d)

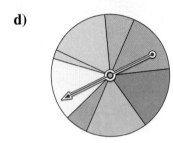

Tree Diagrams

A tree diagram is a visual representation of the number of outcomes of an event.

A coin is tossed twice in a row. Use a tree diagram to show how many different possible outcomes there are.

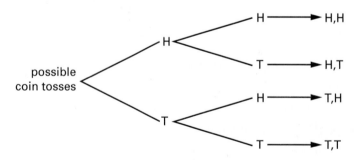

The last "branch" on the tree diagram shows how many different possible outcomes there are. In this example, there are four possible outcomes.

2. a) Draw a tree diagram for rolling a die twice in a row.

  b) How many possible outcomes are there?

Probability

Probability is the measure of how likely an event is to occur.

3. State which of the two events has the greater probability of occurring.

   a) rolling a 5 on a die or tossing Heads on a coin

   b) winning a national lottery or winning a draw at your school

   c) drawing a king from a deck of cards or drawing the 3 of spades from a deck of cards

   d) getting an A on a math test after studying or getting an A on a math test without studying

4. Order these events from most probable to least probable.

   A. The first baby born next year is a girl.

   B. The Sun will set today.

   C. You will live to be 100 years old.

   D. A classmate will live to be 100 years old.

   E. Your best friend will live to be 90 years old.

5. Suppose that you have four red cards, three blue cards, and two white cards in a box.

   a) How many possible outcomes are there for selecting a card from the box?

   b) What is the probability of selecting a red card from the box?

   c) What is the probability of selecting a blue card from the box?

   d) What is the probability of selecting a yellow card from the box?

6. You have 12 blank cards. On each card, you write a different number from 1 to 12. Find the probability of selecting each number below.

   a) the number 1   **d)** a prime number

   b) any number except 5   **e)** a composite number

   c) an odd number   **f)** a multiple of 3

7. A purse has five loose keys in it. Two keys are house keys, and three keys are car keys. One key is taken from the purse.

   a) What is the probability that the key taken is a house key?

   b) What is the probability that the key taken is a car key?

Glossary

Instructional Words

C

calculate: Figure out the number that answers a question; compute

clarify: Make a statement easier to understand; provide an example

classify: Put things into groups according to a rule and label the groups; organize into categories

compare: Look at two or more objects or numbers and identify how they are the same and how they are different (e.g., Compare the numbers 6.5 and 5.6. Compare the size of the students' feet. Compare two shapes.)

conclude: Judge or decide after reflection or after considering data

construct: Make or build a model; draw an accurate geometric shape (e.g., Use a ruler and a protractor to construct an angle.)

create: Make your own example

D

describe: Tell, draw, or write about what something is or what something looks like; tell about a process in a step-by-step way

determine: Decide with certainty as a result of calculation, experiment, or exploration

draw: 1. Show something in picture form (e.g., Draw a diagram.)
2. Pull or select an object (e.g., Draw a card from the deck. Draw a tile from the bag.)

E

estimate: Use your knowledge to make a sensible decision about an amount; make a reasonable guess (e.g., Estimate how long it takes to cycle from your home to school. Estimate how many leaves are on a tree. What is your estimate of 3210 + 789?)

evaluate: 1. Determine if something makes sense; judge
2. Calculate the value as a number

explain: Tell what you did; show your mathematical thinking at every stage; show how you know

explore: Investigate a problem by questioning, brainstorming, and trying new ideas

extend: 1. In patterning, continue the pattern
2. In problem solving, create a new problem that takes the idea of the original problem further

J

justify: Give convincing reasons for a prediction, an estimate, or a solution; tell why you think your answer is correct

M

measure: Use a tool to describe an object or determine an amount (e.g., Use a ruler to measure the height or distance around something. Use a protractor to measure an angle. Use balance scales to measure mass. Use a measuring cup to measure capacity. Use a stopwatch to measure the time in seconds or minutes.)

model: Show or demonstrate an idea using objects and/or pictures (e.g., Model addition of integers using red and blue counters.)

P

predict: Use what you know to work out what is going to happen (e.g., Predict the next number in the pattern 1, 2, 4, 7, ….)

R

reason: Develop ideas and relate them to the purpose of the task and to each other; analyze relevant information to show understanding

relate: Describe how two or more objects, drawings, ideas, or numbers are similar

represent: Show information or an idea in a different way that makes it easier to understand (e.g., Draw a graph. Make a model. Create a rhyme.)

S

show (your work): Record all calculations, drawings, numbers, words, or symbols that make up the solution

sketch: Make a rough drawing (e.g., Sketch a picture of the field with dimensions.)

solve: Develop and carry out a process for finding a solution to a problem

sort: Separate a set of objects, drawings, ideas, or numbers according to an attribute (e.g., Sort 2-D shapes by the number of sides.)

V

validate: Check an idea by showing that it works

verify: Work out an answer or solution again, usually in another way; show evidence of

visualize: Form a picture in your head of what something is like; imagine

Mathematical Words

A

algebraic equation: An **equation** that includes algebraic expressions (e.g., $3x + 5 = 8$)

algebraic expression: The result of applying arithmetic operations to numbers and variables (e.g., In a formula for the perimeter of a rectangle, $P = 2 \times (l + w)$, the algebraic expression $2 \times (l + w)$ shows the calculation.)

area: The number of square units needed to cover a surface

array: An arrangement of objects in equal rows

attribute: A characteristic or quality, usually of a pattern or geometric shape (e.g., Some common attributes of shapes are size, colour, texture, and number of edges.)

B

bar graph: A **graph** that uses bars of different heights to represent different quantities

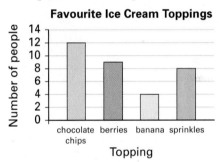

base: 1. In a 3-D **shape**, the face on which it is resting
2. In a prism, the face that determines the number of edges
3. In a 2-D shape, the line segment that is **perpendicular** to the height
4. The number that is used as a factor in a **power**

BEDMAS: A made-up word used to recall the **order of operations**, standing for **B**rackets, **E**xponents, **D**ivision, **M**ultiplication, **A**ddition, **S**ubtraction

biased results: When the results of a survey of one group are not likely to apply to another group selected from the same population

C

capacity: The amount, in millilitres or litres, that a container will hold when filled

Cartesian coordinate system: A method for describing a location by identifying its distances from two intersecting axes, one horizontal (the *x*-axis) and one vertical (the *y*-axis); the location is represented by an ordered pair of coordinates, (*x*, *y*)

cell: The intersection of a column and a row, where individual data entries are stored (e.g., Cell B2 shows the entry in row 2 and column B.)

census: The counting of an entire population

centimetre (cm): A unit of measurement for **length**; one-hundredth of a metre (e.g., A fingertip is about 1 cm wide.); 1 cm = 10 mm; 100 cm = 1 m

centre of rotation: A fixed point around which other points in a shape rotate in a clockwise (cw) or counterclockwise (ccw) direction; the centre of rotation may be inside or outside the shape

circle graph: A **graph** in which data are shown in a circle (Also called a *pie chart*.)

My Day

common denominator: A common multiple of the denominators of two or more fractions (e.g., 12 is a common denominator of $\frac{1}{2}$ and $\frac{1}{3}$.)

common factor: A number that divides into two or more other numbers with no remainder

common multiple: A number that is a **multiple** of two or more given numbers (e.g., 12, 24, and 36 are common multiples of 4 and 6.)

composite number: A number with more than two factors (e.g., 12 is a composite number with the factors 1, 2, 3, 4, 6, and 12.)

congruent: Identical in size and shape

These shapes are congruent.

coordinate grid: A grid that has data points named as ordered pairs of numbers (e.g., (4, 3))

coordinates: An ordered pair, used to describe a location on a grid labelled with an *x*-axis and a *y*-axis (e.g., The coordinates (1, 3) describe a location on a grid found by moving 1 unit horizontally from the origin and then moving 3 units vertically.)

cube: A 3-D **shape** with six congruent square faces

D

data: Information gathered in a survey, in an experiment, or by observing (e.g., Data can be in words like a list of students' names, in numbers like quiz marks, or in pictures like drawings of favourite pets. Note that the word *data* is plural, not singular.)

database: An organized and sorted list of facts or information, usually generated by a computer; made up of individual entries from records and sorted by fields

decimal: A way of writing a fraction or mixed number when the denominator is 10, 100, 1000, …

denominator: The number in a **fraction** that tells how many are in the whole set, or how many parts the whole set has been divided into (Also see **numerator**.) (e.g., In $\frac{3}{4}$, the fractional unit is fourths.)

$$\frac{3}{4} \longleftarrow \text{denominator}$$

dimension: A way to describe how an object can be measured (e.g., A line has one dimension, length. Area has two dimensions, length and width (2-D). Volume has three dimensions, length, width, and height (3-D).)

divisibility rule: A way to determine if one number is a **factor** of another number without actually dividing

E

entry: A single piece of information or data that is entered into a field (e.g., In a database, it is the contents of one cell.)

equation: A mathematical statement that two **expressions** are equal (e.g., $4 + 2 = 6$, $5x + 3 = 8$)

equivalent rates: Rates that represent the same comparison (e.g., $\frac{90 \text{ km}}{2 \text{ h}}$ and 45 km/h)

equivalent ratios: Two or more ratios that represent the same comparison (e.g., $1:3$, $2:6$, and $3:9$)

event: A set of one or more outcomes for a probability experiment (e.g., If you roll a cube with the numbers 1 to 6, the event of rolling an even number has the outcomes 2, 4, or 6.)

experimental probability: The observed probability of an event based on **data** from an experiment; calculated using the expression

$$\frac{\text{number of trials in which desired event was observed}}{\text{total number of trails in experiment}}$$

exponent: The number that tells how many equal **factors** are in a **power**

expression: A **variable** or combination of variables, numbers, and symbols that represents a mathematical relationship (e.g., x, $5 + 3x$, $6^2 + 3^2$, $\sqrt{25}$)

F

factor: One of the numbers you multiply in a multiplication operation

$$2 \quad \times \quad 6 \quad = \quad 12$$

factor   factor

favourable outcome: The desired result in a probability experiment (e.g., If you spin a coloured spinner to see how often the red section comes up, then red is the favourable outcome.)

Fibonacci sequence: A special series of numbers in which each number is the sum of the two numbers before it: 1, 1, 2, 3, 5, 8, 13, …

field: A section or category of a database that requires one specific type of data; a field may be numeric, text, date, or memo; in a database, a field is a single column

formula: 1. A rule represented by symbols, numbers, or letters, often in the form of an equation (e.g., The formula for the area of a rectangle is $A = l \times w$.) 2. Calculations made within a spreadsheet cell, using data from more than one cell (e.g., Cell B8 has the formula = sum (B2−B7), which tells the program to total the orders in column B starting in row 2 and ending in row 7); formulas may look different in different spreadsheet programs

fraction: Numbers used to name part of a whole or part of a set

(e.g., $\frac{3}{4}$ is a **proper fraction**;

$\frac{4}{3}$ is an **improper fraction**;

0.2 is a **decimal fraction**;

$5\frac{1}{2}$ is a **mixed number**.

Also see **numerator** and **denominator**.)

frequency: The number of times that an event or item occurs

frequency table: A count of each item, organized by categories or intervals

G

gram (g): A unit of measurement for **mass** (e.g., 1 mL of water has a mass of 1 g.); 1000 g = 1 kg

graph: A way of showing information so that it is more easily understood; a graph can be concrete (e.g., boys in one line and girls in another), pictorial (e.g., pictures of boys in one row and girls in another row), or abstract (e.g., two bars on a bar graph to show how many students are boys and how many are girls); types of abstract graphs include **bar graphs** and **circle graphs**

greater than (>): A sign used when comparing two numbers (e.g., 10 is greater than 5, or $10 > 5$.)

greatest common factor (GCF): The greatest **whole number** that divides into two or more other whole numbers with no remainder (e.g., 4 is the greatest common factor of 8 and 12.)

H

height of a triangle: The perpendicular distance from one side of a triangle (the base) to the opposite **vertex**; since a triangle has three sides, it has three heights

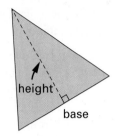

horizontal: Parallel to the horizon; straight from left to right (e.g., In a **coordinate grid**, the x-axis is a horizontal line.)

I

improper fraction: A fraction in which the **numerator** is greater than the **denominator** (e.g., $\frac{4}{3}$)

integers: All positive and negative **whole numbers**, including zero …; $-3, -2, -1, 0, 1, 2, 3, …$

interval: The space between two values (e.g., 0–10 represents the interval from 0 to 10, including 0 and 10.)

isometric: Any **transformation** of a geometric figure that leaves the distance between the points in the figure unchanged

isometric drawing: A drawing that seems 3-D; in an isometric drawing of a cube, each side of the cube has the same length on the diagram

isosceles trapezoid: A trapezoid where the non-parallel sides have equal lengths

K

kilogram (kg): A unit of measurement for **mass** (e.g., A math textbook has a mass of about 1 kg.); 1 kg = 1000 g

kilometre (km): A unit of measurement for **length**; one thousand metres; 1 km = 1000 m

L

least common multiple (LCM): The least **whole number** that has two or more given numbers as factors (e.g., 12 is the least common multiple of 4 and 6.)

length: The distance from one end of a line segment to the other end

The length of this line segment is 2 cm.

less than (<): A sign used when comparing two numbers (e.g., 5 is less than 10, or $5 < 10$.)

litre (L): A unit of measurement for **capacity** (e.g., A thousands cube from a set of base ten blocks has a capacity of 1 L.); 1 L = 1000 mL

lowest common denominator (LCD): The smallest common multiple of the denominators of two or more fractions (e.g., The LCD of $\frac{3}{4}$ and $\frac{1}{6}$ is 12.)

lowest terms: A fraction whose numerator and denominator have no common factor greater than 1 (e.g., $\frac{3}{9}$ is $\frac{1}{3}$ in lowest terms.)

M

mass: The amount of matter in an object (e.g., Common units of measurement are grams (g) and kilograms (kg).)

mean: The average; the sum of a set of numbers divided by the number of numbers in the set

median: The middle value in a set of ordered data (e.g., When there is an odd number of numbers, the median is the middle number. When there is an even number of numbers, the median is the mean of the two middle numbers.)

metre (m): A unit of measurement for **length** (e.g., 1 m is about the distance from a doorknob to the floor.); 1000 mm = 1 m; 100 cm = 1 m; 1000 m = 1 km

millilitre (mL): A unit of measurement for **capacity** (e.g., A centimetre cube from a set of base ten blocks has a capacity of 1 mL.); 1000 mL = 1 L

mixed number: A number made up of a **whole number** and a **fraction** (e.g., $5\frac{1}{2}$)

mode: The number that occurs most often in a set of data; there can be more than one mode or there might be no mode

multiple: The product of a **whole number** when multiplied by any other whole number (e.g., When you multiply 10 by the whole numbers 0 to 4, you get the multiples 0, 10, 20, 30, and 40.)

N

net: A 2-D pattern you can fold to create a 3-D **shape**

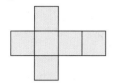

This is a net for a cube.

number line: A diagram that shows ordered numbers or points on a line; positive numbers go from 0 to the right and negative numbers go from 0 to the left

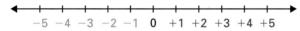

numerator: The number in a **fraction** that shows how many parts of a given size the fraction represents (Also see **denominator**.)

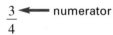

O

opposite integers: Two **integers** that are the same distance away from zero (e.g., +6 and −6 are opposite integers.)

order of operations: Rules describing what sequence to use when evaluating an expression
1. Evaluate within brackets.
2. Calculate powers and square roots.
3. Multiply or divide from left to right.
4. Add or subtract from left to right.

orientation: The direction that a shape or an object is facing (e.g., m ABC and m A′B′C′ have the same orientation.)

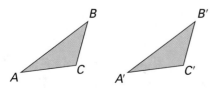

P

palindrome: A number or word that reads the same backwards as forwards (e.g., 1331 and MOM are palindromic.)

parallel: Always the same distance apart (e.g., Railway tracks are parallel to each other.)

parallelogram: A quadrilateral with equal and **parallel** opposite sides (e.g., A **rhombus**, **rectangle**, and **square** are all types of parallelograms.)

percent: A special ratio that compares a number to 100 using the symbol %

perfect square: The product of a **whole number** multiplied by itself (e.g., 81 is a perfect square because it is 9×9.)

perimeter: The distance around a figure

perpendicular: At right angles (e.g., The base of a triangle is perpendicular to the height of the triangle.)

polygon: A closed 2-D shape with sides made from straight line segments

population: The total number of individuals or items

possible outcome: A single result that can occur in a probability experiment (e.g., Getting Heads when tossing a coin is a possible outcome.)

power: A numerical expression that shows repeated multiplication (e.g., The power 4^3 is a shorter way of writing $4 \times 4 \times 4$.)

3 is the exponent of the power.

$$4^3 = 64$$

4 is the base of the power.

primary data: Information that is collected directly

prime number: A number with only two **factors**, 1 and itself (e.g., 17 is a prime number since its only factors are 1 and 17.)

prism: A 3-D **shape** with opposite congruent bases; the other faces are parallelograms (e.g., A triangular-based prism is shown.)

triangular base

triangular base

parallelogram face

proper fraction: A fraction in which the **denominator** is greater than the numerator (e.g., $\frac{1}{2}, \frac{5}{6}, \frac{2}{7}$)

proportion: A number sentence that shows two equivalent ratios (e.g., $1:3 = 4:12$ or $\frac{1}{3} = \frac{4}{12}$)

pyramid: A 3-D shape with a polygon for a base; the other faces are triangles that meet at a single **vertex** (e.g., A rectangular-based pyramid is shown below.)

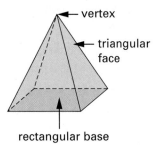

Q

quadrilateral: A polygon with four straight sides (Also see **parallelogram**, **rectangle**, **rhombus**, **square**, **trapezoid**.)

R

rate: A comparison of two quantities measured in different types of units; unlike ratios, rates include units

ratio: A way to compare two or more numbers (e.g., In a group of 5 boys and 7 girls, the ratio of boys to girls is $5:7$ or 5 to 7 or $\frac{5}{7}$.

record: All the data that are listed in each field for a single person, item, or company; in a database, a record is a single row

rectangle: A parallelogram with four square corners

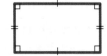

rectangular prism: A 3-D **shape** that has congruent rectangular bases (Also see **prism**.)

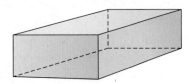

reflection: The result of a flip of a 2-D **shape**; each point in a 2-D shape flips to the opposite side of the line of reflection, but stays the same distance from the line (Also see **transformation**.)

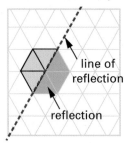

reflection symmetry: A line that divides a shape into two parts that can be matched by folding the shape in half

regroup: To change the order of terms in an arithmetic expression to form groups

regular polygon: A polygon with all sides equal and all angles equal

rhombus: A **parallelogram** with four equal sides

rotation: A transformation in which each point in a shape moves about a fixed point through the same angle

rotational symmetry: A type of *symmetry* that exists when a figure fits over its own outline as it turns around its centre; the order of rotational symmetry is the number of times a shape fits onto itself after a turn that is less than a full turn (e.g., A square has a rotational symmetry of order 4 because it matches its original orientation after each of 4 turns: $\frac{1}{4}$ turn, $\frac{1}{2}$ turn, $\frac{3}{4}$ turn, and full turn.)

S

sample: A part of a population that is used to make predictions about the whole population

scale factor: A number that you can multiply or divide each term in a ratio by to get the equivalent terms in another ratio; can be a whole number or a decimal

scatter plot: A graph that attempts to show a relationship between two variables by means of points plotted on a coordinate grid

secondary data: Information that is collected by someone else

sequence: A set of numbers or things arranged in order, one after the other, according to some rule; for example, the sequence of numbers 1, 1, 2, 3, 5, …

shape: 1. A geometric object (e.g., A square is a 2-D shape. A cube is a 3-D shape.)
2. The attribute that describes the form of a geometric object (e.g., Circles and spheres both have a round shape.)

similar: Identical in shape, but not necessarily the same size

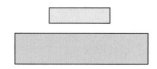

These rectangles are similar.

solution: The answer to a problem, especially any value of a **variable** that makes an **equation** true (e.g., A solution for $5x = 10$ is $x = 2$.)

sort: Put the information in order, from greatest to least or least to greatest; a database can be sorted by fields

spreadsheet: An orderly arrangement of numerical data using a grid of rows and columns; computerized spreadsheets can use hidden formulas to perform calculations with the data

square: A parallelogram with four equal sides and four square corners

square centimetre: A unit of measurement for **area**; the area covered by a square with sides that are 1 cm long

square metre: A unit of measurement for **area**; the area covered by a square with sides that are 1 m long

square root: A number when multiplied by itself equals the original number (e.g., The square root of 81 is represented as $\sqrt{81}$ and is equal to 9 because 9×9 or $9^2 = 81$.)

stem-and-leaf plot: An organization of numerical data into categories based on place values; the most significant digits are the stems and the least significant digits are the leaves (e.g., The circled leaf in this stem-and-leaf plot represents the number 258.)

Stem	Leaves
24	1 5 8
25	2 2 3 4 7 (8) 9
26	0 3
27	
28	8

surface area: The area of the surface of a 3-D object, including the base

The surface area of this box is 42 square units.

T

table of values: An orderly arrangement of facts set out for easy reference; for example, an arrangement of numerical values in vertical and horizontal columns

tally: A way to keep track of data using marks (e.g., ⦀⦀ ⦀⦀ ⦀)

term: 1. Each number or item in a sequence (e.g., In the sequence 1, 3, 5, 7, …, the third term is 5.)
2. The number that represents a quantity in a ratio
3. A number, **variable**, product, or quotient in an expression; not a sum or difference (e.g., In $x + 5x + 3 - \frac{x}{3}$, there are four terms: x, $5x$, 3, and $\frac{x}{3}$. $5x + 3$ is not a term.)

tessellation: An arrangement of plane figures that are the same shape and size to cover a plane (in all directions), without gaps or overlapping

theoretical probability: How likely an event is to occur, expressed as a number from 0 (will never happen) to 1 (certain to happen); calculated using the expression $\dfrac{\text{number of favourable outcomes}}{\text{total number of possible outcomes}}$ (e.g., P(rolling a 4 on a six-sided die) $= \dfrac{1}{6}$)

transformation: The result of moving a shape according to a rule; transformations include **translations**, **rotations**, and **reflections**

translation: The result of a slide; the slide must be along straight lines, left or right, up or down (Also see **transformation**.)

trapezoid: A **quadrilateral** with only one pair of parallel sides

tree diagram: A way to record and count all combinations of **events** (e.g., This tree diagram shows all the three-digit numbers that can be made from the digits 1, 2, and 3, if 1 must be the first digit and each digit is used only once.)

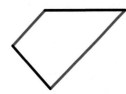

truncated: Having the point of a figure removed by slicing through the solid with a plane cut

U

unit fraction: A fraction with a numerator of 1 (e.g., $\dfrac{1}{2}$)

V

variable: A letter or symbol, such as a, b, or x, that represents a number (e.g., In the formula for the area of a rectangle, the variables A, l, and w represent the area, length, and width of the rectangle.)

vertex (plural is **vertices**): The point at the corner of an angle or **shape** (e.g., A cube has eight vertices. A triangle has three vertices. An angle has one vertex.)

vertical: At a right angle to the horizon; straight up and down (e.g., In a coordinate grid, the y-axis is a vertical line.)

vertices: Plural of **vertex**

volume: The amount of space occupied by an object

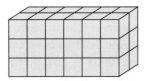

The volume of this box is **36** cubes.

W

whole numbers: The counting numbers that begin at 0 and continue forever; 0, 1, 2, 3, …

Z

zero principle: Two **opposite integers**, when added, give a sum of zero (e.g., $(-1) + (+1) = 0$)

Answers

Chapter 1, p. 1

Getting Started, p. 3

1. a) e.g, 24 marbles   **b)** 12   **c)** 36

2. a) 5 lengths of 5 black cubes

   b) 4 lengths of 5 black cubes or 10 lengths of 2 red cubes

   c) 6 lengths of 3 white cubes or 9 lengths of 2 red cubes

   d) not possible

   e) 6 lengths of 5 black cubes or 10 lengths of 3 white cubes or 15 lengths of 2 red cubes

   f) 16 lengths of 2 red cubes

3. a) 1 cm by 12 cm, 2 cm by 6 cm, 3 cm by 4 cm

   b) 1 cm by 17 cm

   c) 1 cm by 20 cm, 2 cm by 10 cm, 4 cm by 5 cm

   d) 1 cm by 24 cm, 2 cm by 12 cm, 3 cm by 8 cm, 4 cm by 6 cm

4. 2 is a factor because it divides 6 with no remainder. 3 is another factor since it divides 6 with no remainder.

5. 16 is a multiple of 1 because all numbers are multiples of 1. 16 is a multiple of 4 because any number that is a multiple of 8 must also be a multiple of 4. 16 is a multiple of 16 because every number is a multiple of itself.

6. a) 5, 7, 11, 13   **b)** 2 factors   **c)** 1, 853

7. e.g., 4, 6, and 9

8. a) composite   **b)** prime   **c)** composite

   d) prime   **e)** prime   **f)** composite

9. a)

5 cm

5 cm

b)

6 cm

6 cm

c)

7 cm

7 cm

d)

10 cm

10 cm

10. a) 133   **b)** 57   **c)** 84   **d)** 80

1.1 Using Multiples, pp. 6–7

3. a) 2, 4, 6, 8, 10   **b)** 5, 10, 15, 20, 25

   c) 6, 12, 18, 24, 30

4. 30

5. a) e.g., 8, 16, 24; LCM = 8

   b) e.g., 15, 30, 45; LCM = 15

   c) e.g., 36, 72, 108; LCM = 36

8. 12 packages of patties and 16 packages of buns

14. a) e.g., 10 packages of wieners and 15 packages of buns

   b) e.g., 26 packages of wieners and 39 packages of buns

15. Every six minutes, 10:06, 10:12, 10:18, and so on.

16. a) 8   **b)** 12th car

17. the other number

18. Since 12 is a factor of 156, then 156 is a multiple of 12. Also, 156 is a multiple of 156. So, the LCM of 12 and 156 is 156.

19. a) 42 and 48   **b)** 120 and 126

   c) 6000, 6006, 6012, 6018

20. yes

1.3 Factoring, pp. 12–13

4. a) $1 \times 28, 2 \times 14, 4 \times 7$

   b) $1 \times 32, 2 \times 16, 4 \times 8$

   c) $1 \times 64, 2 \times 32, 4 \times 16, 8 \times 8$

   d) $1 \times 120, 2 \times 60, 3 \times 40, 4 \times 30, 5 \times 24,$ $6 \times 20, 8 \times 15, 10 \times 12$

5. a) 1, 2, 4; GCF = 4   **b)** 1, 2, 4; GCF = 4

   c) 1, 2, 4; GCF = 4

   d) 1, 2, 4, 8, 16, 32; GCF = 32

14. b) 36 or 48 players

17. 1, 2, and 4

18. a) true   **b)** false   **c)** true   **d)** true

   e) false

19. 12 and 24

20. a) 2   **b)** the even number itself

   c) twice the odd number   **d)** 1   **e)** 1

1.5 Powers, pp. 18–19

5. a) base 9, exponent 4   **b)** $9 \times 9 \times 9 \times 9$
  c) 6561

6. a) $\$3^{12}$ **b)** $\$531\ 441$ **c)** Option 2
  d) March 13

13. $2^5 > 5^2$ **14. a)** $128 = 2^7$ **b)** 2^{10}

16. a) 49 cats   **b)** 343 kittens

17. a) 8^4 **b)** $\$1024$

18. a) 1   **b)** 0   **c)** 16   **d)** 1000
  e) 1 000 000

19. a) e.g., $2^4, 4^2$ **b)** e.g., $3^4, 9^2$
  c) e.g., $2^8, 16^2$ **d)** e.g., $2^{10}, 32^2$

20. a) increase by powers of 2
    $(2, 2^2, 2^3, 2^4, 2^5; 5 = 3 + 2, 9 = 5 + 2^2$, and so on)
  b) 3, 5, 9, 17, 33, 65   **c)** 513, 1025

Mid-Chapter Review, p. 21

1. If 12 is a multiple, then multiples of 12 are also multiples of the original number; e.g., 24, 36, and 48 are three other multiples.

2. a) 21, 42, 84; LCM = 21
  b) 30, 60, 90; LCM = 30
  c) 60, 120, 180; LCM = 60
  d) 48, 96, 144; LCM = 48

3. a) e.g., 60   **b)** e.g., 7

4. 2 and 5

5. a) $1 \times 100, 2 \times 50, 4 \times 25, 10 \times 10$
  b) $1 \times 27, 3 \times 9$
  c) $1 \times 45, 3 \times 15, 5 \times 9$
  d) $1 \times 1000, 2 \times 500, 4 \times 250, 5 \times 200,$
    $8 \times 125, 10 \times 100, 20 \times 50, 25 \times 40$

6. a) factors of 100: **1**, 2, 4, **5**, 10, 20, **25**, 50, 100; factors of 25: **1**, **5**, **25**; common factors: 1, 5, 25

  b) factors of 27: **1**, **3**, 9, 27; factors of 60: **1**, 2, **3**, 4, 5, 6, 10, 12, 15, 20, 30, 60; common factors: 1, 3

  c) factors of 40: **1**, **2**, **4**, **5**, 8, **10**, **20**, 40; factors of 60: **1**, **2**, 3, **4**, **5**, 6, **10**, 12, 15, **20**, 30, 60; common factors: 1, 2, 4, 5, 10, 20

  d) factors of 75: **1**, **3**, **5**, **15**, 25, 75; factors of 135: **1**, **3**, **5**, 9, **15**, 27, 45, 135; common factors: 1, 3, 5, 15

7.

Drink package size	Number of drinks to buy	Number of cookie packages (12 cookies/package)	Number of cookies/drink
3	4	1	12
4	3	1	12
5	12	5	60
6	2	1	12
8	3	2	24
10	6	5	60
12	1	1	12

8. e.g., Divide 1001 by 7 and 13 to get factors: $1001 \div 7 = 143$ and $1001 \div 13 = 77$, so 143 and 77 are factors.

9. 135   **10.** 2070

11. a) $2^{10} > 10^2$ **b)** $3^2 > 2^3$ **c)** $7^3 < 3^7$ **d)** $4^5 > 5^4$

12. 100 is greater than 75, so 2^{100} must be greater than 2^{75}.

1.6 Square Roots, pp. 24–25

4. b) 6 m by 6 m, 7 m by 7 m, and 8 m by 8 m

5. a) 2   **b)** 4   **c)** 9

6. 32   **7.** 121

12. a) 144, 169, 196   **b)** 1024

15. a) 31   **b)** 431   **c)** 17   **d)** 43

16. 32 m by 32 m

17. e.g., 26 m by 26 m

18. a) i) 10 **ii)** 100 **iii)** 1000
  b) The number of zeros in each square root is half the number of zeros under the square root sign.
  c) 4 zeros in the square root

19. If you square a positive number and then take the square root of the result, you end up with the original number. Or, if take the square root of a number and then square the result, you end up with the original result.

1.7 Order of Operations, pp. 28–29

5. a) no   **b)** 154

6. d)

7. a) 49   **b)** 35   **c)** 20
  d) 12   **e)** 7   **f)** 5

8. a) correct   **b)** incorrect   **c)** incorrect
    d) correct   **e)** correct   **f)** correct

9. e.g., 3.2 L   **10. c)** 20.5

11. a) 1   **b)** 22   **c)** 49

12. a) 3015   **b)** 10   **c)** 17 302
    d) 97   **e)** 1

13. a), c), and e) do not need brackets.
    a) 75   **b)** 46   **c)** 25
    d) 2352   **e)** 4

14. The second formula is correct.

17. a) $(5 + 8)^2 \times 6$ **b)** $(5 + 8^2) \times 6 - 3$
    c) $(10 - 2) \times 4 \div 2$ **d)** $10 \div 2 \times 3 - 1$

18. $((\underline{3 + 2})^2)^2 = (\underline{5^2})^2 = 25^2 = 625$

19. a) 5   **b)** 8   **c)** 10

20. b) e.g., $4 + 4 - 4 - \sqrt{4} = 2$;
      $((4 \times 4) - 4\,) \div 4 = 3$;
      $\sqrt{4 + 4 + 4 + 4} = 4$;
      $\sqrt{4} + \sqrt{4} + 4 \div 4 = 5$;
      $4 + (4 + 4) \div 4 = 6$; $4 + (4 - 4 \div 4) = 7$;
      $4 + 4 + 4 - 4 = 8$; $4 + 4 + (4 \div 4) = 9$;
      $(44 - 4) \div 4 = 10$

21. a) $12 + 3 \times 2 = 18$ **b)** $100 \div (3^2 + 1) = 10$
    c) $(3^2 + 1) \div 2 + 1 = 6$
    d) $2 \times (\sqrt{4} + 1) - 1 = 5$

22. a) $500 + (10^2 \times 25)$ **b)** $3000

1.8 Solve Problems by Using Power Patterns, p. 32

5. a) 6   **b)** 4   **c)** 2   **d)** 6

Chapter Self-Test, p. 33

1. She missed 4, 24, and 144.

2. common factors: 1, 2, and 4; multiples: 168 and 336

3. a) 1 km, 5 km, or 25 km
    b) For Anthony: 175 days at 1 km per day, 35 days at 5 km per day, and 7 days at 25 km per day. For Samantha: 250 days at 1 km per day, 50 days at 5 km per day, and 10 days at 25 km per day.

4. a) 6×4 photo
    b) 216 photos

5. a) i) yes   **ii)** no   **iii)** yes   **iv)** no   **b)** yes

6. a) 4^3 **b)** 64   **c)** 256

7. a) 13 060 694 016   **b)** 362 797 056

8. a) 2 and 4 are both factors of 8, so they must also be factors of any number of which 8 is a factor.

b) $8^{10} = 8$ multiplied by itself 10 times. This is a multiple of 8, so 8 is a factor of 8^{10}.

9. a) 5   **b)** 9   **c)** 10

10. a) 21 m by 21 m   **b)** 11 m by 11 m
     c) 30 cm by 30 cm   **d)** 100 cm by 100 cm

11. a) 4   **b)** 7

Chapter Review, p. 35

1. e.g., 2 and 6

2. a) $2, $5, $20; LCM = 20   **b)** 18

3. common factors: 1, 2, 4, 8, 16, 32, 64

4. Since 272 is a multiple of 16, then 16 must be a factor of 272. The greatest factor of 16 is 16, so the GCF of 16 and 272 must be 16.

5. a) Factors are **i)** 1, 2, 4   **ii)** 1, 3, 9   **iii)** 1, 2, 4, 8, 16   **iv)** 1, 5, 25   **v)** 1, 2, 3, 4, 6, 9, 12, 18, 36
    b) e.g., 81

6. a) e.g., 600   **b)** 97   **c)** e.g., 80

7. a) Keep doubling, 2, 4, 8, 16, 32
    b) $2^{10} = 2^9 \times 2$, so 2^{10} is 2 times 2^9.

8. a) 3 is a factor of 9, so any multiple of 9 is also a multiple of 3. If 9 is a factor of some number, then 3 is also a factor, since the number must be a multiple of both 9 and 3.
    b) Since $3^{10} = 3$ multiplied by itself 10 times, it is a multiple of 3, and so 3 is a factor of 3^{10}.
    c) 177 147

9. a) 64 people   **b)** 4^3 **c)** 16 384 e-mails

10. 81   **11.** 40 m by 40 m

12. a) 115   **b)** 36   **c)** 19

13. 5

Chapter 2, p. 37

Getting Started, p. 39

1. 4 and 25, 5 and 20, 10 and 10

2. a) 8:7   **b)** 6:8   **c)** 2:1   **d)** 6:24
    e) 24:6

3. 16:20, 12:15, 32:40

4. a) $\frac{5}{10}$ **b)** $\frac{3}{4}$ **c)** $\frac{1}{4}$ **d)** $\frac{4}{4}$

5. a)

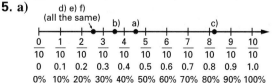

6. 25%, 0.25, $\dfrac{1}{4}$

7. a) i) $7:3$ **ii)** $53:47$ **b) i)** $\dfrac{7}{10}$ **ii)** $\dfrac{53}{100}$

   c) i) 0.7, 70% **ii)** 0.53, 53%

8. a) 23.5 **b)** 2.35 **c)** 8.76 **d)** 0.876

9. a) 99 **b)** 0.99 **c)** 208 **d)** 20.8

10. a) 2 **b)** 1.5 **c)** 1.5 **d)** 1.6

   e) 0.01 **f)** 0.01

11. a) 6.7 **b)** 6.7 **c)** 10 **d)** 10

   e) 100 **f)** 100

12. a) 1 **b)** 1 **c)** 5 **d)** 5.5

2.2 Solving Ratio Problems, pp. 44–45

5. a) i) $2:4 = 5:$ ▨ **ii)** $8:2 =$ ▨$:5$

   iii) $4:3 =$ ▨$:9$ **iv)** $3:$ ▨$= 5:15$

   b) i) 5 **ii)** 5 **iii)** 3 **iv)** 3

   c) i) 10 **ii)** 20 **iii)** 12 **iv)** 9

6. a) 6 **b)** 15

11. a) e.g., 180 cm long

   b) e.g., 300 cm tall **c)** e.g., 2 cm long

14. a) $30:420$ **b)** $150:2100$ **c)** 10 h

15. a) $50:49$ **b)** $16:12$ **c)** $25:74$

18. a) 310

   b) 114 vanilla, 190 chocolate, 76 strawberry

2.3 Solving Rate Problems, pp. 48–49

5. a) $\dfrac{15 \text{ mm}}{3 \text{ days}}$ **b)** $\dfrac{4 \text{ chocolate bars}}{\$2.20}$

   c) $\dfrac{\$14.00}{1 \text{ week}}$ **d)** $\dfrac{12 \text{ cm}}{4 \text{ months}}$

6. a) e.g., $\dfrac{1 \text{ goal}}{2 \text{ games}}$ and $\dfrac{10 \text{ goals}}{20 \text{ games}}$

   b) e.g., $\dfrac{5 \text{ km}}{30 \text{ min}}$ and $\dfrac{1 \text{ km}}{6 \text{ min}}$

   c) e.g., $\dfrac{12 \text{ pizzas}}{60 \text{ min}}$ and $\dfrac{3 \text{ pizzas}}{15 \text{ min}}$

   d) e.g., $\dfrac{2 \text{ penalties}}{5 \text{ games}}$ and $\dfrac{20 \text{ penalties}}{50 \text{ games}}$

18. 11 min **20.** 2.125 h

21. a) Honda Civic

   b) Chevrolet Corvette and Porsche 911 Turbo

2.4 Communicating about Ratio and Rate Problems, p. 52

3. Yes, Marlene can run 6 km in 45 min.

4. Yes, Chris is correct.

5. 1800 flyers

6. No, his reasoning is incorrect.

7. No, the photocopier can make 30 copies in 1 min.

8. No, the iceberg is 81 m from top to bottom.

9. Buy peanuts at bulk-food store.

10. a) Save $0.55 by buying cheaper raisins.

   b) 550 g of golden raisins or 660 g of dark raisins

Mid-Chapter Review, p. 55

1. a) e.g., 10 to 12, 15 to 18, and 20 to 24

   b) e.g., $2:3$, $6:9$, and $56:84$

   c) e.g., $\dfrac{4}{5}$, $\dfrac{20}{25}$, and $\dfrac{108}{35}$

2. a) $40:300$ **b)** $100:120$ **c)** $50:1000$

3. a) 12 **b)** 4 **c)** 32

   d) 30 **e)** 5 **f)** 78

4. 3000 cm **5.** 3 m **6.** 350 passes

7. a) ratio **b)** rate **c)** rate

8. $2400 **9.** 3 h **10.** $22.50

11. 35 coaches

2.5 Ratios as Percents, pp. 58–59

5. a) 44% **b)** 36% **6.** 90%

7. a) 50% **b)** $\dfrac{2}{5}$, 0.4 **c)** $21:100$, 0.21, 21%

   d) $\dfrac{3}{10}$, $3:10$, 30% **e)** $\dfrac{1}{4}$, $1:4$, 0.25

11. 60%

14. a) 8¢ **b)** 7¢ **c)** 6¢

15. 95%, 92%, 90%, 68% (Mohammed, Andrew, Ian, Tyrone)

19. 40%

20. a) $5951:8848$ **b)** 0.67 **c)** 67%

21. a) e.g., You could use the proportion $\dfrac{1}{3} = \dfrac{▨}{100}$ to calculate $\dfrac{1}{3}$ as a percent. Since the scale factor is not a whole number, and since the percent $= 1 \times$ scale factor, the percent will not be a whole number.

   b) e.g., $\dfrac{1}{7}$

2.6 Solving Percent Problems, pp. 62–63

5. a) 9 **b)** $\dfrac{15}{100}$, $\dfrac{▨}{60}$, 9 **c)** 0.15×60, 9

6. a) 15 **b)** 4 **c)** 225 **d)** 450

11. a) $1500 **b)** $7500

14. 39% **15. a)** $200 **b)** $160

16. The jeans are cheaper at Jane's Jean Shop.

17. e.g., sleep 8 h, school 6 h, job 2 h 24 min, homework 1.7 h, meals 2 h, entertainment 4 h

18. 30 cm   **19.** 250 units   **20.** $20 100
21. 300%

2.7 Decimal Multiplication, pp. 66–67

5. a) 0.24   **b)** 0.14
6. a) e.g., about 5   **b)** e.g., about 24
7. a) 0.68   **b)** 6.072
11. $8.7 \times 0.6 = 5.22$
14. $7.56   **19.** $191.21
20. Miguel 1.785 m, Romona 1.712 m
21. about 329 297 births
22. 24 775 673, to the nearest person
23. $5760   **24.** $1.3 \times 1.3 = 1.69$

2.8 Decimal Division, pp. 70–71

5. a) 3   **b)** 20   **c)** 31
6. a) e.g., 4   **b)** e.g., 1
7. a) 90   **b)** 5.1   **c)** 6.25
10. a) 115 dimes **b)** 230 nickels **c)** 46 quarters
  d) 1150 pennies
13. a) 14 pieces plus some rope left over
  b) 14 pieces   **c)** 8 pieces   **d)** 16 pieces
14. about 2 h   **b)** about 3 h
15. 4 glasses can be filled, with some water left
  in the bottle.
16. 23 L   **17.** 22.5 h
18. about 112 km/h
19. a) about 1.5 m tall   **b)** about 1.7 m tall
20. 8 min
21. a) 13.08 cm³   **b)** 91.7 cm³ of water
22. about 400 days
23. 1.75 h (airplane), 1.8 h (bullet train),
  and 2.27 h (car)

Chapter Self-Test, p. 73

1. a) 4:12   **b)** $\frac{4}{16}$ **c)** 75%

2. e.g., 10:18, 15:27, and 50:90
3. a) 10   **b)** 14   **c)** 12   **d)** 10
4. 54 m   **5.** $18   **6.** $96   **7.** 80%
8. a) $\frac{21}{60}$, 0.35, 35% **b)** $\frac{18}{100}$, 18:100, 18%

  c) $\frac{82}{100}$, 82:100, 0.82

9. a) 14   **b)** 70%   **c)** 250
10. 1170 fans   **11.** $15.00
12. $22.80   **13. a)** 55% **b)** 45%

14. e.g., The decimal equivalent of 10% is 0.10, and
  the decimal equivalent of 17% is 0.17. You can
  multiply an amount by 0.10 in your head, but
  that's not easy to do with 0.17.
15. e.g., The decimal equivalent of 1% is 0.01, and
  the decimal equivalent of 10% is 0.10. It is easy
  to multiply an amount by both 0.01 and 0.10 in
  your head.
16. e.g., To calculate 50% of 212, you might take
  half of 212 (divide it by 2). To calculate 42% of
  212, you would probably use a proportion.
17. a) e.g., 1.4 is about 1.5, and 3.5 is a little more
  than 3. $3 \times 1.5 = 4.5$. My answer should be
  a little more than that, which is about 5.
  b) 0.7 is more than half but less than 1, so
  0.7×7.34 is a bit less than halfway between
  7.34 and half of $7.34 = 3.67$. This is about 5.
  c) e.g., 65.2 is about 60, and 12.9 is about 12.
  $60 \div 12 = 5$, so the answer is about 5.
  d) 0.53 is about 0.5, and dividing by 0.5 is the
  same as multiplying by 2. 2.46 is about 2.5,
  and $2.5 \times 2 = 5$.
18. 119.35

Chapter Review, p. 75

1. a) e.g., 18:40 and 27:60
  b) e.g., $\frac{16}{20}$ and $\frac{24}{30}$ **c)** e.g., 42 to 6 and 63 to 9
2. a) $\frac{85}{120}$; e.g., $\frac{17}{24}$ **b)** $\frac{24\,000}{80}$; e.g., $\frac{300}{1}$
3. a) 6   **b)** 18   **c)** 7   **d)** 8
4. 8.5 cm   **5.** 300 km
6. a) $\frac{24}{30}$, 0.80, 80%   **b)** $\frac{36}{100}$, 36:100, 36%
  c) $\frac{86}{100}$, 86:100, 0.86   **d)** 3:3, 1.0, 100%
  e) $\frac{5}{100}$, 5:100, 5%
7. No, she would have to calculate the percent for
  $\frac{6}{14}$.
8. a) crust $\frac{50}{100}$, cheese $\frac{25}{100}$, tomato sauce $\frac{12}{100}$,
  sausage $\frac{8}{100}$, mushrooms $\frac{5}{100}$
  b) crust 0.5, cheese 0.25, tomato sauce 0.12,
  sausage 0.08, mushrooms 0.05
  c) No, the shape would not change.
9. a) 21   **b)** 48   **c)** 5%

10. 330 tickets   **11.** Raj

12. a) 1.8   **b)** 8

13. a) 7.42   **b)** 2.952   **c)** 3   **d)** 21

Math in Action: Rock Band Manager, pp. 77–78

1. $2000   **2.** $1338.75   **3.** 15 points

4. e.g., advantage: could make a lot of money; disadvantage: uncertain income

5. e.g., $15 \times 1000 - 0.2(15 \times 1000 - 0.1(15 \times 1000)) - 0.10(15 \times 1000)$

6. Without order of operations, different answers would be possible; e.g., amount earned = $0.8 \times 0.9 \times$ (cost of one ticket $\times$ number of tickets sold)

7. $10 800, $2700, $1500

8. band $14 440, manager $3600, agent $2000

9. band $12 960, manager $3240, agent $1800

10. $54 000

11. e.g., money from T-shirts = $0.8 \times 0.33 \times 0.85 \times$ total sales

12. for $7500: hall $1125, GST $525, PST $600, merchandisers $4271.25, manager $420.75, band $1683; for $12 750: hall $1912.50, GST $892.50, PST $1020, merchandisers $7261.10, manager $715.28, band $2861.10

13. $25 000, $25 000

14. $31 250

Chapter 3, p. 79

Getting Started, p. 81

1. a) Time of day   **b)** Commuters

c) one million   **d)** 4 p.m. to 6 p.m.

2. a) 200   **b)** 24   **c)** 23%   **d)** 43%

3. soccer 5; swimming ̶H̶H̶H̶ I, $\frac{6}{20}$, 30%; hockey IIII, 4, $\frac{4}{20}$; baseball III, 3, 15%; cycling 2, $\frac{2}{20}$, 10%

4. a) $11.50   **b)** 5   **c)** 5

3.2 Avoiding Bias in Data Collection, pp. 85–87

5. a) yes   **b)** yes   **c)** yes

6. a) primary data   **b)** secondary data

c) secondary data

10. a) would not   **b)** would not

c) would   **d)** would

11. Surveys will represent entire population.

13. e.g., b), a), f), c), d), e)

3.3 Using a Database, pp. 90–91

4. a) Yan, Ye   **b)** Goring, Mandy   **c)** Adams, Mark

d) Goring, Mandy

5. a) text   **b)** numeric   **c)** numeric   **d)** memo

e) date   **f)** text

6. a) numeric: Code, Year Released, Running time (min); text: Title, Studio; date: (none); memo: Notes

b) Casper

7. a) by country   **b)** China

c) population   **d)** population density

e) Canada, China, U.S.A., Brazil, Australia

10. a) Date published   **b)** Status

c) Call #   **d)** Author

11. e.g., **a)** research organ donors

b) research new medical treatment and medicines

c) search for missing people, check laws

d) find homes for sale by price or location

3.4 Using a Spreadsheet, pp. 94–95

4. a) I4, J4, E9, I9, J9

b) I4 = 37, J4 = 296.00, E9 = 32

c) Column J = Column I * 8.5

5. a) i) reptiles   **ii)** France's information

iii) number of pet reptiles in France

iv) population of Canada   **b)** U.S.

6. b) Christine's total hours watching TV

c) $-$ sum(D2:D7)/6   **e)** Friday

7. b) Wednesday   **c)** $3010

9. a) 40

10. a) 560:280; Order 600 hot dogs.

b) 560:140; He will sell 75 chips.

Mid-Chapter Review, p. 97

1. e.g., ask a government health office, inquire with restaurant owners, research from Statistics Canada, inquire with bread and cereal companies

2. a) sample   **b)** yes

3. e.g., **a)** Order #   **b)** Last name   **c)** Order date

d) Notes

4. Order # 3321

5. e.g., First Name, Price, Has Been Paid

6. 3323   **7.** e.g., Click on column heading to sort.

8. a) Quantity   **b)** swordtail

c) number of swordtail fish   **d)** e.g., Tax=0.15*D

9. a) e.g.,

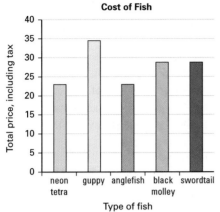

Cost of Fish

(Bar graph: Total price, including tax (y-axis, 0 to 40) vs. Type of fish (x-axis): neon tetra ≈23, guppy ≈34.5, anglefish ≈23, black molley ≈29, swordtail ≈29)

b) Use bar or pie graph.

10. From left to right: platyfish, 10, $0.75, $7.50, $1.13, $8.63

11. D − 0.50*B

3.5 Frequency Tables and Stem-and-Leaf Plots, pp. 100–101

5. a) 0–199, 200–399, 400–599, 600–799, 800–999

b) 120–121, 122–123, 124–125, 126–127, 128–129

c) 0–999, 1000–1999, 2000–2999, 3000–3999, 4000–4999

d) 0–99, 100–199, 200–299, 300–399, 400–499

6. a) frequency **b)** either **c)** stem-and-leaf

8. a) 37 **b)** 26 **c)** 10 **d)** about 27%

9. absences secondary, heights secondary, textbooks primary

10. a)

Rosa's Scores	
Stem	**Leaf**
9	4 4 6
10	4 4 4 6 6 8
11	2 6 8 8 8
12	0 2 2 4
13	0 2

b) 38 **c)** 55%

11. a)

Number of Potato Chips per Bag	
Stem	**Leaf**
9	6 9
10	7 8 8
11	7 9 9
12	2 2 3 3 3 5 6 7 7 7 8 9
13	2 2 3 4 5 7 7
14	2 3 4 5
15	0 0 1 4
18	8

12.

Passengers	Frequency
95–99	3
100–104	3
105–109	3
110–114	3
115–119	4
120–124	2
125–129	6
130–134	3
135–139	2
140–144	4
145–149	1
150–154	2

13.

Points Scored by Team		
Points for leaf	**Stem**	**Points against leaf**
	8	8
4 7 9	9	0 7
7 8 8 8	10	1 2 9 9
4 7	11	0 2 3 9
4 5 7 9	12	2 7
2	13	2

3.6 Mean, Median, and Mode, pp. 104–105

3. a) 7, 7.5, 8 **b)** 5.875, 6, 7

c) 19.857…, 18, 18 and 22 **d)** 5.714…, 5, 5

4. a) 48.333… **b)** 221.1666… **c)** 6.778 333…

d) 70% **e)** 70.333…

5. a) 3 **b)** 54 **c)** 76%

d) 268 **e)** 2.295

6. a) 8 and 9 **b)** 18 **c)** 7.1

d) F and G **e)** 93%

9. a) mean 90.45, median 89, mode 93

10. b) mean 17.2, median 18.5, mode 19
11. a) mean 9.666…, median 10, mode 11
　　b) mean 9.2857 …, median 9.5, mode 7
12. no, e.g., 40, 42, 44, 52, 54, 56
13. a) mean $64 375, median $42 500,
　　　mode $35 000
　　b) mean 　　**c)** mode
　　d) mode, more factory workers than high-paying
　　　executives
14. a) mean 　　**b)** mode 　　**c)** mode 　　**d)** mean
15. 74 　　　**16.** 12 　　　**17.** 4, 8, 8

3.7 Communicating about Graphs, p. 110

4. Second sentence is valid, but does not follow
　　from first sentence. Conclusions are
　　unreasonable, not justified, and unconvincing.
　　Too many drinks would be ordered.
5. Serve twice as many apple and cherry pies than
　　any other type.
6. a) Graph is three-dimensional with Pants and
　　　months on horizontal axes and Number of
　　　pants on vertical axis.

Chapter Self-Test, pp. 111–112

1. a) A, B, D 　　**b)** A, B 　　**c)** C
2. e.g., **a)** only surveys students of same age
　　b) only surveys people who prefer that restaurant
　　c) only surveys shoppers of that store
　　d) only surveys football players
3. a) i) Stable Structures 　**ii)** Intro to Integers
　　　iii) Art Is for All
　　b) e.g., **i)** Catalogue # 　**ii)** Title 　**iii)** 7
　　　iv) all data from row 1
4. a) 5788 　　**b)** E2 = C2*D2 　　**c)** $5.88
　　d) A7: 4898, B7: paint, acrylic, blue, 2L,
　　　C7: $15.23, D7: 4, E7: $60.92, F7: $70.06
5. ability to enter formula directly within cell
6. a) 120–129, 130–139, 140–149, 150–159,
　　　160–169
　　b)

Wingspan (cm)	Frequency
120–129	7
130–139	7
140–149	6
150–159	2
160–169	2

c) 120–129 and 130–139 　　**d)** 122
7. a) median 25.5, mean 25.5333…, mode 27
　　　and 33
　　b) median, only a measure of one time during day
　　c) mode, people can expect a mode temperature
8. a) More than half of newspaper involves
　　　non-advertising sections.
　　b) More space is devoted to advertising than any
　　　other section.

Chapter Review, pp. 114–115

1. a) census 　　**b)** survey 　　**c)** census
　　d) questionnaire
2. a) asking only already ill people
　　b) bias toward a greater number of sick people
　　　in community
3. a) descending order of Date and time
　　b) Triman 　　**c)** Lizzaboo
4. a) Gonzer 　　**b)** E7 　　**c)** 3 　　**d)** C/B
5. a)

Play rating	Frequency
E	11
G	11
S	8
P	6

b) at least good

6. a)

Stem	Leaf
0	4 8 8
1	4 5 7 7
2	4 5 5 6 7 7
3	1 3 7 8 8
4	2 5

b) e.g., Advertise to people of most common
　　age group, stock up on products appropriate
　　to most common age group, sell products in
　　proportion to various age groups.
c)

Age Groups

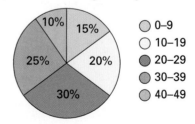

- 0–9
- 10–19
- 20–29
- 30–39
- 40–49

7. a) 20–39, 40–59, 60–79, 80–99
b) 440–441, 442–443, 444–445, 446–447
8. a) mean 5.7, median 4, mode 4
b) mean 104.25, median 106.5, mode 115
9. mean; mode doesn't change, median changes slightly
10. a)

Stem	Leaf
8	1 3
9	0 0 0 2 2 2 2 6
10	
11	2 2 6 7 8
12	0

b) mean 99.6, median 92, mode 92
c) mean; The other two exaggerate his abilities because they do not give enough emphasis to the fact that he often scores in the 110s range.
11. a) bar graph; we're comparing different objects
b)

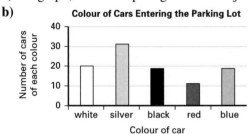

Colour of Cars Entering the Parking Lot

12. a) 8 mm   **b)** between 2 p.m. and 3 p.m.
c) between 5 p.m. and 6 p.m.   **d)** 2 mm
e) 1.33 mm   **f)** 4 mm

Cumulative Review: Chapters 1–3, pp. 117–118

1. D	**2.** D	**3.** C
4. C	**5.** D	**6.** D
7. D	**8.** A	**9.** D
10. A	**11.** B	**12.** C

Chapter 4, p. 119

Getting Started, p. 121

1. a) 13, 16, 19; rule: Add 3.
b) fifteen, eighteen, twenty-one; rule: Write out every third number.
c) n, q, t; rule: Write every third letter.
d) 55, 66, 77; rule: Add 11.
e) 100 001, 1 000 001, 10 000 001; rule: Add a 0 between the ones.
f) 31, 41, 52; rule: The difference between each number increases by 1.
2. a)

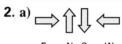

E   N   S   W

b) The arrow points east, north, south, west, and then it repeats.
3. a) 28, 20, 12; pattern: subtract 8
b) 15, 26, 25; pattern: add the next prime number, subtract 1
c) 8, 11, 9; pattern: subtract 2, add 3
4.

Input	Output
1	7
2	13
3	19
4	25
5	31
6	37
7	43
8	49
9	55
10	61

5. a) 6
b) e.g., I labelled 2, 4, 6, and 8 as 1, 2, 3, and 4. I know that every fourth number after 3 will be a 6. Since 23 will be in that sequence of numbers, I know that the 23rd number is 6.
6. a)

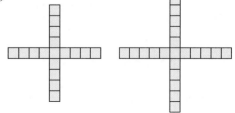

b) In each new figure, 4 cubes are added to the ends of the cross.

c) 25 cubes

7. a)

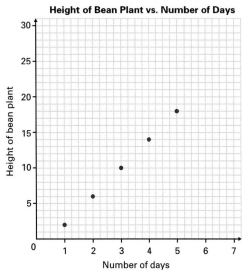

Height of Bean Plant vs. Number of Days

b) 22 mm

4.2 Applying Pattern Rules, p. 126

4. a) 16, 26, 42   **b)** 24, 39, 63

5. a) pattern rule: add 8; 40, 48, 56

b) pattern rule: add 4; 24, 28, 32

c) pattern rule: add 3; 17, 20, 23

d) pattern rule: multiply by 10; 10 000, 100 000, 1 000 000

14. a) 67, 66, 69   **b)** 311, 310, 313

15. a) term value = sum of the 3 previous term values where the first 3 terms are 0, 1, 2

b) 125, 230, 423

4.3 Using a Table of Values to Represent a Sequence, pp. 130–131

4. a)

Term number (figure number)	Picture	Term value (number of stars)
⋮		
4		9
5		11

b) Start with 3 and add 2 each time.

c) Term value = term number + next sequential number

d) 17

5. They are both right; e.g., Since the term numbers are sequential numbers from 1 to 5, then starting with 3 and adding 3 to each term value is the same as multiplying the term number by 3.

12. 1275

Mid-Chapter Review, p. 133

1. e.g., The 2nd diagonal has the pattern "Start at 4, add 2 each time."
The 5th row has the pattern "Start at 5, add 5 each time, and stop at 25."
The 4th diagonal has the pattern, "Start at 16, add 4 each time."

2. 8, 18, 38, 78, 158

3. e.g., pattern rule: term value = (term number + 1)^2
6th term value = 49

4. 25

5. a) e.g., The term values increase by 4 each time. Also, term value = (3 × term number) + previous term number.

b)

Term number	Term value
⋮	
4	15
5	19
6	23

6. 22

7. a)

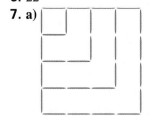

b) 70

8. a)

Term number	Rule	Term value
1	1 × 5 − 3	2
2	2 × 5 − 3	7
3	3 × 5 − 3	12
4	4 × 5 − 3	17

b) term value = previous term value + 5

c) term value = (term number × 5) − 3

d) 47

9. $54

4.4 Solve Problems Using a Table of Values, p. 137

5. 76; e.g., For each figure number, I added 12 to the number of toothpicks needed for the previous figure number.

6. 72 games for 9 teams and 90 games for 10 teams.

7. 96   **8.** 91   **9.** 7th day   **10.** yes

11. a) 27   **c)** 5

12. a) 110   **b)** 5th day

4.5 Using a Scatter Plot to Represent a Sequence, pp. 140–141

5. 49 posts

6. a)

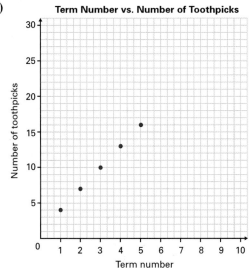

b) 28 toothpicks   **c)** 7th term

12. 11 triangle stones long

Chapter Self-Test, p. 143

1. a) 60, 72, 84; pattern rule: add 12

b) 110, 222, 446; pattern rule: start by adding 7 and then double the difference each time

c) 64, 104, 168; pattern rule: each number is the sum of the 2 previous numbers

d) 29, 37, 46; pattern rule: the difference increases by 1 each time

2. a)

Term number	Term value
1	5
2	10
3	15
4	20
5	25
6	30
7	35

b) term value = term number × 5

3. a)

Input	Output
⋮	
5	26
6	32
7	38
8	44
9	50
10	56

b) e.g., You could find a pattern in the points of the scatter plot and estimate what output would match an input of 30.

c) output = (input × 6) − 4

4. a)

Term number	Term value
1	4
2	6
3	8
4	10
5	12
6	14
7	16

b) You can calculate the term value using the previous term or by using a relationship between the term number and the term value.

term value = previous term value + 2
term value = (2 × term number) + 2

5. 44 guests   **6.** 72 games

7. a) 48   **b)** $240; yes

Chapter Review, p. 145

1. a) 14, 17, 20; pattern rule: add 3

b) 80, 70, 58; pattern rule: the difference increases by 2 each time

c) 48, 78, 126; pattern rule: add the 2 previous terms

2. a)

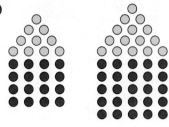

b)

Figure number	Number of green counters	Number of red counters	Total number of counters
1	1	1	2
2	3	4	7
3	6	9	15
4	10	16	26
5	15	25	40

c) The difference between the number of green counters increases by 1 each time.
The number of red counters is equal to the figure number squared.
The difference between the total number of counters increases by 3 each time.

d) 6th figure: 57 counters; 10th figure: 155 counters

3. 105 line segments

4. Mike will read 235 pages; he is right

5. a)

Term number	Term value
1	5
2	6
3	7
4	8
5	9
6	10
7	11
8	12

b) I could extend the line on the scatter plot until it passes through a term number of 20.

6. a)

Figure	Blue	Purple	Green
1	2	6	2
2	4	8	2
3	6	10	2
4	8	12	2

b)

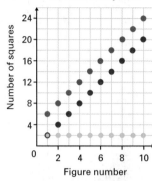

Figure Number vs. Number of Squares

Math in Action: Games Designer, pp. 147–148

1. Start with a 2, then alternate 7 and 4.

2. e.g., below 2 in "20," increase by 3: 2, 5, 8, 11

3. 2, 4, 6, 8 gets bonus, $2 + 4 + 6 + 8 = 20$, $20 \div 4 = 5$ no remainder

4. e.g., 1, 2, 3

5. a) Increase by 1.   **b)** Increase by 4.
c) Increase by 3 for 2 turns, then decrease by 3 for 2 turns.
d) Divide by 5.

6. a) 32, 36   **b)** 36, 39   **c)** 7776, 46 656
d) 4, 1

7. a) alternate 0 and 1, decrease by 3 and increase by 6, decrease by 3
b) no

Chapter 5, p. 149

Getting Started, p. 151

1. a) >   **b)** =   **c)** <   **d)** >
2. a) m   **b)** cm^2 **c)** m^2 **d)** km
e) mm
3. a) length 30 cm, perimeter 84 cm
b) length 7 m, width 7 m, perimeter 28 m
c) width 6 mm, perimeter 78 mm
d) width 10 mm, perimeter 176 mm
4. a) 56 m^2 **b)** 3600 cm^2
5. a) 6.8 L of paint   **b)** $25.50
6. a) 65.5   **b)** 439.3   **c)** 21.4   **d)** 26.6
7. yellow shape regular hexagon, red shape trapezoid, orange shape square, green shape equilateral triangle, blue and beige shapes rhombuses

5.1 Area of a Parallelogram, pp. 154–155

5.

Parallelogram	Base (cm)	Height (cm)
A	3	3
B	10	6
C	16	9

6. e.g., **a)** A: base 3 cm, height 2 cm, B: base 6 cm, height 2 cm, C: base 6 cm, height 1 cm
b) A is 6 squares, B is 12 squares, C is 6 squares
c) A 6 cm^2, B 12 cm^2, C 6 cm^2

7. 32.0 m^2

8. a) 20.6 mm^2 **b)** 9.9 dm^2 **c)** 15.00 m^2

11. e.g., rectangles with length 36 cm and width 1 cm, length 12 cm and width 3 cm, and length 9 cm and width 4 cm

12. a) 7 m   **b)** 220 cm^2 **c)** 7 cm   **d)** 4.42 dm^2
e) 7 m   **f)** 896.5 mm^2

15. a) e.g., Divide height of original parallelogram by 2.
b) e.g., Multiply height of original parallelogram by 2.
c) e.g., Multiply height of original parallelogram by $\frac{3}{4}$.

16.

Number of parking spaces	Area (m²)	Total cost ($)
1	14	21.50
5	70	107.50
10	140	215.00
50	700	1075.00

5.2 Area of a Triangle, pp. 158–159

3. a) triangle 24 cm^2, parallelogram 48 cm^2
b) triangle 12 cm^2, parallelogram 24 cm^2
c) triangle 8 cm^2, parallelogram 16 cm^2

4. a) e.g., triangles with base 6 cm and height 4 cm and with base 8 cm and height 3 cm
b) e.g., triangles with base 9 cm and height 4 cm and with base 6 cm and height 6 cm
c) e.g., triangles with base 10 cm and height 2 cm and with base 5 cm and height 4 cm
d) e.g., triangles with base 8 cm and height 2 cm and with base 4 cm and height 4 cm

5. A 2 cm^2, B 3.75 cm^2, C 2 cm^2, D 2 cm^2

11. a) 24 cm   **b)** 3.4 m   **c)** 18 cm^2 **d)** 50 cm

15. a) 6 cm   **b)** 12 cm^2

16. a) 27.0 cm   **b)** 4 mm

5.4 Area of a Trapezoid, pp. 164–165

4. 45 cm^2 **5.** 5.04 cm^2 **6.** 14 cm^2

7. a) the side that is 4.8 cm
b) area 20.2 cm^2, perimeter 19.6 cm

9. a) 15 cm^2 **b)** 180 cm^2

10. c)   **11.** 3.0 cm

12. a) 612 cm^2 **b)** 1050 cm^2

13. 8 cm   **14.** 6.25 m^2

15. 10.2 cm^2 **16.** 750 mm^2

Mid-Chapter Review, p. 168

1. A parallelogram with base 48 m and height 1 m.

2. a) 14 cm   **b)** 7.0 km^2 **c)** 4.90 cm   **d)** 2.0 mm

3. a) 12 cm^2 **b)** 2.8 m^2 **c)** 26 cm^2

4. a) 8 units^2 **b)** 9 units^2 **c)** 12 units^2
d) 12 units^2 **e)** 10.5 units^2 **f)** 6 units^2

5. a) e.g., triangle with base 6 cm and height 9 cm
b) e.g., parallelogram with base 8 mm and height 8 mm
c) e.g., trapezoid with bases 2 cm and 6 cm and height 4 cm
d) e.g., trapezoid with bases 4 cm and 10 cm and height 3 cm

6. $1620

5.6 Calculating the Area of a Complex Shape, pp. 174–175

4. a) 26 m^2 **b)** 36.0 cm^2

6. a) 60 cm^2 **b)** 60 cm^2 **c)** 94.5 cm^2

7. a) e.g., rectangles that are 60 m by 60 m, 2 m by 1800 m, 1 m by 3600 m, 4 m by 900 m, and 3 m by 1200 m
b) track with a side length of 3600 m
c) track with side lengths of 60 m

8. $438   **9.** 6 units^2

10. a) 14.5 m^2 **b)** 86.4 cm^2

11. 302 sides

5.7 Communicating about Measurement, p. 178

3. The perimeter is 8 cm and the area is 4 cm^2.

4. Possible rectangles: 1 m by 36 m, 2 m by 18 m, 3 m by 12 m, 4 m by 9 m, and 6 m by 6 m

5. 15 cm^2

6. They all have the same area, 5.25 cm^2; for each triangle, the base and height are the same.

7. a) 50 cm^2 **8. a)** 58.88 m^2

1. a) perimeter 22.5 cm, area 22.5 cm²
   b) perimeter 19.0 m, area 21.0 m²
2. a) 7.5 cm² b) 42 cm²
3. a) 30 cm² b) 8.75 cm² c) 30 cm²
4. a) 16 cm² b) 20 cm²
5. 12 cm²
6. a) 140 m² b) 48 m

Chapter Review, p. 181

1. a) e.g., parallelograms with base 9 cm and
      height 4 cm, base 12 cm and height 3 cm,
      and base 6 cm and height 6 cm
   b) e.g., triangles with base 12 cm and height
      4 cm, base 8 cm and height 6 cm, and
      base 16 cm and height 3 cm
2. no
3. a) 1.5 cm² b) 1.5 cm² c) 2.25 cm²
   d) 2.5 cm²
4. 1134.08 m²   5. 4 units²
6. a) Design 1 $469, Design 2 $522.60,
      Design 3 $318.25
   b) e.g., Design 2 because it has the largest area.

Chapter 6, p. 183

Getting Started, p. 185

1. a) >   b) <   c) >   d) <
2. $a < b$
3. a) 31   b) 77   c) 113   d) 154
4. a) 16   b) 0   c) 68   d) 57
5. c) −1°C
   d) −6°C < −4°C because it is lower on the
      thermometer than −4°C.
   e) 3°C colder
6. a) +25   b) −2   c) −20   d) −38°C
   e) +41°C
7. −4 T minus four seconds, +4 T plus four
   seconds, 0 takeoff
8. a) e.g., Positive is greater than zero. Negative
      is less than zero.
   b) e.g., positive: not negative or greater than zero;
      negative: not positive or smaller than zero

6.1 Comparing Positive and Negative Numbers, pp. 187–189

4. a) >   b) >   c) >   d) >
5. a)

Number line from −10 to +10 with points at −7, −5, −2, 0, +1, +8

   b) −7, −5, −2, 0, +1, +8
6. a) <   b) >   c) >
19. a) true   b) false   c) true   d) true
20. −14

6.3 Adding Integers Using the Zero Principle, pp. 194–195

3. a) −1   b) +2   c) −1   d) +2
   e) −6   f) −10
4. a) 0   b) 0
15. b)

+2	+1	+6
+7	+3	−1
0	+5	+4

c)

+3	−4	+1
−2	0	+2
−1	+4	−3

d) e.g.,

−3	−10	−5
−8	−6	−4
−7	−2	−9

16. a) true   b) true   c) false
17. a) +4, −4, +5   b) −13, −21, −34
18. 0   19. a) 8°C   b) −4°C   c) 17°C

6.4 Adding Integers That Are Far from Zero, pp. 198–199

5. a) 0   b) −27   c) −27   d) +7   e) +7
6. a) 39 red, 26 blue
   b) e.g., 26 red counters and 26 blue
      counters will add to 0 since each red
      counter represents (+1) and each blue
      counter represents (−1).
   c) 13 red   d) +13
7. a) −70   b) −30   c) +30   d) −70
11. a) 10 more b) 20 more
17.

−8	+6	+5	−5
+3	−3	−2	0
−1	+1	+2	−4
+4	−6	−7	+7

18. e.g., Both processes involve moving to the
   left on a number line; e.g., (+25) + (−15) =
   +10 is the same as 25 − 10 = 15. If the move
   does not cross 0, that is, if you are left with a
   positive number, then adding a negative number
   to a positive number is like subtracting.

6.5 Integer Addition Strategies, pp. 202–203

4. a) -30 **b)** $+2$ **c)** -73
10. $+15$
11. Samantha sold the stock in Week 5.
12. 345 m deep
13. a)

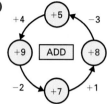

b) e.g., The final sum is $+5$ because the 4 integers that were added $(+4, -2, +1, -3)$ have a sum of 0.
c) e.g.,

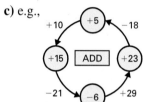

Mid-Chapter Review, p. 205

1. a) e.g., A profit is a positive gain for a company. A loss is negative since it leads to a debt for the company.
b) e.g., $(+5)°C$ is warmer than $(-5)°C$ because it is above zero.
c) e.g., A hockey player on the ice earns a negative point when there is a goal against her team and a positive point when there is a goal for her team.
d) e.g., A driver travelling at 50 km/h wants to accelerate by 5 km/h. The driver adds $(+5)$ km/h to the speed. If the driver wants to decelerate, the speed would decrease by 5 km/h. This is like adding (-5) to the speed.
2. a) -2 **b)** -6 **c)** -9 **d)** -2
3. a) $>$ **b)** $<$ **c)** $<$ **d)** $<$
4. a) -7 **b)** -3 **c)** -2 **d)** 0
5. a) $+3$ **b)** -7 **c)** $+1$ **d)** 0
6. a) $+12$ **b)** 0
7. a) e.g., The sum of two integers is positive when the sign of the integer with the largest magnitude is positive.
b) e.g., The sum of two integers is negative when the sign of the integer with the largest magnitude is negative.
8. 55 m above sea level
9. a) -100 **b)** -130 **10.** -36
11. a) -30 **b)** -41 **12.** \$1154

6.6 Using Counters to Subtract Integers, pp. 210–211

5. a) $(-4) - (+2) = (-6)$
b) $(+3) - (+2) = (+1)$
c) $(+3) - (-2) = (+5)$
d) $(-3) - (-2) = (-1)$
e) $(-2) - (-3) = (+1)$
6. a) a), c), e)
b) a) -6 b) $+1$ c) $+5$ d) -1 e) $+1$
21. a) -235 **b)** -50 **c)** $+350$
22. a) $+3$ **b)** -5 **c)** $+1$
23. a) e.g., $(+1) + (-1) - (+10) = (-10)$
b) e.g., $0 + (-5) - (+5) = (-10)$, $(+2) + (-12) - 0 = (-10)$

6.7 Using Number Lines to Subtract Integers, pp. 214–215

4. & 5. a) $(+10) - (-5) = +15$
b) $(-10) - (+15) = -25$
c) $(-55) - (-20) = -35$
d) $(+50) - (-100) = +150$
e) $(+90) - (+200) = -110$
6. a) $+40$ **b)** -35 **c)** $(-35) - (+40) = -75$
16. a) -2355 **b)** -1613 **c)** $+87$
17. a) $+30$ **b)** -60 **c)** $+112$ **d)** $+200$
e) -110 **f)** $+353$
18. a) $(+15) - (-9) = (+15) + (+9) = (+24)$

$(+15) + (+9) = (+24)$

b) $(+15) - (-20) = (+15) + (+20) = (+35)$

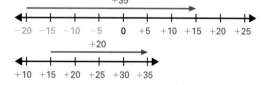

c) $+50$

6.8 Solve Problems by Working Backwards, p. 218

3. Yes. The original number is always one more than the result.

4. 14   **5.** 28   **9.** 64 kg

10. 128 grapes   **11.** $320.00

14. Referring to the discs in order from smallest to largest as A, B, C, D, and E, and the posts from left to right as 1, 2, and 3, one solution is A to 3, B to 2, A to 2, C to 3, A to 1, B to 3, A to 3, D to 2, A to 2, B to 1, A to 1, C to 2, A to 3, B to 2, A to 2, E to 3, A to 1, B to 3, A to 3, C to 1, A to 2, B to 1, A to 1, D to 3, A to 3, B to 2, A to 1, C to 3, A to 1, B to 3, A to 3.

Chapter Self-Test, p. 219

1. a) $+4$ **b)** -1

2. a) $>$ **b)** $>$ **c)** $<$

3. a) $(+5) + (-4) = (+1)$
   b) $(-3) + (+1) = (-2)$
   c) $(-2) + (-3) = (-5)$
   d) $(-5) + (+7) = (+2)$
   e) $(+20) + (-35) = (-15)$

4. $+1$ **5.** $+1$ **6.** -7

7. a) $(-3) - (+2) = (-5)$
   b) $(+30) - (-10) = (+40)$
   c) $(+3) - (-2) = (+5)$
   d) $(-1) - (-2) = (+1)$
   e) $(-4) - (-2) - (-2)$

8. a) 148   **b)** 206

9. a) -3 **b)** -13 **c)** -9 **d)** -4
   e) -3 **f)** -10

10. loss of $343

Chapter Review, p. 221

1. a) $+4$ **b)** $+2$ **c)** 0

2. a) $>$ **b)** $>$ **c)** $>$

3. yes, $+9$ and -9

4. e.g., $(+4) + (-6) = (-2)$;
   $(-5) + (+3) = (-2)$;
   $(-7) - (-5) = (-2)$

5.

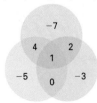

6. up $3   **7. a)** -231 **b)** -569

8. a) $+16$ **b)** -17 **c)** $+130$ **d)** $+380$

9. a) ▨ $+ 1$

10. a) $+2$ **b)** -11 (least)
   c) $+9$ (greatest)   **d)** $+2$

11. e.g., $(+4) - (-2) - (-18) + (-11) + (-3)$
   $= 10$

12. e.g., $(-1) + (-1) + (-2); (-4) + (+4) + (-4);$
   $(-10) + (+5) + (+1)$

13. e.g., Subtracting an integer is like adding the integer with the opposite sign. When subtracting, you find the space between two integers. When adding, you start at one integer and then move a distance that corresponds to the other integer.

14. a) $+5$ **b)** -5 **c)** -1 **d)** -2

15. $-4, -9$

16. a) -413 **b)** -666 **c)** -314 **d)** -881

Cumulative Review: Chapters 4–6, pp. 223–224

1. D   **2.** A   **3.** D
4. C   **5.** C   **6.** D
7. C   **8.** A   **9.** B

Chapter 7, p. 225

Getting Started, p. 227

1.

2. $-7, -4, -2, +3, +5$

3. a) C   **b)** D   **c)** B   **d)** A
   e) E

4. a) A   **b)** C   **c)** D   **d)** B

5. trapezoid

6. Top shape is a reflection; middle shape is a translation; bottom shape is a rotation.

7. 6

8. a) no   **b)** yes   **c)** yes
   d) yes   **e)** yes

9. a) reflection symmetry about a vertical line
   b) reflection symmetry about a horizontal, vertical, or diagonal line, as well as rotational symmetry

c) reflection symmetry about a diagonal line

d) not symmetrical

e) reflection symmetry about a diagonal line

f) rotational symmetry

7.1 Comparing Positions on a Grid, pp. 230–231

4. $A(-2, 3)$, $B(2, 2)$, $C(1, -4)$, $D(-2, 0)$, $E(-3, -4)$

5.

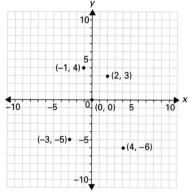

6. a) right angle scalene triangle   **b)** trapezoid

7. a) $-5, 6$ **b)** $-2, -8$

8. a) For the box, choose any number less than 5.
(e.g., $(5, -3)$ is to the right of $(-2, 2)$ because $5 > -2$.)

b) For the box, choose any number greater than -6. (e.g., $(2, -6)$ is below $(-2, 1)$ because $-6 < 1$.)

10. a) obtuse scalene triangle   **b)** rhombus

c) parallelogram   **d)** pentagon

21. a) (c, f) is above and to the left of (d, g).

b) not possible to say

22. $(3, 2)$, $(3, -2)$, $(-3, 2)$, $(-3, -2)$

23. e.g., triangles with coordinates $(-10, 2)$, $(0, 5)$, $(10, 2)$ and $(-6, 2)$, $(-6, 4)$, $(4, 10)$

7.2 Translations, pp. 234–235

3. $(14, 7)$ **4.** 4 units to the left

5. C **6.** $A'(3, 1)$, $B'(5, 4)$, $C'(5, -2)$

14. 2 blocks east

15. c) It does not matter in which order the translations are applied.

16. $A(4, 1)$, $B(6, 0)$, $C(5, 3)$

17. $A'(2, 1)$, $A''(5, -1)$, $B(4, -1)$, $B''(9, -2)$, $C(1, 1)$, $C'(3, 2)$

7.3 Reflections, pp. 237–239

5. A

6. a) $A'(5, 3)$, $B'(-2, 4)$, $C'(0, -2)$, $D'(4, 0)$

b) $R'(-3, -2)$, $S'(4, -1)$, $T'(2, 2)$

17. a) $A'(2, 2)$, $B'(5, 1)$, $C'(5, 3)$

c) $A''(3, 2)$, $B''(0, 1)$, $C''(0, 3)$

d) translation 5 units to the right

e) Two reflections in parallel lines can always be represented by a translation.

18. a) $A''(0, 2)$, $B''(0, 0)$, $C''(3, 0)$

b) $A''(0, -2)$, $B''(0, 0)$, $C''(-3, 0)$

c) The order in which reflections are applied does matter.

7.4 Rotations, pp. 242–243

4. a) figures 3 and 4   **b)** figure 5

5. 10:30, $180°$

6. B'

7. $A'(-3, 2)$, $B'(-5, 1)$, $C'(-1, -1)$

8. $Q'(3, 5)$, $R'(3, 1)$, $S'(1, 1)$, $T'(1, 5)$

18. a) $A'(0, 0)$, $B'(3, 0)$, $C'(3, -1)$

b) $A''(0, 0)$, $B''(0, -3)$, $C''(-1, -3)$

c) Two $90°$ rotations in the same direction about point A make a rotation of $180°$ about point A.

19. a) $A'(5, 2)$, $B'(2, 3)$, $C'(2, 1)$

c) $A''(5, -2)$, $B''(-3, 2)$, $C''(2, -1)$

d) $180°$ about the origin

Mid-Chapter Review, p. 247

1. e.g., **a)** $(-1, -2)$ to the right of $(-2, 0)$; $-1 > -2$

b) $(0, 0)$ below $(-4, 3)$; $0 < 3$

2. b) $C'(6, 1)$, $D'(8, 2)$, $E'(7, 4)$, $F'(5, 3)$

3. b) $A'(-6, 2)$, $B'(-2, 2)$, $C'(-6, 5)$

4. b) $D'(-5, 7)$, $E'(-3, 3)$, $F'(-12, 3)$

5. can't be sure

6. b) $A'(4, 8)$, $B'(8, 3)$, $C'(3, 3)$

7. WW' 18 units, XX' 10 units, YY' 6 units, ZZ' 20 units

8. b) $L'(5, 2)$, $M'(-8, 0)$, $N'(-10, 3)$

9. b) $A'(-2, 4)$, $B'(3, 3)$, $C'(2, 1)$

10. a) reflection about y-axis, $90°$ rotation about point $(0, 3)$

b) $180°$ rotation about point $(0, 0)$, translation of 8 to the right and 6 down

11. They are similar but may not be congruent.

7.7 Communicating about Geometric Patterns, p. 252

3. a) e.g., There are two similar rectangles whose bases are 7 cm apart. The rectangle on the left has a base of 8 cm and a height of 4 cm. The rectangle on the right has a base of 4 cm and a height of 2 cm and is a reduction of the larger rectangle by a scale factor of $\frac{1}{2}$.

4. a) e.g., There are four congruent right-angle triangles. The four right angles meet at a point. The top-right triangle has base 6 cm and height 3 cm. The top-left triangle has base 3 cm and height 6 cm. The bottom right triangle has base 3 cm and height 6 cm. The bottom-left triangle has base 6 cm and height 3 cm.

7.9 Tessellating Designs, p. 260

4. The angles at A add up to $360°$.

5. a) yes   **b)**

6. b) e.g.,

c) yes

9. a) a rectangle

b) rotations, reflections, transformations

c) He changed the shape as in James's solution by rotating the changed part $180°$ to the other half of the same side.

Chapter Self-Test, p. 262

1. Translate the triangle 9 cm along the base DE.

2. a), b), c) e.g.,

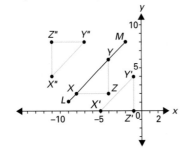

3. b) $D'(2, -2), E'(4, 1), F'(-1, 2), G'(-2, -3)$

4. a)

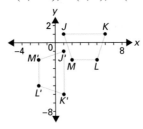

b) clockwise rotation of $90°$ about the origin or a counterclockwise rotation of $270°$ about the origin

5. a) congruent   **b)** similar   **c)** neither

6. a) yes   **b)** yes   **c)** no

7. The design has four congruent right-angle triangles, each with height 6 cm and base 4 cm. The smallest angle in each triangle meet at a common vertex such that the triangle on the right is lying on its height. The triangle on top is the image of a $90°$ counterclockwise rotation of the first triangle about the common vertex. The two remaining triangles can be found by performing two more $90°$ counterclockwise rotations about the vertex.

8. a) yes

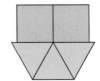

b) The vertices of each square are each $90°$, while the vertices of each equilateral triangle are each $60°$. One vertex each from two squares and three equilateral triangles will sum to $360°$.

Chapter Review, pp. 264–265

1. a) $(14, 0)$ **b)** $(-14, 0)$ **c)** $(0, -14)$ **d)** $(0, 14)$

2. c) $P'(-7, 7), Q'(-6, 5), R'(-4, 8)$

3. B

4. c) $A'(0, 2), B'(-3, 1), C'(-3, -2), D'(1, -1)$
   e) $A''(-2, 3), B''(-5, 2), C''(-5, -1), D''(-1, 0)$

5. e.g., reflection in the base of the triangle, or a rotation of $180°$ about the centre of the base, or a reflection in a horizontal line followed by a translation

6. reflection in a vertical line

7. Reflect A in a vertical line.

8. e.g., rotation of 90° about the centre of the shape, followed by a reflection in a vertical or horizontal line; or a reflection in a vertical or horizontal line, followed by a rotation of 90° about the centre of the shape; or a reflection in a diagonal line and translation

9. b) $A'(-5, 1)$, $B'(-2, 0)$, $C'(-1, 3)$

10. a)

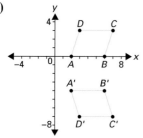

b) e.g., reflection in the *x*-axis, followed by a translation 4 units down

11. same as coordinates of $\triangle PQR$; a clockwise rotation of 360°

12. The only possible transformation is rotation.

13. a) e.g., The red diagonally placed rectangle was translated to get the green one, and the yellow diagonally placed rectangle was translated to get the blue one.

b) e.g., The green triangle was reflected to get the red one, and the red diagonally placed rectangle was reflected to get the green one.

c) e.g., The blue triangle was rotated to get the yellow one, and the blue diagonally placed rectangle was rotated to get the yellow one.

14. congruent

15. right-angle isosceles triangle

16. e.g., The two orientations are 180° rotations of one another.

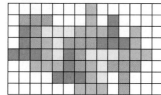

17. a) 6　　　**b)** 360°

18. a) $\frac{1}{3}$　　　**b)** three rhombuses

c) Rotate rhombus 60° cw about top vertex of rhombus, or rotate 60° ccw about bottom vertex of rhombus to move it to the bottom left position.

Math in Action: Computer Animator, pp. 267–268

4. A computer can produce exact copies of objects and allow for precise rotations. It is also helpful to be able to undo mistakes.

6. translation　　　**7.** rotation

Chapter 8, p. 269

Getting Started, p. 271

1. a) 99　　　**b)** 17

2. a) 20, 24, 28; term value = 4 × term number

b) 17, 20, 23; term value = 3 × term number + 2

3. a)

Figure number	Number of blocks
1	7
2	11
3	15
4	19
5	23

b) Start with seven square blocks arranged in the shape of an "H." Then add four more blocks, one to each arm of the "H," each time.

c) 83 blocks

4. a) =　　**b)** =　　**c)** ≠　　**d)** ≠　　**e)** =

5. a) 7　　**b)** 7　　**c)** 6　　**d)** 9　　**e)** 72

f) 15

6. a) Each five-metre increase in distance results in a one-second increase in time.

b)

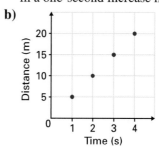

c) 50 m

8.2 Using Variables to Write Pattern Rules, pp. 276–277

4. $n + 4$

5. a) There is only one yellow block in each shape. The number of blue blocks changes.

b) The first shape has three blocks, and every shape after that has two more blocks than the shape before it.

c) $1 + 2n$

10. a)

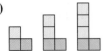

b)

c)

d)

13. a)   **b)**

14. a) yes   **b)** no

8.3 Creating and Evaluating Expressions, pp. 280–281

4. c)

5. a) 30   **b)** 24   **c)** 6   **d)** 21

6. $12h + 35$

8. a) 9   **b)** 40   **c)** 27   **d)** 8
e) 11   **f)** 14   **g)** 32   **h)** 40

9. 21   **11. c)** $170

12. c) $59   **13. c)** $41

16. a) $(4 + 2)s + 200$ or $6s + 200$
b) $4(a + p)$ or $4a + 4p$ **c)** $6c + 8m$

17. a) $0.75p + 1.25d$ **b)** November
c) November $172.25, February $158.75, April $144.75

Mid-Chapter Review, p. 283

1. a) 6, 7, 8
b) The number of squares is always three more than the figure number.
c) $n + 3$ **d)** n
e) n represents the figure number.
f) The 3 does not change.

2. a) $1 + 2(n + 1)$ **b)** $3 + 2n$

3. 33 toothpicks

4. a) D   **b)** C   **c)** A   **d)** E   **e)** B

5. a) $5n$ involves multiplication, while $5 + n$ involves addition.
b) $5n$ is likely to have a greater result since adding 5 to itself a certain number of times is likely to result in a greater value than simply adding 5 to that number.

6. a) $a + 5$ **b)** $r + 11$ **c)** $4s$ **d)** $b + 9$

7. 35

8.4 Solving Equations by Inspection, pp. 286–287

3. a) 7   **b)** 33   **c)** 7   **d)** 6   **e)** 3

4. a) $s = 4n + 3$ **b)** $23 = 4n + 3$ **c)** $n = 5$
d) 23

14. 10, 11, 12   **15.** 33 squares

16. 44 triangles, with one toothpick left over

8.5 Solving Equations by Systematic Trial, pp. 290–291

4. a) 4   **b)** 24   **c)** 8   **d)** 15

5. a) $n = 134$ **b)** $b = 21$ **c)** $x = 16$

6. a) 20   **b)** $p = 25$

7. c) 69 counters

10. 85 and 86   **15.** $50   **16.** D

17. 4 m   **18.** $n = 5$

19. a) yes   **b)** no

8.6 Communicating the Solution for an Equation, pp. 293–294

4. Divide both sides by two; both sides still balance. Solve for c.

11. a) $x = 2$ **b)** $x = 5$ **c)** $x = 8$ **d)** $x = 11$

Chapter Self-Test, p. 297

1. In the first pattern, each figure has three blocks on the bottom and $2 \times$ the figure number is added. This is $3 + 2n$ or $2n + 3$. In the second pattern, each figure has $2 \times$ the figure number of blocks, with two blocks on top. This is $2n + 2$. There is one block added on the right. This makes it $(2n + 2) + 1$.

2. 12   **3. a)** $2n$ **b)** $5x - 6$ **c)** $4s + 9$

4. a) $8p$ **b)** $10 + 6p$
c) $14 cheaper for members

5. 52   **6.** $x = 5$

7. $n = 12$ **8.** 12

9. a) $2b = 10$ **b)** $b = 5$
  c) Each bag contains 5 counters.

Chapter Review, p. 299

1. a)

Figure number	Number of toothpicks
1	5
2	7
3	9
4	11
5	13

b) The number of toothpicks is three more than twice the figure number.
  c) $t = 3 + 2s$
2. a) $n - 38$ **b)** $83 + x$ **c)** $4p$ **d)** $2s + 79$
3. 15
4. a) $n = 11$ **b)** $w = 1$ **c)** $r = 45$ **d)** $y = 4$
5. a) $x = 260$ **b)** $g = 218$ **c)** $z = 30$ **d)** $h = 5$
6. a) B, $n = 16$ **b)** C, $n = 10$ **c)** A, $n = 44$
7. a) 12   **b)** 88   **c)** 26   **d)** 12
8. $m = 12$ **9.** $y = 8$

Math in Action: Business, pp. 301–302

1. cookies $2.96, crackers $1.49, butter $1.61, bread $1.34, eggs $1.78, flour $4.30
2. $466.55   **3.** $511.06   **4.** $0.96
5. a) price of shirt = $2 \times$ cost   **b)** $29.98
  c) $14.99
6. $8.00   **7. a)** $4300   **b)** $499
8. selling price = cost + expenses + profit
9. $298   **10.** $20.50

Chapter 9, p. 303

Getting Started, p. 305

1. a) $\frac{1}{10}$ **b)** 20%   **c)** $\frac{1}{4}$ **d)** 40%
  e) $\frac{1}{2}$ **f)** 75%
2. a) Z   **b)** X   **c)** W   **d)** Y
3. a)
  b)

c)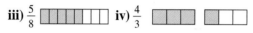
d)

4. a) $=$ **b)** $<$ **c)** $>$ **d)** $<$
5. a) hexagon: 1, rhombus: $\frac{1}{3}$, triangle: $\frac{1}{6}$,
  trapezoid: $\frac{1}{2}$
  b) e.g., **i)** 5 triangles **ii)** rhombus **iii)** hexagon and a trapezoid
  c) $\frac{5}{6}$
6. a) i) $1\frac{1}{2}$ **ii)** $1\frac{2}{3}$ **iii)** $2\frac{2}{3}$ **iv)** $1\frac{1}{4}$ **b)** $\frac{8}{3}$
7. a) i) $\frac{8}{5}$ **ii)** $\frac{14}{3}$ **iii)** $\frac{23}{4}$ **iv)** $\frac{23}{9}$ **b)** $1\frac{3}{5}$
8. a) i) $\frac{2}{3}$ 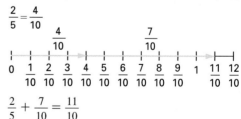 **ii)** $\frac{4}{5}$

  iii) $\frac{5}{8}$ **iv)** $\frac{4}{3}$
  b) e.g., **i)** $\frac{4}{6}, \frac{8}{12}$ **ii)** $\frac{12}{15}, \frac{16}{20}$ **iii)** $\frac{10}{16}, \frac{20}{32}$ **iv)** $\frac{8}{6}, \frac{12}{9}$
9. a) 4, 8, 12, 16, 20   **b)** 5, 10, 15, 20, 25
  c) 6, 12, 18, 24, 30   **d)** 8, 16, 24, 32, 40
  e) 12, 24, 36, 48, 60

9.2 Adding Fractions with Models, pp. 310–311

4. a) $\frac{1}{6}$ is less than $\frac{1}{4}$, and $\frac{3}{4} + \frac{1}{4} = 1$, so $\frac{3}{4} + \frac{1}{6}$ must be less than 1.
  b) $\frac{11}{12}$
5.

$$\frac{2}{5} = \frac{4}{10}$$

$$\frac{2}{5} + \frac{7}{10} = \frac{11}{10}$$

6. a) about $\frac{1}{2}$ **b)** $\frac{9}{20}$
13. a) $\frac{69}{100}$ **b)** 69%
16. a) no   **b)** $1\frac{1}{12}$ c.
17. a) i) $\frac{7}{12}$ **ii)** $\frac{9}{20}$ **iii)** $\frac{11}{30}$ **iv)** $\frac{13}{42}$

b) The numerator of the sum is the sum of the denominators. The denominator of the sum is the product of the denominators. e.g.,
$$\frac{1}{7} + \frac{1}{8} = \frac{7+8}{7 \times 8} = \frac{15}{56}$$

18. a) $6, 8, \frac{1}{5}, \frac{1}{10}$ **b)** $\frac{3}{40}$

9.3 Multiplying a Whole Number by a Fraction, pp. 314–315

4. $4\frac{2}{3}$ c.

5. a)

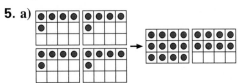

b) $\frac{20}{12}, 1\frac{8}{12}$

6. a) $\frac{3}{4} + \frac{3}{4} + \frac{3}{4} + \frac{3}{4} + \frac{3}{4}$

b)

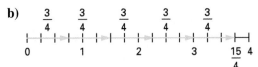

c) $\frac{15}{4}, 3\frac{3}{4}$

16. any multiple of 8

17. a)

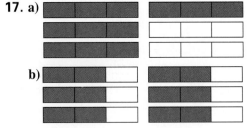

b)

18. c) $\frac{1}{2} \times 4 = 2$, so $\frac{1}{2} \times \frac{4}{5} = \frac{2}{5}$

     d) e.g., double the answer; $2 \times \frac{2}{5} = \frac{2}{5} + \frac{2}{5} = \frac{4}{5}$

9.4 Subtracting Fractions with Models, pp. 318–319

5. b) $\frac{7}{15}$ **6.** $\frac{5}{6}$ **7. a)** $\frac{13}{30}$ **b)** $\frac{2}{5}$

14. yes   **15. a)** $\frac{1}{12}$ **b)** $\frac{2}{12}$ **c)** $\frac{1}{12}$

16. a) i) $\frac{1}{12}$ **ii)** $\frac{1}{20}$ **iii)** $\frac{1}{30}$ **iv)** $\frac{1}{42}$

The numerator in the difference is the difference in the denominators. The denominator in the difference is the product of the denominators.

b) $\frac{1}{6}$

18. c) $\frac{7}{12}$ **19. a)** 0   **b)** $\frac{1}{4}$

20. yes   **21.** 6

Mid-Chapter Review, p. 321

1. a) e.g.,

$\frac{1}{3} + \frac{1}{3} = \frac{2}{3}$

b) e.g.,

$\frac{1}{6} + \frac{1}{6} + \frac{1}{6} + \frac{1}{6} + \frac{1}{6} = \frac{5}{6}$

c) e.g.,

$\frac{1}{2} + \frac{1}{2} = 1$

2. a) about $\frac{1}{2}, \frac{10}{20}$ **b)** about $1\frac{1}{4}, \frac{11}{8}$

   c) about $1, \frac{11}{10}$ **d)** about $\frac{3}{4}, \frac{19}{24}$

3. a) $\frac{19}{20}$ **b)** $\frac{38}{24}$

   c) $\frac{7}{8}$ **d)** $\frac{11}{12}$

4. e.g., $\frac{3}{3}, \frac{2}{4}$; e.g., $1\frac{1}{2} = 1 + \frac{1}{2}$

5. 1, 2, 3   **6.** about $\frac{12}{35}$

7. a) $\frac{1}{5} + \frac{1}{5} + \frac{1}{5} + \frac{1}{5} + \frac{1}{5} + \frac{1}{5} = \frac{6}{5}, 1\frac{1}{5}$

   b) $\frac{5}{12} + \frac{5}{12} + \frac{5}{12} + \frac{5}{12} = \frac{20}{12}, 1\frac{2}{3}$

   c) $\frac{3}{5} + \frac{3}{5} + \frac{3}{5} + \frac{3}{5} + \frac{3}{5} + \frac{3}{5} + \frac{3}{5} + \frac{3}{5} = \frac{24}{5}, 4\frac{4}{5}$

   d) $\frac{4}{9} + \frac{4}{9} + \frac{4}{9} + \frac{4}{9} + \frac{4}{9} = \frac{20}{9}, 2\frac{2}{9}$

8. a) $\frac{9}{8}$ **b)** $\frac{10}{9}$ **c)** $\frac{25}{6}$ **d)** $\frac{8}{5}$

9. e.g., 3, 4

10. a) $\frac{5}{21}$ **b)** $\frac{9}{20}$ **c)** $\frac{33}{54}$ **d)** $\frac{23}{20}$

11. $\frac{11}{20}$

12. a) $\frac{13}{30}$ **b)** $\frac{11}{12}$ **c)** $\frac{2}{5}$ **d)** $\frac{7}{3}$

13. $\frac{7}{30}$

9.5 Subtracting Fractions with Grids, pp. 324–325

5. a) $\frac{7}{15}$ **b)** $\frac{14}{24}$

6. She has $\frac{3}{12}$ or $\frac{1}{4}$ left to watch.

7. a) $\frac{2}{15}$ **b)** $\frac{1}{12}$ **c)** $\frac{1}{21}$

d) $\frac{5}{24}$ **e)** $\frac{7}{20}$ **f)** $\frac{7}{20}$

8. a) $\frac{1}{6}$ **b)** $\frac{8}{48}$ or $\frac{1}{6}$

c) $\frac{6}{27}$ or $\frac{2}{9}$ **d)** $\frac{11}{21}$

9. $\frac{5}{12}$ **10.** $\frac{4}{10}$ or $\frac{2}{5}$

11. a) $\frac{5}{12}$ **b)** $\frac{7}{12}$ **c)** $\frac{5}{12}$

12. $\frac{5}{8}$ **14.** $\frac{1}{8}$ **15.** $\frac{4}{15}$ **17.** $\frac{2}{5}$

18. a) yes   **b)** yes   **19.** e.g., two $\frac{1}{16}$ notes

20. a) $\frac{30}{100}$ or $\frac{3}{10}$ **b)** \$30   **22.** $\frac{4}{2}$

23. a) $\frac{1}{2}$ a beat

9.6 Adding and Subtracting Mixed Numbers, pp. 328–329

4. a) $3\frac{1}{3}$ rows   **b)** $3\frac{1}{2}$ rows   **5. a)** $4\frac{1}{8}$

6. a) $2\frac{3}{7}$ **b)** $\frac{1}{3}$ **c)** $1\frac{5}{6}$

7. a) $3\frac{3}{8}$ **b)** $5\frac{1}{15}$ **c)** $4\frac{7}{10}$ **d)** $2\frac{1}{6}$

e) $7\frac{4}{9}$ **f)** $7\frac{5}{6}$

8. a) $5\frac{9}{10}$ **b)** $5\frac{5}{12}$ **c)** $3\frac{1}{4}$ **d)** $6\frac{1}{6}$

e) $4\frac{17}{20}$ **f)** $2\frac{1}{9}$

9. a) $6\frac{1}{5}$ **b)** $6\frac{4}{7}$ **c)** $1\frac{1}{10}$ **d)** $4\frac{1}{8}$

e) $1\frac{1}{3}$ **f)** $3\frac{1}{6}$

10. a) 15 years old   **b)** $14\frac{1}{6}$ years old

11. $4\frac{1}{15}$

12. a) $3\frac{11}{12}$ **b)** $1\frac{1}{3}$

13. a) $4\frac{3}{8}$ **b)** $\frac{1}{4}$ **c)** yes

15. $2\frac{1}{2}$ **16.** $1\frac{5}{6}$

17. a) $1\frac{1}{3}$ pies   **b)** 8 pieces

18. $5 - 1\frac{2}{4} = 3\frac{3}{6}$

19. $1\frac{4}{5}$ **20.** $4\frac{7}{8}$

9.7 Communicating about Estimation Strategies, pp. 332–333

4. A little more than $1\frac{1}{2}$.

5. A little more than $3\frac{1}{2}$.

6. A little more than $\frac{6}{10}$ or $\frac{3}{5}$.

7. A little more than 4.

8. A little less than $2\frac{1}{2}$.   **9.** $1\frac{2}{3}$ boxes

9.8 Adding and Subtracting Using Equivalent Fractions, p. 336

5. a) 12   **b)** 16   **c)** 35   **d)** 36

6. a) $\frac{7}{8}$ **b)** $1\frac{9}{20}$ **7. a)** $\frac{3}{8}$ **b)** $\frac{1}{20}$

8. a) $\frac{17}{20}$ **b)** $1\frac{1}{6}$ **c)** $\frac{13}{24}$ **d)** $\frac{13}{18}$ **e)** $1\frac{3}{20}$ **f)** $\frac{27}{40}$

9. a) $\frac{13}{20}$ **b)** $\frac{1}{2}$ **c)** $\frac{5}{24}$ **d)** $\frac{7}{18}$ **e)** $\frac{7}{20}$ **f)** $\frac{3}{40}$

10. a) $\frac{4}{9}$ **b)** $\frac{1}{2}$ **c)** $\frac{11}{20}$ **d)** $\frac{1}{4}$ **e)** $1\frac{7}{12}$ **f)** $\frac{4}{21}$

12. $\frac{9}{40}$ **13. a)** $\frac{1}{6}$ **b)** $\frac{1}{3}$

15. e.g., $\frac{1}{8} + \frac{1}{4} + \frac{4}{16}$ **16.** e.g., $\frac{2}{4} - \frac{1}{2}$

17. $4\frac{2}{3}$

Chapter Self-Test, p. 338

1. a) bottom half (triangle and trapezoid)   **b)** $\frac{1}{2}$

2. a) i) $\frac{5}{8}$ **ii)** $\frac{23}{20}$

b) i) $\frac{3}{8}$ is about $\frac{1}{2}$. $\frac{1}{2} + \frac{1}{4}$ is $\frac{3}{4}$ or $\frac{6}{8}$.
$\frac{6}{8}$ is close to $\frac{5}{8}$.

ii) $\frac{2}{5}$ is about $\frac{1}{2}$ or $\frac{2}{4}$. $\frac{2}{4} + \frac{3}{4} = \frac{5}{4}$

$\frac{5}{4} = \frac{25}{20}$ is close to $\frac{23}{20}$.

3. $\frac{1}{2} + \frac{1}{4} = \frac{3}{4}$

4. a) i) $\frac{1}{8}$ **ii)** $\frac{1}{20}$

b) i) $\frac{3}{8}$ is about $\frac{1}{2}$. $\frac{1}{2} - \frac{1}{4}$ is $\frac{1}{4}$ or $\frac{2}{8}$.

$\frac{2}{8}$ is close to $\frac{1}{8}$.

ii) $\frac{4}{5}$ is about $\frac{3}{4}$. $\frac{3}{4} - \frac{3}{4} = 0.\ 0$ is close to $\frac{1}{20}$.

5. $\frac{1}{5}$ **6. a)** $1\frac{3}{5}$ **b)** $2\frac{2}{8}$

7. a) $6\frac{23}{36}$ **b)** $3\frac{9}{10}$ **c)** $\frac{4}{5}$ **d)** $4\frac{5}{8}$

8. The fraction parts of the mixed numbers are the same.

9. e.g., The fraction parts of the mixed numbers are not the same.

10. a) $\frac{5}{12}$ **b)** $\frac{1}{8}$

11. $\frac{31}{40}$ **12. a)** $\frac{7}{18}$ **b)** $\frac{13}{40}$

Chapter Review, p. 341

1. 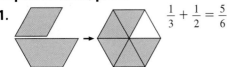 $\frac{1}{3} + \frac{1}{2} = \frac{5}{6}$

2. a) $\frac{7}{8}$ **b)** $\frac{11}{10}$ or $1\frac{1}{10}$

3. a) $3\frac{3}{4}$ **b)** $1\frac{5}{7}$

4. $1\frac{2}{3}$ h

5. a) $\frac{5}{21}$ **b)** $\frac{1}{4}$ **c)** $\frac{11}{18}$ **d)** $\frac{3}{20}$

6. a) $\frac{11}{12}$ of a bag   **b)** $\frac{1}{12}$ **7.** $\frac{4}{45}$

8. a) $2\frac{9}{10}$ **b)** $4\frac{7}{12}$ **c)** $5\frac{2}{9}$

d) $4\frac{1}{6}$ **e)** $3\frac{5}{7}$ **f)** $\frac{2}{9}$

9. $1\frac{1}{6}$ h   **10.** $2\frac{2}{9}$

11. a) yes, $2\frac{3}{4}$ **b)** no, $\frac{41}{42}$ **12. b) i)** $1\frac{8}{21}$ **ii)** $1\frac{5}{28}$

13. a) $\frac{31}{35}$ **b)** $1\frac{5}{9}$ **c)** $\frac{1}{30}$ **d)** $\frac{1}{15}$

Cumulative Review: Chapters 7–9, pp. 343–344

1. D   **2.** C   **3.** B
4. A   **5.** D   **6.** A
7. D   **8.** B   **9.** B

Chapter 10, p. 345

Getting Started, p. 347

1. a)   **b)**

c)   **d)**

e) 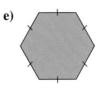   **f)**

2. a) pyramid **b)** prism   **c)** neither
d) prism   **e)** neither   **f)** prism

3. a) pentagonal prism   **b)** octagonal prism
c) rectangular prism

4. a) hexagonal pyramid   **b)** triangular pyramid
c) pentagonal pyramid

5. triangular pyramid

6. a) C   **b)** A   **c)** D

7. a) hexagonal prism   **b)** triangular prism
c) square-based pyramid

8. 4   **9.** 6   **10.** 4

11. A polygon is two-dimensional, while a polyhedron is a three-dimensional shape based on a polygon.

12. a) e.g.,   **b)** e.g.,

c) No, because the sides of the polygon base are not equal in length.

10.2 Building Objects from Nets, pp. 352–353

5. a) square-based pyramid, rectangular prism
  b) 4   **c)** 9

6. a) A   **7.** A, B, and C

8. a) B   **b)** B

9. a) A, B, and C   **b)** A

10. a) hexagonal based prism, hexagonal based pyramid

11. B and C

12. a) Top and bottom are square-based pyramids.

13. a) cube and a truncated square-based pyramid

14. b)

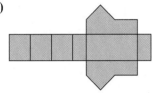

10.3 Top, Front, and Side Views of Cube Structures, pp. 356–357

5. c) A: left-side, B: front, C: top   **d)**

e) the left-side view and the top view

12. b) 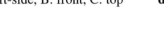   **13. b)**

Mid-Chapter Review, p. 359

1. a) octahedron **b)** square-based pyramid

2. a) No, the bottom-left square interferes with the bottom-right square.

b) Yes, this net forms a cube.

c) No, the large square needs to be divided in half.

d) Yes, this net forms a cube, but the lid for this box will be triangular.

4. A and B

10.4 Top, Front, and Side Views of 3-D Objects, pp. 362–363

4. octagonal prism

10.5 Isometric Drawings of Cube Structures, pp. 366–367

11. left

10.6 Isometric Drawings of 3-D Objects, pp. 370–371

10. c) The drawings represent the same objects, but do not show the same surfaces.

11. b) I chose a view that shows three different surfaces.

12. Isometric drawings are slanted to give an illusion of depth.

10.7 Communicating about Views, p. 374

4. e.g., The shape is a pentagonal prism with a regular pentagon as its base.

Step 1: Cut five 2 cm squares to make the side faces of the prism.

Step 2: Cut out two regular pentagons with side lengths of 2 cm.

Step 3: Tape the pieces together to form a net, as shown, then fold the net to form the final shape.

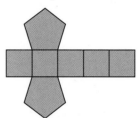

Chapter Self-Test, p. 377

1. a) cube, square-based pyramid

2. B   **3. b)**

7. First, draw one regular hexagon. Then draw six congruent rectangles using each side of the hexagon as a width. At the open width of one of the rectangles, draw a regular hexagon using the width of the rectangle as a side.

Chapter Review, p. 379

1. a)

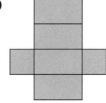

b) If the width of the package matches its height, then the package will have no gaps.

2. C   **3.** C

5. a) hexagonal prism   **b)** triangular prism
6. a) C

Chapter 11, p. 381

Getting Started, p. 383

1. a) e.g.,

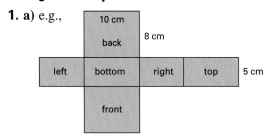

b) area of top and bottom 50 cm², area of front and back 80 cm², area of 2 sides 40 cm²

2. a) centimetres   **b)** square centimetres
   c) centimetres   **d)** cubic centimetres
   e) cubic centimetres   **f)** square centimetres

3. a) e.g.,

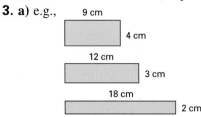

b) e.g., 26 cm, 30 cm, and 40 cm

4. 1500 1 cm ice cubes

5. Volume of smaller linking cube is 1 cm³, and volume of regular linking cube is 8 cm³.

6. a) 27 cm³   **b)** 12 cm³

7. a) B   **b)** D   **c)** C   **d)** A

11.1 Surface Area of a Rectangular Prism, pp. 386–387

5. a) e.g.,

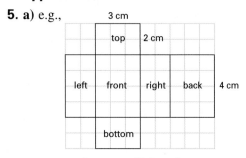

surface area 52.0 cm²

b) 52.0 cm²

6. a) 38.0 cm² **b)** 52.0 cm² **c)** 94.0 cm²

7. b) 126 cm²

8. a) 600 cm² **b)** 216 cm²

9. b) 12 cm by 1 cm by 1 cm
   c) 2 cm by 3 cm by 2 cm

11. a) 128.0 m² **b)** 4 cans of paint

12. c) 60 cm by 120 cm by 90 cm crate

14. b) The new surface area is 4 times greater than the original surface area.

   c) The new surface area is $\frac{1}{4}$ of the original surface area.

11.2 Volume of a Rectangular Prism, pp. 390–391

5. a) Use the orange face for the base. It has dimensions that are easy to multiply together to find its area.

   b) Use the green face for the base. It has dimensions that are easy to multiply together to find its area.

   c) Use the yellow face for the base. It has dimensions that are easy to multiply together to find its area.

6. a) 80.4 m³   **b)** 43 cm³   **c)** 336 cm³

8. a) e.g., 5 cm × 5 cm × 20 cm prism
   b) e.g., 10 cm × 2.5 cm × 20 cm prism

9. a) 512.0 cm³   **b)** 0.5 m³   **c)** 21.0 km³

11. B

14. a) 90 000 L   **b)** 81 000 L

15. 2238.5 cm³

16. A   **17.** Y

18. a) unknown side length 4 cm, volume 64 cm³
   b) unknown side length 8 cm, volume 288 cm³

19. 1705.5 cm³

Mid-Chapter Review, p. 393

1. a) 888 cm²   **b)** 1334.5 cm²
   c) 347.5 cm²   **d)** 40 300 mm²

2. 272 cm²

3. e.g., I would use the following formula: Surface Area = Area of bottom + Area of front + Area of back + Area of left side + Area of right side

4. e.g., It is more because now there are two extra sides.

5. yes   **6.** 19 250 cm³

7. a) 543.375 cm³ **b)** 7320 cm³ **c)** 53 820 cm³

8. small $324.00, medium $609.00, large $1152.00

9. 19 250 mL

10. a) prism B   **b)** prism B

11.3 Solve Problems by Guessing and Testing, pp. 396–397

5. a) e.g., 30 cm by 10 cm by 5 cm and 20 cm by 25 cm by 3 cm

b) e.g., 1000 cm² and 1270 cm²

c) e.g., My first set of dimensions would be better because its surface area is less.

6. a) e.g., 40 cm by 25 cm by 3 cm and 50 cm by 10 cm by 6 cm

b) e.g., 2390 cm² and 1720 cm²

c) e.g., My second set of dimensions would be better because its surface area is less.

7. a) e.g., length 100 cm, width 75 cm; length 90 cm, width 85 cm

b) e.g., 100 cm by 75 cm

8. a) e.g., length 25 cm, width 20 cm; length 30 cm, width 15 cm

b) e.g., 25 cm by 20 cm

9. length 2 m, width 1 m

11. a) 4 cm **b)** 3 cm

11.4 Relating the Dimensions of a Rectangular Prism to Its Volume, pp. 399–400

4. a) e.g.,

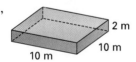

b) 400 m³ **c)** 600 m³ **d)** 400 m³

10. a) 420 cm³ **b)** no   **c)** no

Chapter Self-Test, p. 405

1. a) 354.3 cm²   **b)** 1350 cm² **c)** 553 cm²

2. a) 140 cm³   **b)** 3780 cm³

3. A

4. a) 5 cm   **b)** 5 m

5. a) no   **b)** 31 500 cm³

c) 7050 cm² of fabric

6. 862 cm²   **7.** 4 times

Chapter Review, pp. 406–407

1. a) 166 cm²   **b)** 664 cm²

2. a) 726 m²   **b)** 5.42 cm² **c)** 1080 cm²

3. a) 343 cm³   **b)** 384 cm³ **c)** 360 cm³

4. surface area 291.7 cm², volume 225.25 cm³

5. B

6. a) e.g., 20 m by 10 m by 2 m and 4 m by 25 m by 4 m

b) e.g., 4 m by 25 m by 4 m

7. No. e.g., The dimensions of three different prisms with a volume of 24 m³ are 24 m by 1 m by 1 m, 6 m by 2 m by 2 m, and 3 m by 4 m by 2 m.
Surface area of first is 98 m², surface area of second is 56 m², and surface area of last is 52 m².

8. a) 100 m³   **b)** 150 m³   **c)** 100 m³

d) 200 m³   **e)** 100 m³

9. a)

Area of base (cm²)	Height (cm)	Volume (cm³)	Surface Area (cm²)
4	2	8	24
4	4	16	40
4	8	32	72
4	16	64	136

b) It doubles.

c) The difference increases by a factor of 2.

Chapter 12, p. 409

Getting Started, p. 411

1. a) 7   **b)** 12   **c)** 4 and 5, 6 and 8

d) 9   **e)** 10

2. a) 0.7, 70% **b)** $\frac{1}{10}$, 0.1 **c)** 0.09, 9%

d) $\frac{22}{25}$, 88% **e)** $\frac{11}{20}$, 0.55 **f)** $\frac{17}{50}$, 0.34

g) 0.6, 60% **h)** $\frac{13}{25}$, 52%

3. Enter the numerator; press ⌦ ; enter the denominator; press ⌸ .

4. Heads or Tails

5. a) G, R, B

b) G-G, G-R, G-B, R-G, R-R, R-B, B-G, B-R, B-B

c) G-G-G, G-G-R, G-G-B, G-R-G, G-R-R, G-R-B, G-B-G, G-B-R, G-B-B, R-G-G, R-G-R, R-G-B, R-R-G, R-R-R, R-R-B, R-B-G, R-B-R, R-B-B, B-G-G, B-G-R, B-G-B, B-R-G, B-R-R, B-R-B, B-B-G, B-B-R, B-B-B

6. 6 possible outcomes

7. e.g., **a)** $\frac{5}{20}$ **b)** $\frac{10}{20}$

12.2 Calculating Probability, pp. 416–417

4. a) $\frac{1}{4}$ or 0.25 or 25% **b)** $\frac{1}{4}$ or 0.25 or 25%

   c) $\frac{1}{2}$ or 0.5 or 50% **d)** $\frac{1}{52}$ or about 0.02 or 2%

5. a) $\frac{1}{13}$ **b)** $\frac{1}{4}$ **c)** $\frac{4}{13}$ **d)** $\frac{4}{11}$

13. a) $\frac{3}{16}$ **b)** yes

Mid-Chapter Review, p. 419

1. Y-Y-Y, Y-Y-N, Y-Y-M, Y-N-Y, Y-N-N, Y-N-M,
   Y-M-Y, Y-M-N, Y-M-M, N-Y-Y, N-Y-N, N-Y-M,
   N-N-Y, N-N-N, N-N-M, N-M-Y, N-M-N, N-M-M,
   M-Y-Y, M-Y-N, M-Y-M, M-N-Y, M-N-N,
   M-N-M, M-M-Y, M-M-N, M-M-M

2. $\frac{6}{10}$ or 0.6 or 60%

3. a) B **b)** A **c)** C

4. a) $\frac{1}{2}$ **b)** $\frac{1}{4}$ **c)** $\frac{1}{6}$ **d)** $\frac{1}{12}$

5. a) yes **b)** no **c)** no **d)** yes

6. a) $\frac{1}{4}$ **b)** $\frac{1}{4}$ **c)** $\frac{1}{2}$

7. a) Kyle's **b)** Winnie's **c)** Kyle's

12.3 Solve Problems Using Organized Lists, p. 425

5. a) 16 combinations if order matters;
      10 combinations if order does not matter
   b) 8 combinations if order matters;
      5 combinations if order does not matter
   c) P(with order) $= \frac{1}{2}$, P(without order) $= \frac{1}{2}$

6. Spot's age, Fido's age, Rover's age: 1, 2, 12;
   1, 3, 11; 1, 4, 10; 1, 5, 9; 1, 6, 8; 2, 3, 10; 2, 4, 9;
   2, 5, 8; 2, 6, 7; 3, 4, 8; 3, 5, 7; 4, 5, 6

7. a) 37 combinations **b)** $\frac{27}{37}$

10. b) $\frac{63}{64}$ **13. a)** 13 combinations **b)** $\frac{1}{3}$

12.4 Using Tree Diagrams to Calculate Probability, pp. 428–429

4. $\frac{1}{4}$

5. H-H-H, H-H-T, H-T-H, H-T-T, T-H-H, T-H-T,
   T-T-H, T-T-T

7. b) if order matters, $P = \frac{19}{27}$; if it does not, $P = \frac{3}{5}$

10. a) 64 combinations **b)** 24 combinations

13. $\frac{1}{6}$ **14.** 1587 combinations

12.5 Applying Probabilities, pp. 432–433

3. Derek **7. b)** $69 **c)** $31 **d)** $\frac{1}{10}$

10. a) Bag C **b)** Bag A **c)** $\frac{35}{63}$

11. a) $\frac{1}{6561}$ **b)** $\frac{6560}{6561}$

12. The probability of getting a basket would be $\frac{1}{2}$
   only if both possible outcomes, getting a basket
   and not getting a basket, were equally likely. The
   likelihood of getting a basket varies depending
   on factors such as how far away the shooter is
   away from the basket and the skill of the shooter.

13. A and C

Chapter Self-Test, p. 434

1. a) $\frac{2}{11}$ **b)** $\frac{4}{11}$ **c)** $\frac{4}{11}$ **d)** 0

2. Theoretical probability gives the expected
   likelihood of an event. Experimental probability
   is the likelihood that an event will occur based
   on the number of times the event occurred during
   an experiment. For example, the theoretical
   probability of a coin landing on Heads is $\frac{1}{2}$,
   whereas if a coin is tossed 50 times and it
   lands on Heads 30 times, then the experimental
   probability of landing on Heads would be
   $\frac{30}{50} = \frac{3}{5}$, which is close to the theoretical
   probability, but not equal to it.

3. a) $\frac{1}{12}$ **b)** 0.08

4. a) 75 outcomes **b)** 24 **c)** $\frac{8}{25}$

5.

$2 coins	$1 coins	Quarters
0	0	10
0	1	6
0	2	2
1	0	2

6. a) 6 ways **b)** $\frac{1}{3}$ **7.** C

8. a) 13% **b)** 9 times

Chapter Review, p. 436

1. a) A   **b)** B   **c)** C

2. a) $\frac{1}{2}$ **b)** $\frac{1}{13}$

3. first task, second task, third task, fourth task: piano, walk dog, homework, clean room; piano, walk dog, clean room, homework; piano, homework, walk dog, clean room; piano, homework, clean room, walk dog; piano, clean room, walk dog, homework; piano, clean room, homework, walk dog; walk dog, piano, homework, clean room; walk dog, piano, clean room, homework; walk dog, homework, piano, clean room; walk dog, homework, clean room, piano; walk dog, clean room, piano, homework; walk dog, clean room, homework, piano; homework, walk dog, piano, clean room; homework, walk dog, clean room, piano; homework, piano, walk dog, clean room; homework, piano, clean room, walk dog; homework, clean room, walk dog, piano; homework, clean room, piano, walk dog; clean room, walk dog, homework, piano; clean room, walk dog, piano, homework; clean room, homework, walk dog, piano; clean room, homework, piano, walk dog; clean room, piano, walk dog, homework; clean room, piano, homework, walk dog

4. a) 1ˢᵗ hurdler, 2ⁿᵈ hurdler, 3ʳᵈ hurdler: Dave, Tony, Colin; Dave, Tony, Joel; Dave, Colin, Tony; Dave, Colin, Joel; Dave, Joel, Tony; Dave, Joel, Colin; Tony, Dave, Colin; Tony, Dave, Joel; Tony, Colin, Dave; Tony, Colin, Joel; Tony, Joel, Dave; Tony, Joel, Colin; Colin, Dave, Tony; Colin, Dave, Joel; Colin, Tony, Dave; Colin, Tony, Joel; Colin, Joel, Dave; Colin, Joel, Tony; Joel, Dave, Tony; Joel, Dave, Colin; Joel, Tony, Dave; Joel, Tony, Colin; Joel, Colin, Dave; Joel, Colin, Tony

b) $\frac{3}{4}$

5.

Food	Drink	Outcome
hamburger	juice	hamburger-juice
	water	hamburger-water
	lemonade	hamburger-lemonade
hot dog	juice	hot dog-juice
	water	hot dog-water
	lemonade	hot dog-lemonade

6. a) 18   **b)** 14   **c)** 8

7. $\frac{5}{36}$ **8.** $\frac{1}{8}$

9. Tim   **10.** Panthers

Cumulative Review: Chapters 10–12, pp. 438–439

1. C   **2.** B   **3.** A

4. D   **5.** D   **6.** B

7. B   **8.** A   **9.** C

Review of Essential Skills from Grade 6, p. 440

Chapter 1: Factors and Exponents, pp. 440–441

1. Millions: 1, Hundred thousands: 4, Ten thousands: 3, Thousands: 6, Hundreds: 9, Tens: 2, Ones: 7

2. a) 1, 2, 4, 7, 14, 28   **b)** 1, 2, 4, 8, 16, 32

c) 1, 2, 4, 7, 8, 14, 28, 56   **d)** 1, 29

e) 1, 3, 13, 39   **f)** 1, 3, 17, 51

3. b), c), f)

4. a) 1   **b)** 1, 27   **c)** 2

d) no   **e)** 1, 3, 9, 27

5.

Number	Factors	Number of factors	Prime or composite
1	1	1	neither
2	1, 2	2	prime
3	1, 3	2	prime
4	1, 2, 4	3	composite
5	1, 5	2	prime
6	1, 2, 3, 6	4	composite
7	1, 7	2	prime
8	1, 2, 4, 8	4	composite
9	1, 3, 9	3	composite
10	1, 2, 5, 10	4	composite
11	1, 11	2	prime
12	1, 2, 3, 4, 6, 12	6	composite
13	1, 13	2	prime
14	1, 2, 7, 14	4	composite
15	1, 3, 5, 15	4	composite
16	1, 2, 4, 8, 16	5	composite
17	1, 17	2	prime
18	1, 2, 3, 6, 9, 18	6	composite
19	1, 19	2	prime
20	1, 2, 4, 5, 10, 20	6	composite

6. a) 11   **b)** 6   **c)** 44   **d)** 108
    e) 9   **f)** 25   **g)** 28   **h)** 31
    i) 14
7. a) $<$ **b)** $<$ **c)** $<$ **d)** $>$
    e) $=$ **f)** $=$

Chapter 2: Ratio, Rate, and Percent, pp. 442–444

1. a) 2.5, 2.6, 2.7, 2.8   **b)** 0.07, 0.08, 0.10, 0.11
2. c) 6.35
3. a) 450, 4.5, 0.45, 0.045
    b) 3215, 3.215, 3.215, 0.032 15
    c) 0.4032, 0.040 32, 40.32, 40.32
4. a) 73   **b)** 0.003 365   **c)** 875 600
    d) 8 548 000   **e)** 0.043 15   **f)** 0.007 462
    g) 42.57   **h)** 9800   **i)** 0.076 93
5. 40
6. a) 4 to 9, 4:9, $\frac{4}{9}$ **b)** 5 to 9, 5:9, $\frac{5}{9}$
7. about 26 to 1
8. a) 83%   **b)** 1%   **c)** 50%   **d)** 9%
    e) 100%
9. a) 45%   **b)** 76%   **c)** 40%   **d)** 8%
    e) 80%   **f)** 80%
10. a) 50%   **b)** 75%   **c)** 10%
11. a) 8%   **b)** 70%   **c)** 45%   **d)** 37%
    e) 0.8%   **f)** 55.43%
12. a) 0.1, 10%   **b)** 0.8 80%   **c)** 0.5, 50%
    d) 0.75, 75%   **e)** 0.75, 75%   **f)** 0.625, 62.5%
13. a) 36%   **b)** 64%
14.

Fraction	Decimal number	Precent
$\frac{1}{100}$	0.01	1%
$\frac{1}{10}$	0.1	10%
$\frac{1}{20}$	0.05	5%
$\frac{3}{8}$	0.375	37.5%
$\frac{1}{5}$	0.2	20%
$\frac{1}{4}$	0.25	25%

Chapter 3: Data Management, pp. 445–446

1. scale for y-axis, title for x-axis, and data points
2. a) (2, 3)   **b)** (4, 7)   **c)** (7, 4)   **d)** (0, 5)
    e) (1, 6)
3. a) sleep   **b)** basketball   **c)** about 4 times more

4. 5.88 s
5. a) 8.9   **b)** 9.0   **c)** 8.5, 9.0

Chapter 4: Patterns and Relationships, pp. 447–448

1. a) e.g., 75   **b)** Sums are equal.
    c) Sums are equal for any box.
2. a) Sums are equal.   **b)** Pattern is similar.
    c) e.g., The numbers in each row increase by 1
    while the numbers in each column increase
    by 10.
    d) Sums are equal.   **e)** no
4. a) Same patterns would hold.
5. a) e.g., The sum of the top left number and the
    bottom right number equals the sum of the
    bottom left number and the top right number,
    and equals the sum of the two numbers in the
    middle. For

1	2	3
11	12	13

    $1 + 13 = 14$, $11 + 3 = 14$, $2 + 12 = 14$
    b) Same pattern is true.
6. e.g., The sum of the top left number and the
    bottom right number equals the sum of the
    bottom left and top right number. For

1	2	3	4
11	12	13	14

    $1 + 14 = 15$, $11 + 4 = 15$
7. a) $32 \times 43 = 1376$, $33 \times 42 = 1386$,
    $1386 - 1376 = 10$
    b) difference always 10
8. e.g., For

32	33	34
42	43	44
52	53	54

    a) 40   **b)** 1720
    c) 43, the number in the centre of the square
    d) Yes, the answer in part (c) is different for
    different sets of squares.
9. e.g., For any square box around 16 numbers, the
    products of the corner numbers of the diagonals
    differ by 90.

Chapter 5: 2-D Measurement, pp. 449–451

1.

Millimetres (mm)	Centimetres (cm)	Decimetres (dm)	Metres (m)	Kilometres (km)
4 800	480	48	4.8	0.004 8
90 000	9 000	900	90	0.09
200	20	2	0.2	0.000 2
8 000 000	800 000	80 000	8 000	8
3 000 000	300 000	30 000	3 000	3
764 000	76 400	7 640	764	0.764

2. a) 26 cm **b)** 48 cm **c)** 16 cm **d)** 60 cm
3. a) 10 cm² **b)** 100 cm² **c)** 72 cm²
   d) 30.25 m² **e)** 28 m² **f)** 36 cm² **g)** 144 cm²
4. a) 40 cm² **b)** 63 cm² **c)** 60 cm² **d)** 4 cm²
5. a) 31.5 m² **b)** 10 cm² **c)** 8.1 m²

Chapter 6: Addition and Subtraction of Integers, p. 452

1. a) decrease **b)** forward **c)** plus **d)** lose
   e) up **f)** loss **g)** drop **h)** save
   i) above **j)** negative
2. a) 6°C **b)** 0°C **c)** −5°C **d)** 10°C
   e) −1°C **f)** −3°C
3. a) 7; 5 < 7 **b)** 10; 10 > 0
   c) 19; 19 > 9 **d)** 28; 23 < 28
4. a) 0, 3, 5, 9, 12, 18 **b)** 6, 7, 7, 15, 15, 108
   c) 4, 7, 10, 13, 88, 109
5. a) 7 **b)** 14 **c)** 15 **d)** 20
   e) 12 **f)** 3 **g)** 0 **h)** 4
   i) −5 **j)** 2 **k)** 8 **l)** 2
   m) 4 **n)** 4 **o)** 4 **p)** 2
   q) 2 **r)** 12

Chapter 7: 2-D Geometry, pp. 453-455

1. a) triangle, yes **b)** trapezoid, yes
   c) open rectangle, no **d)** circle, no
   e) cube, no **f)** parallelogram, yes
2. a) 30° **b)** 60° **c)** 120° **d)** 100°
4. a) acute **b)** right **c)** reflex **d)** straight
   e) acute **f)** obtuse
5. a) congruent **b)** similar
   c) similar **d)** congruent
6. a) flip **b)** turn **c)** slide
7. a)

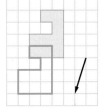

b)

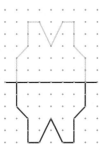

c)

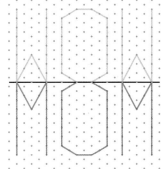

8. a)

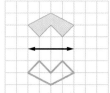

b)

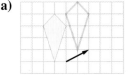

c)

d)

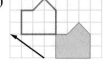

e)

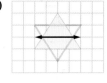

f)

Chapter 8: Variables, Expressions, and Equations, p. 456

1. a) 555-7283
   b) e.g., discount long-distance service
2. a) 555-8463 **b)** e.g., a time-pieces store
3. a) e.g., 555-DEAL
   b) e.g., a second-hand car dealer

4. a) e.g., My company sells apple pies to people in my community.
   b) e.g., 555-PIES
5. the product
6. The bar code scanner can easily register the thickness of lines, but it cannot read numbers.
7. No; I bought the same brand of candy bar from two different stores and they had the same bar code.
8. Each number is represented by two bars of different thickness.

Chapter 9: Fraction Operations, pp. 457–458

1. a) $\frac{5}{6}$ **b)** $\frac{2}{5}$ **c)** $\frac{7}{16}$ **d)** $\frac{3}{10}$

2. a) $\frac{13}{3}$ **b)** $\frac{65}{9}$ **c)** $\frac{11}{6}$ **d)** $\frac{43}{5}$

3. a) $2\frac{1}{3}$ **b)** $1\frac{7}{9}$ **c)** $6\frac{1}{5}$ **d)** $2\frac{1}{6}$

4. a) $\frac{1}{2}$ **b)** $\frac{1}{3}$ **c)** $\frac{1}{5}$ **d)** $1\frac{7}{8}$

5. $\frac{3}{4} = \frac{15}{20}, \frac{4}{5} = \frac{16}{20}; \frac{4}{5}$ is greater

Chapter 10: 3-D Geometry, pp. 459–460

1. a) pentagonal pyramid   **b)** rectangular prism
   c) hexagonal prism   **d)** triangular prism
   e) square-based pyramid
   f) triangular pyramid or tetrahedron

2.

	Faces	Edges	Vertices
a)	6	10	6
b)	6	12	8
c)	8	18	12
d)	5	9	6
e)	5	8	5
f)	4	6	4

3.

Polyhedron	Polygons
triangular prism	2 triangles, 3 rectangles
rectangular prism	6 squares
pentagonal pyramid	1 pentagon, 5 triangles
octagonal prism	2 octagons, 8 squares
square-based pyramid	1 square, 4 triangles

4. a) 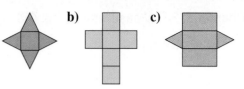   **b)**   **c)**

Chapter 11: Surface Area and Volume, p. 461

1. a) 400   **b)** 500   **c)** 1000
2. a) 60 cm³   **b)** 82.5 m³   **c)** 25.0 cm³
   d) 17.4 m³   **e)** 0.1 cm³
3. 461 700 mm³   **4.** 95
5. a) 100 m³   **b)** 117.37 m³   **c)** 85 m³

Chapter 12: Probability, pp. 462–463

1. a) 8   **b)** 8   **c)** 8   **d)** 8
2. a)

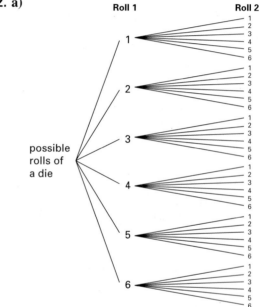

b) 36
3. a) Heads on a coin   **b)** draw at your school
   c) drawing a king   **d)** after studying
4. B, A, E, D, C or B, A, E, C, D
5. a) 3, since the cards of like colour are indistinguishable
   b) $\frac{4}{9}$ **c)** $\frac{3}{9}$ or $\frac{1}{3}$ **d)** 0
6. a) $\frac{1}{12}$ **b)** $\frac{11}{12}$ **c)** $\frac{1}{2}$ **d)** $\frac{5}{12}$
   e) $\frac{6}{12}$ or $\frac{1}{2}$, since 1 is neither prime nor composite
   f) $\frac{4}{12}$ or $\frac{1}{3}$
7. a) $\frac{2}{5}$ **b)** $\frac{3}{5}$

Index

Fuel efficiency, 49

G

Geometrical patterns, communicating about, 250–52
Graphs, 99, 108–10, 445
Greatest common factor (GCF), 11, 20
Grid paper, 150, 152
    drawing views on, 356
    subtracting fractions with, 322–25, 339
Grids, 57, 64, 68
    Cartesian, 228–31
    subtracting fractions with, 322–25, 339
Guessing and testing
    solving equations by, 288–91
    solving measurement problems by, 394–397

H

Halving, 15
Hexagon, 459
Hexagonal pyramids, 459, 460

I

Image, 232
Improper fractions, 305, 457
Inspection, solving equations by, 284–87, 298
Integer addition
    for integers that are far from zero, 196–99
    strategies, 200–3, 204
    zero principle for, 192–95, 204
Integers
    comparing positive and negative, 186–89, 204
    defined, 186
    opposite, 192
    ordering, 187
Integer subtraction
    counters for, 208–11, 220
    number lines for, 212–15, 220
Intervals, 98
Isometric drawings
    of cube structures, 364–67

of 3-D objects, 368–71, 378
Isosceles trapezoid, 170

K

Keyboards, 107

L

Least common multiple (LCM), 5–6, 20
Letter patterns, and algebra, 300
Lists. *See* Organized lists
Lowest common denominator (LCD), 458
Lowest terms, 305, 411

M

Math Games
    About 7, 169
    Alge-Match, 295
    Fishing for Solids, 375
    Fraction Bingo, 337
    Integro, 206
    Nim, 123
    Rolling Powers, 25
    Target Mean, 106
    Transformational Golf, 261
    Turn Up the Volume!, 401
    Unlucky Ones, 421
    Wastepaper Basketball, 72
Maze, 427
Mean, 102–3, 113, 446
Measurements
    communicating about, 176–78
    division with, 70
    2-D, 149–82
Measuring cups, 304
Median, 102–3, 113, 446
Memo field, 88, 96
Metric staircase, 171
Missing term, 43–44, 54
Mixed numbers, 305, 457–58
    adding and subtracting, 326–29, 339, 340
Mode, 102–3, 113, 446
Multiples, 3, 4–7, 20, 305
Multiplication
    decimal, 64–67, 74, 307

Credits

This page constitutes an extension of the copyright page. We have made every effort to trace the ownership of all copyrighted material and to secure permission from copyright holders. In the event of any question arising as to the use of any material, we will be pleased to make the necessary corrections in future printings. Thanks are due to the following authors, publishers, and agents for permission to use the material indicated.

Cover: Martin Barraud/Stone/Getty Images

Chapter 1 Opener Page 1: David Sailors/First Light; Page 7: Digital Vision/Getty Images; Page 8 top right: © Rodney Hyett; Elizabeth Whiting and Associates/Corbis/Magma, top left: Ryan McVay/Photodisc Red/Getty Images, centre right: Albert J. Copley/Photodisc Green/Getty Images, bottom right: Alanie/LifeFile/Photodisc Green/Getty Images; Page 10: © Wolfgang Kaehler/Corbis/Magma; Page 13: © Gary Conner/Index Stock; Page 17: Steve Pope/Associated Press; Page 18: Corel; Page 22 top right: Ryan McVay/Photodisc/Getty Images; Page 24: © Dimitri Lundt/Corbis/Magma: Page 35: © Gail Mooney/Corbis/Magma; Page 36: AP Photo/Kevork Djansezian/CP Picture Archive

Chapter 2 Opener Page 37: © Jason Hawkes/Corbis/Magma; Page 38: Brian Sytnyk/Masterfile; Page 42: Richard Price/Taxi/Getty Images; Page 46: © Najlah Feanny/Corbis/Magma; Page 48: © Michael Newman/Photo Edit; Page 49 top left: © Dennis MacDonald/Photo Edit Inc., top right: Corbis/Magma; Page 50: © Gunter Marx Photography/Corbis/Magma; Page 52: Digital Vision/Getty Images; Page 53 left to right: Jessie Parker/First Light, (c) Phil Schermeister/Corbis/Magma, © Inga Spence/Index Stock, © Zefa Visual Media - Germany/Index Stock; Page 55: Jules Frazier/Photodisc Green/Getty Images; Page 56: www.firstlight.ca; Page 60: Digital Vision/Picture Quest; Page 71 top: Thomas Nord/Shutterstock, centre: Corel, bottom: Corbis/ Magma; Page 77: Associated Press/CP Picture Archive; Page 78 centre: © Neal Preston/Corbis/Magma, bottom: Frank Gunn/CP Picture Archive

Chapter 3 Opener Page 81: © David Stoecklein/Corbis/Magma; Page 82: Mark Tomalty/Masterfile; Page 84: Comstock Images/Getty Images; Page 89: © Joseph Sohm; ChromoSohm Inc./Corbis/Magma; Page 97: © Clive Druett; Papilio/Corbis/Magma; Page 98: © Dennis MacDonald/Photo Edit Inc.; Page 105: A. Ramey/Photo Edit Inc.; Page 112: PhotoLink/Photodisc Red/Getty Images; Page 115: D2 Productions/Index Stock Imagery; Page 118: www.firstlight.ca

Chapter 4 Opener Page 119: Roy Ooms/Masterfile; Page 137: Ryan McVay/Photodisc Green/Getty Images; Page 141: © Kelley-Mooney Photography/Corbis/Magma; Page 142 top left: Beata Kroll Myhill, top right: © Steven Emery/Index Stock Imagery, centre: © Corbis/Magma, bottom left: C Squared Studios/Photodisc Green/Getty Images

Chapter 5 Opener Page 149: © Bonnie Kamin/Photo Edit; Page 159: © Steve Crise/Corbis/Magma; Page 160: © Scott T. Smith/Corbis/Magma

Chapter 6 Opener Page 183: Ryan McVay/The Image Bank/Getty Images; Page 185: © Roger Ressmeyer/Corbis/Magma; Page 200: Kevin Fitzgerald/The Image Bank/Getty Images; Page 205: Zoran Milich/Masterfile; Page 212: Corel; Page 224: Mark Lewis/Stone/Getty Images

Chapter 7 Opener Page 225: Alanie/Life File/Getty Images; Page 257: The M.C. Escher Company; Page 260: The M.C. Escher Company; Page 266: © Araldo de Luca/Corbis/Magma

Chapter 8 Opener Page 269: © Neal Preston/Corbis/Magma; Page 271: © Jeff Greenberg/Photo Edit Inc.; Page 278: Jim Cochrane/First Light; Page 280: © Carl and Ann Purcell/Corbis/Magma; Page 301: TRBfoto/Photodisc Red/Getty Images; Page 302: Photolink/Photodisc Green/Getty Images

Chapter 9 Opener Page 303: Rick Souders/First Light; Page 319: Steve Mason/Photodisc Green/Getty Images; Page 322: Lisa Peardon/Taxi/Getty Images; Page 324: © Strauss/Curtis/Corbis/Magma; Page 342: © Lester Lefkowitz/Corbis/Magma

Chapter 10 Opener Page 345: © Barry Winiker/Index Stock Imagery; Page 363 top: © Macduff Everton/Corbis/Magma, bottom: Dick Hemingway; Page 368: Corel; Page 376 left: © Burstein Collection/Corbis/Magma, right: MS Kaufmann A422, fol.6r, Library of the Academy of Sciences, Budapest

Chapter 11 Opener Page 381: Frank Hudec/First Light; Page 391: Lloyd Sutton/Masterfile; Page 397: © North Carolina Museum of Art/Corbis/Magma; Page 398: Comstock Images/Getty Images; Page 405: Kristiina Paul Bowering; Page 407: Corbis/Magma

Chapter 12 Opener Page 409: John Lund/The Image Bank/Getty Images; Page 424: © Spencer Grant/Photo Edit Inc.; Page 426: © Ed Bock/Corbis/Magma; Page 436: Chris Cole/The Image Bank; Page 438: Jeff Haynes/AFP/Getty Images